Electronic Packaging
Science and Technology

# Electronic Packaging Science and Technology

*King-Ning Tu*
*National Chiao Tung University, Hsinchu, Taiwan, ROC*

*Chih Chen*
*Department of Materials Science and Engineering, National Yang Ming Chiao Tung University, Hsinchu, Taiwan, ROC*

*Hung-Ming Chen*
*Institute of Electronics, National Yang Ming Chiao Tung University, Hsinchu, Taiwan, ROC*

WILEY

This edition first published 2022


*Registered Office*
John Wiley & Sons, Inc., 111 River Street, Hoboken, NJ 07030, USA

*Editorial Office*
111 River Street, Hoboken, NJ 07030, USA

For details of our global editorial offices, customer services, and more information about Wiley products visit us at www.wiley.com.

Wiley also publishes its books in a variety of electronic formats and by print-on-demand. Some content that appears in standard print versions of this book may not be available in other formats.



*Library of Congress Cataloging-in-Publication Data*

Names: Tu, K. N. (King-Ning), 1937– author. | Chen, Chih, 1970– author. | Chen, Hung-Ming, author.
Title: Electronic packaging science and technology / King-Ning Tu, University of California, Los Angeles, United States of America; Chih Chen, National Chiao Tung University, Hsinchu, Taiwain; Hung-Ming Chen, National Chiao Tung University, Hsinchu, Taiwain.
Description: 1st edition. | Hoboken, NJ : Wiley, 2022. | Includes bibliographical references and index.
Identifiers: LCCN 2021039114 (print) | LCCN 2021039115 (ebook) | ISBN 9781119418313 (cloth) | ISBN 9781119418320 (adobe pdf) | ISBN 9781119418337 (epub)
Classification: LCC TK7870.15 .T8 2022 (print) | LCC TK7870.15 (ebook) | DDC 621.381/046–dc23
LC record available at https://lccn.loc.gov/2021039114
LC ebook record available at https://lccn.loc.gov/2021039115

Cover image: © Andriy Onufriyenko/Getty Images
Cover design by Wiley

MIX
Paper from responsible sources
FSC® C013604
www.fsc.org

# Contents

# Preface

As we enter the big data era, mobile devices are ubiquitous. Internet of things (IoT) is everywhere, and we have man-to-man, man-to-machine, and machine-to-machine communications. Furthermore, in the Covid-19 virus pandemic period, the trend of distance teaching, distance medicine, home office, and on-line meeting has increased greatly the need of advanced consumer electronic products, demanding smaller form factor, larger memory, more functions, faster and larger data collection and transmission, cheaper cost, and superb reliability. At the same time, 5G advanced communication technology and 3D IC devices have begun their impact to our society, and many new artificial intelligence (AI) applications have been invented.

With the perceived slowing down of Moore's law of miniaturization of Si chip technology, microelectronic industry is searching for alternative ways to sustain Moore's law. 3D IC is most promising in achieving more-than-Moore, wherein the up-scale of packaging technology is critical. Indeed, new advanced packaging factories are being built worldwide. We ask what will be the technical innovations in electronic packaging for 3D IC devices in order to enhance performance and reliability? Or, what are the challenging issues in electronic packaging technology that are essential in the near future development of semiconductor technology?

The goal of this book to present the science and engineering of advanced electronic packaging technology for a deeper understanding of the essence in development and manufacturing of the more-than-Moore technology. Especially, what is new in this book are the subjects of Cu-to-Cu direct bonding by using the (111) uni-directionally oriented nanotwin Cu, innovative 3D IC systems in packaging integration for high performance of wide bandwidth and low power devices, and the analysis of mean-time-to-failure equations based on entropy production.

After the introduction chapter, the following chapters will be divided into three parts. In Part I, the history of bonding technology will be covered in

Chapter 2, starting from wire-bonding, tab-automated bonding, flip chip C-4 solder joint bonding, micro-bump bonding, Cu-to-Cu direct bonding, and hybrid bonding. The microstructure, properties, and applications of randomly oriented and (111) uni-directionally oriented nano-twin Cu will be covered in Chapter 3. Then, Chapter 4 and Chapter 5 will be dedicated to chemical reactions and kinetic processes in Cu-Sn reactions for solder joint formation. Chapter 4 will review solid-liquid interfacial diffusion (SLID) reactions between liquid solder and Cu. Chapter 5 will review solid-solid reactions between solid solder and Cu. The kinetics of growth of intermetallic compound (IMC), which is a stoichiometric compound and has no composition gradient, has been an outstanding problem in solid-solid reactions. We introduce Wagner's diffusivity to overcome it.

Part II consists of chapters on electric circuit integration in packaging technology. The emphasis is on the design of low power devices and intelligent integration. The technical issues related to the need for faster rates and increased amounts of data transport in 2.5D/3D IC are discussed. It is explained how to increase the I/O density and the bandwidth in packaging technology.

Part III is a collection of chapters on reliability science. It begins with a chapter on irreversible processes of atomic flow, heat flow, and charge flow in open systems. The most important issue of Joule heating will be analyzed. The topics of electromigration, thermomigration, stress migration, and failure analysis will be covered. Equations of mean-time-to-failure (MTTF) will be reanalyzed on the basis of entropy production.

Finally, in Chapter 14, a brief discussion on how to use artificial intelligence to accelerate reliability testing will be presented. We propose an x-ray based graphic processing unit (X-GPU) to analyze early reliability failure before it occurs in any newly developed 3D IC device for mass production. The goal of AI here is to change the time-dependent and time-consuming reliability tests to time-independent tests. The basic idea of mean microstructure-change to failure (MMTF) will be introduced, so that we can link MTTF to MMTF.

We appreciate the capable help of Mrs. Jody Lee and Mr. John Wu at NCTU in preparing the book.

Hsinchu, December 2020

*King-Ning Tu*
*Chih Chen*
*Hung-Ming Chen*

# 1

# Introduction

## 1.1 Introduction

As we enter the big data era, mobile devices are ubiquitous. On hardware, nearly everyone has a cell phone. On software, internet of things (IOT) reaches everywhere. We have man-to-man, man-to-machine, and machine-to-machine communications. Furthermore, during the Covid-19 virus pandemic, the trend of distance teaching, distance medicine, home office, and online meeting has increased greatly the need of advanced consumer electronic products, demanding smaller form factor, larger memory, more function, cheaper cost, faster and greater rate of data transmission, and superb reliability. Actually, the advanced 5G communication technology and 3-dimensional integration of circuits (3D IC) have already begun their impact to our society. No doubt the world around us is changing rapidly. In human history, this is the second time of a fundamental revolution.

In eighteenth century, we had industrial revolution when steam engine was invented. It developed machine power to replace human power and animal power. The activities in civilization were changing from agriculture to industry. We had railroad trains, ocean liners, automobiles, airplanes, and electricity. While industrial production has transformed human society from feudal to democratic, it was accompanied by capitalism, then communism, and then socialism. Indeed, the impact to human society was huge in the last two to three hundred years.

In twentieth century, after the invention of transistor, very-large-scale integration of transistor circuits, and mobile technology, we have data power to enhance machine power. What is coming is to have artificial intelligence (AI) revolution. We have robots, and human-less vehicles and

*Electronic Packaging Science and Technology*, First Edition. King-Ning Tu, Chih Chen, and Hung-Ming Chen.

aircrafts to serve us. Mobile technology supported by mobile internet will have a long way to go in the near future. Accompanying the rapid progress, however, Moore's law of miniaturization in Si chip technology is near ending, so people wonder whether the rapid progress can be sustained.

If we look back to the last 10–20 years, semiconductor industry has had some interesting events. Namely, Japan has lost the leadership in semiconductor technology. The large European countries such as England, France, and Russia have no presence in microelectronics. On the other hand, Taiwan and Korea have gained the leadership due to the success of Taiwan Semiconductor Manufacturing Corporation (TSMC) and Samsung. Today, China has identified semiconductor device development and manufacturing to be a national goal and will spend a large sum of money to achieve it. One of the key reasons behind the trade war between United States and China is due to the competition in advanced semiconductor technology and in manufacturing of consumer electronic products. At the same time, 5G communication technology and AI applications are with us and they exert an unlimited impact and change to our society.

There were many reasons of Japan's failure, such as the critical change in currency exchange rate between yuan and dollar, and also the heavy punishments of Fujitsu by United States. However, Japan has recently joined TSMC to form an advanced electronic packaging factory in Japan to maintain its impact. The weaker and weaker presence of England and France is because of the insufficient financial support on semiconductor technology due to socialism. They tended to support rather uniformly the popular needs of society, but did not concentrate on supporting semiconductor technology. Today, to build a Si-based transistor factory will take 3–5 billion US dollars, so fewer and fewer countries can afford it. Russia is simply poor. Samsung in Korea has had the government's focused support to do so. TSMC in Taiwan is unique that it takes orders from all over the world to make VLSI devices, so it knows the major trend of the technology. Thus, it can spend the money it earned to keep improving the technology with innovation. For example, the technique of immersion lithography has enabled TSMC to lead the manufacturing of nanoscale semiconductor devices for the past five generations. Now, we are having the 5 nm node of nanotechnology, and 3 nm and 2 nm nodes are coming.

Some experts in China commented that China might take 10 years to catch up. No reason was given why it will take so long! On the other hand, China has been very successful in high-speed train and satellite technology. When China can make transistors as cheap as those made in United States, it is a sign that China has caught up. A plausible reason is given below.

A Chinese classic philosopher, Wang Yang-Ming, said "Knowing is hard, but practice is easy." Semiconductor technology is completely opposite that "Knowing is easy, but practice is hard." Take an example of teaching a child to play violin or piano; we can find the best teacher to teach the child all the technique and tricks of how to play well. Even a very talented child, he/she still has to take 10–20 years of practice to be good. It is the "experience" that is needed to do well. Indeed, "experience" cannot be taught quickly and cannot be copied or cannot be steal. Furthermore, in semiconductor manufacturing, the experience is not just of one person, it is of an entire industry.

## 1.2 Impact of Moore's Law on Si Technology

Moore's law states that the transistor density per chip area is doubling every 18–24 months without the increase of production cost. Figure 1.1 depicts the achievements of 2D IC of Si technology according to Moore's law. From 1970 to 1985, the density has increased from 1K to 1M per chip, which is an increase of 1000 times. From 1985 to 2005, the increase was from 1M to 1G, which is another increase of 1000 times. It is the most successful event of sustainability for about 40 years of any human activity.

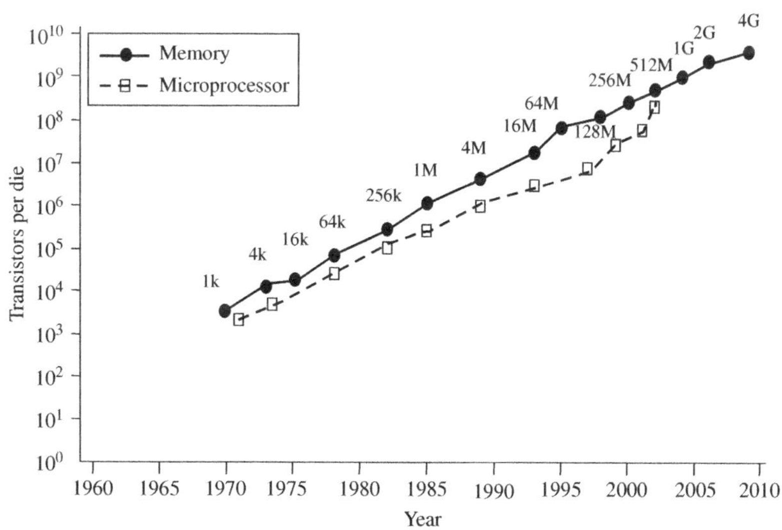


**Figure 1.1** A sketch of the achievements of 2D IC of Si technology according to Moore's law.

Physical size reduction

There are 110 Si chips in the module of a mainframe Computer in 2002.

IBM J. of R&D., November, 2002.

If we replace the 110 chips by 1 chip, we can make the mobile phone now!

Also replace key-board by finger-touch.

**Figure 1.2** An image of part of the device structure of the mainframe computer in 2002. The size of the module is about 10 × 10 cm.

The law has had two significant impacts. First, the price of one transistor is now cheaper than the printing of one alphabet on a newspaper, which is a significant consequence of the Moore's law that circuit density can be doubled without the increase of production cost. Hence, we can have very low cost use of transistors. Second, it enables the reduction of physical size of a computer so that handheld and mobile devices can be made. Figure 1.2 shows part of a mainframe computer in 2002. The module size is about 10×10 cm, and between two of the ceramic plates in the module, there are $10 \times 11 = 110$ pieces of Si chips. According to Moore's law shown in Figure 1.1, the circuit density per chip in 2002 is about 256 M. If we can integrate the central processing unit (CPU) and the memory on the 110 chips into a single chip, we can build a mobile device or a mobile computer by using a single chip! No doubt, we also need to reduce the packaging structure, as well as to replace a keyboard by the finger-touching technique. It is worth mentioning that in Figure 1.2, while we do not see the chips but we see the electronic packaging structure, wherein the bright solder joints are everywhere. It shows how important is solder joint technology in electronic device manufacturing. This is because solder joint is still the best way to join two Cu wires, even two nanowires of Cu.

## 1.3 5G Technology and AI Applications

Those advances mentioned in the above have changed internet to mobile internet. Internet means computer-to-computer communication. Mobile internet means mobile computer (cell phone)-to-mobile computer communication. At this moment, the rapid advances in 5G and AI require Moore's law to keep going ahead for at the least another 10–20 years, but Moore's law is ending.

In human civilization, the technique of communication has advanced step-by-step slowly from language, written words, printing, telephone and telegraph, television, internet, and now to mobile internet. Today, the advanced communication technology behind mobile internet is defined as 5G, with a standard of certain required performance. Figure 1.3 shows the drawing of a flower, which has six petals. There are two sets of petals: the inner brighter ones and the outer darker ones. The latter represents 5G technology and the former represents 4G technology. Each petal defines a specific technical requirement, as shown in Table 1.1. For example, on

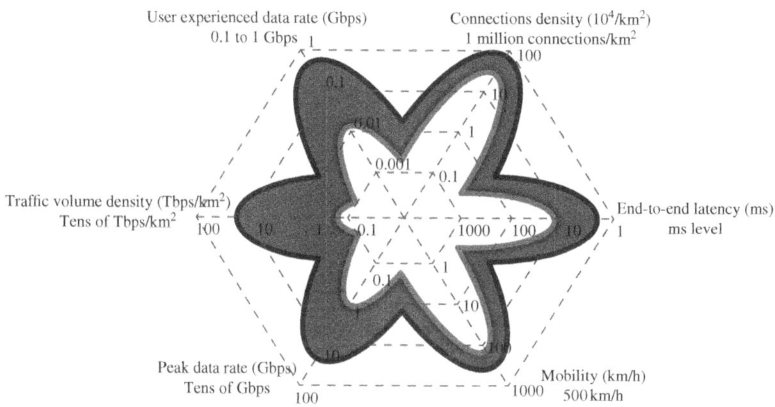


**Figure 1.3** The petals of flower, which shows the required functions in 4G and 5G communication technology.

**Table 1.1** 5G technical requirements.

- Primary requirements:
  End-to-end latency (mille sec)
  Peak data rate (mobility, km/h)
  Traffic volume density (tens of Tbps/km$^2$)
  Density of base stations ($10^4$/km$^2$)
  Internet of everything
  Energy efficiency
- Secondary requirements:
  Security and reliability
  Low cost

point-to-point latency of signals, it was 10 ms in 4G and about 1 ms in 5G. If we can advance the required performance beyond the petals of 5G, it will be 6G.

In 2014, it was already predicted that around 2020, the 5G technology will be widely available. However, the company, Huawei in China, announced that its 5G technology is ready in 2019, ahead of everyone. It became a critical issue in the trade war between China and United States. The latter has put Huawei on its entity list or ban list. The hardware in 5G consists of satellites, servers, optical fiber lines, base stations, cell phones, and terminal sensors, and the software consists of various mobile internets, such as Android of Google. Huawei is especially good at making base stations.

Data and signals can be transmitted via satellites in sky, but satellites can be shot down easily in the beginning of a war. It can be transmitted via optical fibers underground and across oceans. The making of optical fiber requires the doping of rare-earth elements, so the strategic importance of rare-earth elements can be appreciated. Also it can be transmitted via base stations on land surface, so the construction of a large number of base stations is of national security concern. It is worth mentioning that Taiwan has successfully used its network of base stations to detect and control the motion and contact of people with Covid virus.

In the late nineties, dot-com was developed but soon burst because cell phones were not available. After cell phone becomes popular, Apple, Microsoft, Amazon, etc. are now the biggest companies in the world, no more GE, IBM, and Exxon, because of the wide applications of mobile technology.

About the global standard of requirements in 5G technology, see Table 1.1, the first is point-to-point latency of signals, which is only a few milli second. Latency means the total time spent to send out a signal and to receive it back. In a chain of moving cars, if the first car stops suddenly, the second car must stop within the time of latency, otherwise an accident would occur. If we consider a human-less vehicle, the LiDAR (not radar) on top of the car should be able to detect a sudden appearance of a pedestrian or a car, so that it can stop to avoid an accident.

Moreover, LiDAR and radar are line-of-sight techniques. Yet, we need to have a network of vehicle-to-everything in order to have non-line-of-sight awareness to know what is behind a stationary or a moving object. Also it should have an ultrafast rate of transport of data in order to show clearly the change of images of the surrounding of a car moving at high speed. Besides, it can download or upload instantly news or weather reports. To download a movie will now take only three seconds in 5G technology. For the success of 5G, it must have a very large number of base stations, so that information can be received and transmitted continuously from place to place. It is no need to explain other standard requirements in Table 1.1, they should be clear. For example, any device being used under the hood of a car should have a high reliability because of heat. Then, low cost is important for the use of internet of everything in our home and office.

No doubt, 5G technology will enhance AI applications. At the end of this book, in Chapter 14, we shall discuss the need of using AI to accelerate the study of reliability, so that we may change it from a time-dependent event to a time-independent event. Another area for the use of AI will be the biomedical and health applications. For example, Chinese medicine has been based on big data for many years, and the technique of acupuncture could be improved with modern microelectronic devices. The link between microelectronics and biomedical applications will be the most important advanced technology in the future.

## 1.4 3D IC Packaging Technology

As the trend of miniaturization in Si technology slows down, microelectronics industry has been looking for ways to keep the downsizing momentum going, meaning to go to more-than-Moore! [1–3] The critical feature size in Si devices has already reached nanoscale, below 10 nm. Hence, it is harder and harder to make transistor circuits on a Si chip smaller and smaller without a large cost increase. At present, the most promising way to extend Moore's law is to go from 2D IC to 3D IC. Actually, the paradigm change has occurred more than 10 years ago, but 3D IC is not in mass production, because of cost and reliability.

In semiconductor manufacturing, because the product quantity is extremely large, so high yield and high reliability are critically important. Low yield will increase cost, and poor reliability will lead to recall; one example is the battery failure of cell phones. For any consumer electronic product in mass production, the concern of reliability is critical, especially the electronic packaging in 3D IC for advanced consumer electronic products, which are widely used now for distance teaching and home office.

In this introductory chapter, we explain what is electronic packaging? Also, what are the science and engineering in it, especially those relate to reliability? If we want to add more functions to hand-held devices, the operations of memory, logic, and special functions must be increased. At the same time, power as well as battery capacity must be increased too. A larger size battery will squeeze the volume of the rest of the device, which makes the heating problem worse. To remove heat, we must have a temperature gradient. If we consider a temperature difference of 1 °C across a microstructure of 10 μm in diameter, the temperature gradient is 1000 °C/cm, which will induce thermomigration. In turn, Joule heating will enhance electromigration, and thermo-stress will induce stress-migration. While these are time-dependent events, they are of major reliability concern.

Figure 1.4 is a scanning electron microscopy (SEM) image of the cross-section of a 2.5D IC test device. It has only two pieces of Si chips stacking on a polymer board. Electrically, they are interconnected by three sets of solder joints. At the bottom or on the outside of the polymer board is the set of the largest solder balls of diameter up to 760 μm, which is called the ball-grid-array (BGA). These balls allow the test device to be connected to the circuits on a printed circuit board. Within the polymer board, there are Cu wirings, as well as Cu plated-through-holes, which are not shown in the image. On top of the polymer board, there is the second set of flip chip solder balls of diameter about 100 μm, the so-called C-4 (controlled collapse chip connection) solder balls, connecting the board to the first Si chip, which is the "interposer." In this test device, there is no transistor on the interposer, which is passive and serves only as a substrate without introducing thermal stress to the active Si chip on the top. Often this test device is called 2.5D IC due to the fact that the interposer has no transistors. If the interposer has transistors, it becomes 3D IC.

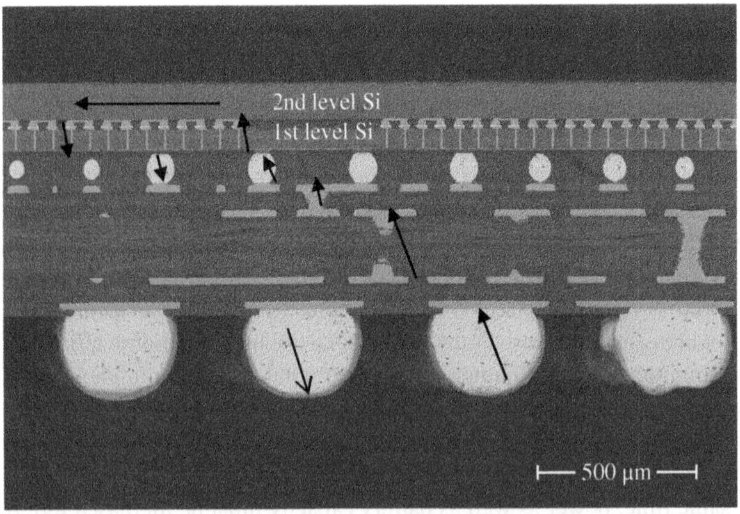


**Figure 1.4** Scanning electron microscopy (SEM) image of the cross-section of a 3D IC test device. It has only two pieces of Si chips stacking on a polymer board.

In the interposer, there are arrays of vertical through-Si-vias (TSV) plated with Cu, making connections to the third arrays of solder joints of diameter about 10–20 μm, the so-called micro-bumps or μ-bumps, which join the interposer to the top Si chip. The top Si chip is an active device chip, so it has transistors. The thickness of the device in Figure 1.4 is about that of a US penny. The thinness of the device is a critical requirement due to the limit of form factor of mobile consumer electronic products. Consequently, the thickness of Si chips is thin too. The thickness of the Si interposer is about 50 μm, which is much thinner than that of a convention Si chip of 200 μm in thickness. The thin interposer has caused the warpage problem, as well as the heat conduction issue, to be discussed in the later chapters. The diameter of the TSV in the interposer is about 5 μm, so the aspect ratio of the TSV is 10.

In the above example, besides the active Si chip, the rest, which includes the interposer, can be regarded as electronic packaging. The packaging enables the Si chip to function, as well as to allow us, to interact with the outside world. In the packaging, it is worth mentioning that between two sets of solder joints of different sizes, there should be a redistribution layer (RDL) structure for circuit fan-out. It increases the number of input–output (I/O) contacts of a circuit in going from a low density of solder joints to a high density of solder joints. The higher the density of I/O, the better the resolution of frequency of a digital electromagnetic wave, because each I/O is designed to transmit a small width of the wave.

At the moment, there are two critically important challenges in electronic packaging technology. The first is the need of denser and denser I/O, which means the diameter of micro-bump and the pitch between them has to be reduced. As to be shown in Chapter 3, hybrid-bonds consist of Cu-to-Cu bonds together with dielectric-to-dielectric bonds are being developed. The second is Joule heating and heat dissipation, which will be discussed in Chapter 9.

About the increase of I/O, from BGA to C-4 joints, there is a RDL of Cu wires in the upper part of the polymer board. From C-4 joints to μ-bumps, there is an RDL of Cu wires at the lower part of the interposer chip. This second RDL is invisible in the figure, but it is new in 3D IC because it does not exist in 2D IC devices, where typically there are only two levels of solder joints. The failure of the new RDL is of concern.

Figure 1.5a and b show synchrotron radiation tomographic images of a 3D IC and part of a 2.5D IC device, respectively. The latter has a length about 4 mm, and a thickness and a height of about 0.5 mm. Due to the weak absorption of X-ray, the two Si chips and the polymer substrate become invisible. We can see the vertical TSV pillars. Also, the solder balls and the

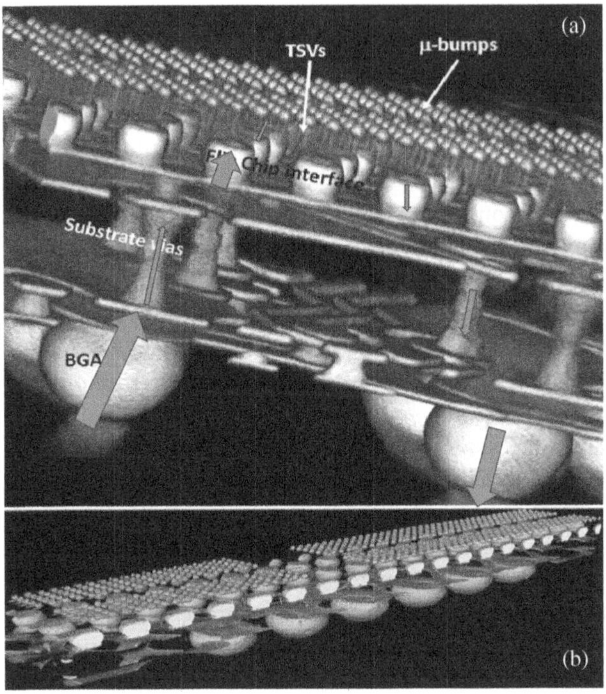


**Figure 1.5** (a) Synchrotron radiation tomographic images of a similar device as shown in Figure 1.1. Due to the weak absorption of X-ray, the two Si chips and the polymer substrate become invisible. But the solder balls and the Cu wires are shown clearly. (b) Synchrotron radiation tomographic image of a 2.5D IC device, having a length about 4 mm, and a thickness and a height of about 0.5 mm. The vertical TSV pillars can be seen.

Cu wires are shown clearly. In Figure 1.5a, by using a pair of the BGA balls as the cathode and the anode and by passing 50 mA at 100 °C, following the arrows which indicate the conduction path, we can study time-dependent failures caused by electromigration and Joule heating, to be discussed in Chapter 10.

Why do we emphasize electromigration and Joule heating? This is because electronic devices are current–voltage (I-V) devices, so the applied electric current goes in and out of the devices in an open system. It causes Joule heating and electromigration, which are of key reliability concern. Figure 1.6 is a schematic diagram of the cross-section of a typical 3D IC device. The structure, in essence, is the same as that shown in Figure 1.4, except that on the right-hand side, there is a stack of memory chips on a logic chip as the CPU. If we replace the stack by an optical or

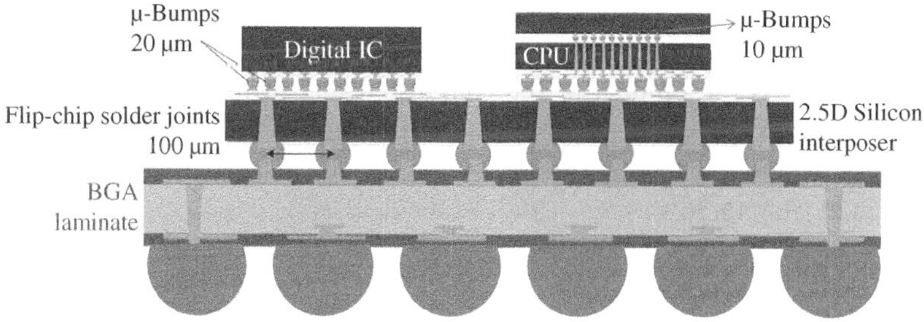


**Figure 1.6** Schematic diagram of the cross-section of a typical 3D IC device.

compound semiconductor or MEMS chip, it becomes heterogeneous integration.

In comparing the structure of 3D IC to that of 2D IC, the difference is the stacking of multilayer of chips and the interconnects using TSV and μ-bumps. On processing TSV, the thinner the chip, the easier the drilling of vias. On making μ-bumps, its melting point should be lower than that of C-4 joints, so that the latter will not melt upon the melting of the former. Thus, the basic challenges are that the wafer is thinner and the processing temperature is lower.

From the viewpoint of packaging technology, we may say that the essence or the major challenge in 3D IC is to scale down the dimension of packaging structures so that it can match those in the chip technology. There is no Moore's law in packaging technology, so it has room to shrink.

What are the key functions of electronic packaging? The cell phone held in our hands is a movable electronic packaging product or a mobile computer, which enables us to compute and to communicate with the world around us. The set of chips in the cell phone can be arranged horizontally, side by side, but it takes space. Or they can be arranged vertically, one on top of the other, this is called 3D IC, and it reduces the form factor and takes less space. However, heat dissipation in 3D IC is harder because the packing is denser. When over-heat occurs, it induces reliability problems. Over all, the product should be electrically, mechanically, chemically, and thermally stable.

## 1.5 Reliability Science and Engineering

An electronic device in operation is an open system because electrical charges flow in and out of the device. While the number of charges in transport is conserved, entropy production is not. The waste heat in entropy production is Joule heating on the basis of irreversible processes. [4, 5] For electrical conduction, Onsager's Eq. (1.1) below shows that entropy production is the product of the conjugated flux of j (current density = coulomb/$cm^2$-sec) and the conjugated driving force of E (electric field E = jρ, where ρ is resistivity). Derivation of the Onsager equation will be given in Chapter 9.

$$\frac{TdS}{Vdt} = jE = j^2\rho \tag{1.1}$$

where $T$ is the temperature, $V$ is the volume of sample, $dS/dt$ is the entropy production rate, and $j^2\rho$ is the Joule heating per unit volume per unit time. Typically, the power from Joule heating is written as $P = I^2R = j^2\rho V$, where $I$ is the applied current and $R$ is the resistance of the sample. Thus, $j^2\rho$ is power density or Joule heating per unit volume per unit time of the sample, in units of Watt/$cm^3$, and $I^2R$ is Joule heating per unit time for the entire sample, in units of Watt. Clearly, this is the reason why we need low-power devices or low entropy production devices.

While the cost of production of 3D IC can be reduced when it is in mass production, the problem of reliability due to over-heating has to be solved fundamentally by a smart system design or by design-for-reliability (DfR) and by a critical selection in materials integration. To put it simply, we need to design low-power devices, and also we need to understand heat production (Joule heating) in irreversible processes and heat dissipation in the device structure. [6] Hence, the science and engineering of electronic packaging come into focus.

Entropy production is the most relevant understanding of failure induced by electromigration, thermomigration, and stress-migration in irreversible processes. [7] Statistical analysis of failure requires the knowing of mean-time-to-failure (MTTF). An example is Black's equation of MTTF for elctromigration. In Chapter 13, we shall present a unified model of MTTF for electromigration, thermomigration, and stress-migration on the basis of entropy production.

Figure 1.7 shows an example of electromigration-induced damage in Cu interconnects. The high current density in the interconnect has induced a flow of atoms along the electron flow direction, going from the cathode to the anode, leading to vacancy accumulation and void formation in the cathode region. Resistance of the interconnect increases gradually until an opening in the circuit occurs, where the resistance increases dramatically.

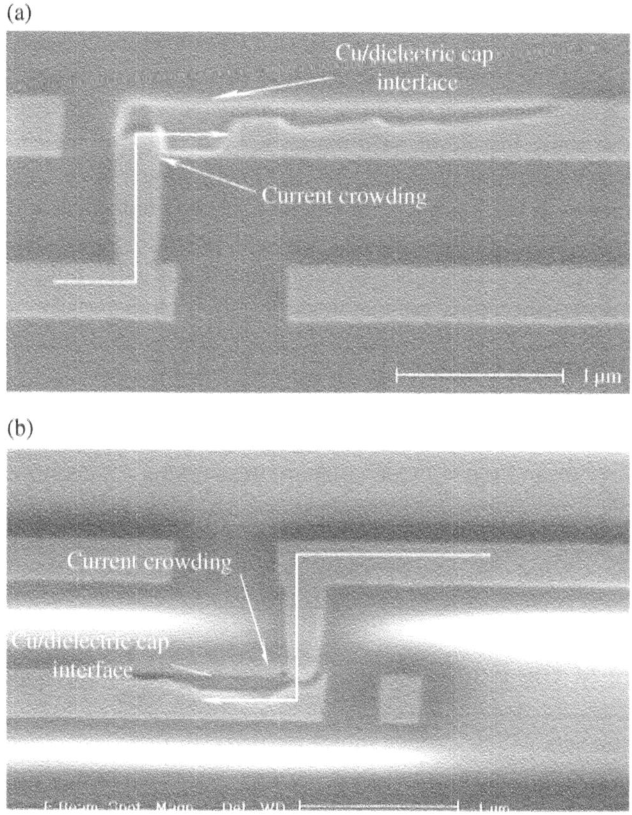


**Figure 1.7** An example of electromigration electromigration-induced failure of void formation in Cu interconnects. (a) Electrons drifted from the bottom Cu line to the top Cu line. (b) Electrons drifted from the top Cu line to the bottom Cu line.

## 1.6 The Future of Electronic Packaging Technology

In the near future, the R&D of packaging technology for 3D IC devices will be the focus. The role of electronic packaging in Si-based microelectronic technology is getting more and more important. In turn, reliability will be of major concern. The introduction of AI to facilitate the applications to new 3D IC devices and the reduction of the time-consuming reliability tests will demand our concentrated effort. In the long run, the applications of electronic packaging technology to biomedical devices will be important. For example, as people lives longer, diabetes is common in old age. To determine oxygen content and sugar content in our blood, we use invasive method to obtain a drop of blood from our finger for measurement. It is very unpleasant! If we can invent a noninvasive method, for example by wearing a mobile device around our finger or arm, it will greatly reduce the unpleasant feeling in our daily live if we have diabetes. Even better, if we can implant a small device in our body to perform the blood testing function, we will need to understand the interfacial interaction between biological and nonbiological materials. In other words, we will need to study biocompactable materials, as well as the chemical reactions in body fluid at body temperature. Furthermore, we may need to join a biomaterial to a non-biomaterial. To do so, we may need a low-temperature solder or binder which can decompose slowly in the body. However, the melting point of the common Pb-free solder, the eutectic SnAg, is over 200 °C, to be discussed in Chapter 4 on solder joint reactions. We may need a new solder whose wetting temperature is around 100 °C, which is above the working temperature and near the body temperature, for biomedical devices. While these issues are beyond the scope of this book, we note that the future of electronic packaging technology has a long way to go.

## 1.7 Outline of the Book

The following chapters will be divided into three parts. In Part I, we cover briefly the history of bonding technology in Chapter 2, starting from wire-bonding, tab-automated bonding (TAB), flip chip C-4 solder joint bonding, micro-bump bonding, Cu-to-Cu direct bonding, and hybrid bonding. In Chapter 3, we shall cover the structure, properties, and applications of randomly oriented and (111) unidirectionally oriented nano-twin Cu. Then, Chapters 4 and 5 will be dedicated to chemical reactions and kinetic processes in solder joint formation. Chapter 4 will review solid–liquid interfacial diffusion (SLID) reactions between liquid solder and Cu. Chapter 5 will review solid–solid reactions between solid solder and Cu upon annealing. The kinetics of growth of IMC, which is a stoichiometric compound without composition gradient, has been an outstanding problem in the kinetic analysis of layered interfacial reactions. We introduce Wagner's diffusivity to overcome it.

Part II consists of three chapters related to electric circuitry in electronic packaging. The emphasis is about the design of low-power devices and high intelligent integration. The technical issues of the need of faster rate and larger amount of data transport are discussed. How to increase the I/O density and the bandwidth in packaging technology are explained.

Part III is a collection of chapters on reliability science. It begins with a chapter on irreversible processes of atomic flow, heat flow, and charge flow in interconnects. The topics of electromigration, thermomigration, stress migration, and failure analysis will be covered. The topic of men-time-to-failure (MTTF) will be reviewed on the basis of entropy production.

In the last chapter, Chapter 14, a discussion on how to use AI to accelerate the solving of reliability problems will be explored. We propose an X-ray-based graphic processing unit (X-GPU) to analyze reliability failure distribution in any newly developed 3D IC device for mass production. The goal is to change the time-dependent and time-consuming reliability tests to time-independent tests.

## References

1 Chen, K.-N. and Tu, K.N. (2015). Materials challenges in three-dimensional integrated circuits. *MRS Bulletin* 40: 219–222.

2 Iyer, S. (2015). Three-dimensional integration: an industry perspective. *MRS Bulletin* 40: 225–232.

3 Chen, C., Yu, D., and Chen, K.-N. (2015). Vertical interconnects of microbumps in 3D integration. *MRS Bulletin* 40: 257–263.

4 Prigogine, I. (1967). *Introduction to Thermodynamics of Irreversible Processes*, 3e. New York: Wiley-Interscience.

5 Tu, K.N. (2011). Chapter 10 on "Irreversible processes in interconnect and packaging technology". In: *Electronic Thin-Film Reliability* (ed. K.N. Tu). Cambridge, UK: Cambridge University Press.

6 Tu, K.N., Liu, Y., and Li, M. (2017). Effect of Joule heating and current crowding on electromigration in mobile technology. *Applied Physics Reviews* 4: 011101.

7 Tu, K.N. and Gusak, A.M. (2019). A unified model of mean-time-to-failure for electromigration, thermomigration, and stress-migration based on entropy production. *Journal of Applied Physics* 126: 075109.

# Part I

# 2

# Cu-to-Cu and Other Bonding Technologies in Electronic Packaging

## 2.1 Introduction

Manufacturing of semiconductor devices can be divided into three portions: the front-end-of-line (FEOL), the back-end-of-line (BEOL), and the packaging. The FEOL deals with manufacturing of transistors, capacitors, and resistors. The BEOL refers to the fabrication processes after the deposition of the first metal layer, and there are more than 10 metal layers deposited by the dual-damascene technique for Cu interconnects in high-end devices. For the present 7 nm node, the first metal layer is Cu. The purpose for the BEOL is to connect the individual transistors, capacitors, and resistors. The final stage in semiconductor device packaging is to connect a chip or multiple chips to metal lead-frames by wire bonding, to polyimide tapes by tape automatic bonding, to organic/ceramic substrates by solder joints, or to Si interposor or another chip by Cu-to-Cu bonds in the state-of-the-art 3D integration of circuits (IC). Figure 2.1 shows the schematic drawing for the first-level packaging on lead-frame. The lead-frames or the substrates with the chips are jointed to a print circuit board (PCB), which is called second-level packaging. And then the PCB with chips are jointed to a mother board (third level packaging) if necessary in order to complete the packaging process. The packaging in movable hand-held consumer products is about the same. The purposes for packaging are to provide signal passage, power distribution, heat dissipation, and physical chip protection and prevention of corrosion.

Generally speaking, first-level packaging can be classified into five generations, including wire bonding, tape-automated bonding, flip-chip technology (C4), micro bumps, and Cu-to-Cu bonding.

*Electronic Packaging Science and Technology*, First Edition. King-Ning Tu, Chih Chen, and Hung-Ming Chen.

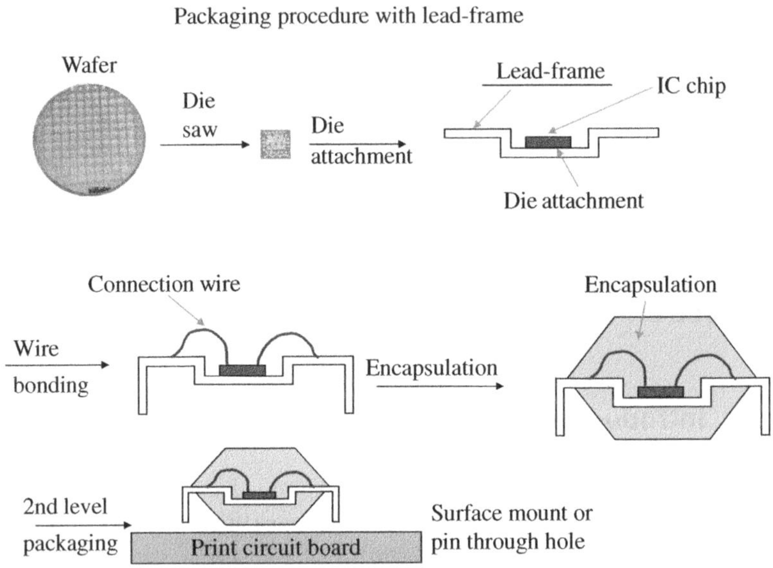


**Figure 2.1** General process of packaging procedure with lead-frame.

## 2.2 Wire Bonding

After the separation of chips or dies from a wafer, they are attached to the central lead-frame plates. A lead-frame is an alloy frame consisting of paddles and packaging leads, as shown in Figure 2.2. A silicon die is attached on the paddle, and the electrical leads are connected to the die with wire bonds.

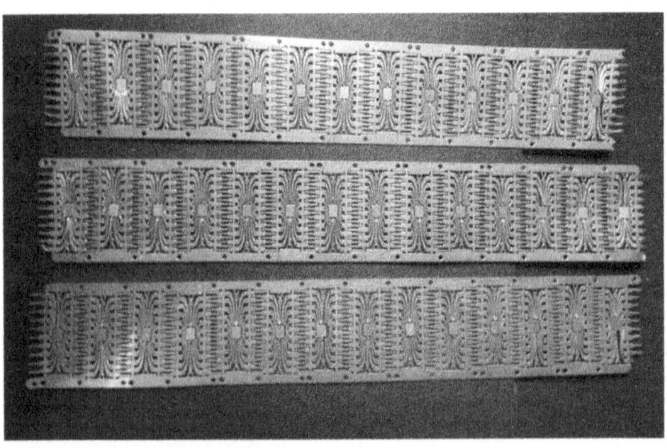

**Figure 2.2** Photo for lead-frames.

The lead-frame serves to provide reliable bonding and good heat conduction. A usual process consists of position reading, chip picking, alignment, die placement, and heat or mechanical treatment to form bonding. Epoxy-based polymer adhesive is added to provide metal particles to form a thermally and electrically conductive path between the chip and the lead-frame. The epoxy must be capable in taking thermal strains during temperature changes and curing temperatures of 60–350 °C. The packaging procedure for lead-frame is below: after wafer manufacturing, the separated die is attached to a lead-frame, which is called die attachment, and wire bonding connects the central die to the lead-frame and is encapsulated for second-level packaging, which will be discussed below. Figure 2.1 presents the schematic drawing for the structure and process of a wire-bonded chip on a lead-frame.

Wire bonding uses methods such as ultrasonic bonding, thermo-compression bonding, and thermos-sonic bonding to attach thin metal wires from pads on top of the lead-frames or printed circuit boards (PCBs) to bond-pads on the IC chips.

Ultrasonic bonding utilizes wedges to guide metal wires for secure fastening onto the bond-pads, which is the first-level bonding. Then the wire needs to be bonded to the substrate, which is the second-level bonding. A 20–60 kHz frequency and 20–200 μm amplitude ultrasonic wave is applied to achieve a cold welding effect to complete the bonding. The application of ultrasonic waves not only removes oxides and contaminants from the pad surface but also causes an acoustic weakening effect to drive a dynamic recovery and recrystallization on the bonding interface to complete the bonding process. Ultrasonic bonding has the advantages such as lower bonding temperature, smaller bonding dimensions, and a lower wire winding profile, which is ideal for bond-pads on small chips.

During a thermal-compression bonding process, the wire passes through a high thermal resistivity bonding capillary, then the bonding equipment guides the wire to the first bonding point where thermal compression forms a ball bond. After that, the wire is then guided to the second bonding point on the substrate, where the release of the wire forms a crescent bond, as illustrated in Figure 2.3. The whole process is operated at 300–400 °C for the first bonding point and 150–250 °C for the second bonding point.

As for the thermosonic bonding technology, it combines both the ultrasonic and thermal compression methods to effectively control the bonding process temperature at 150–250 °C while being able to avoid the formation of thick intermetallic compounds and the deterioration of circuit boards at high temperatures.

Materials used in wire bonding greatly depends on the mechanical strength or on the break strength and elongation required in the bonding

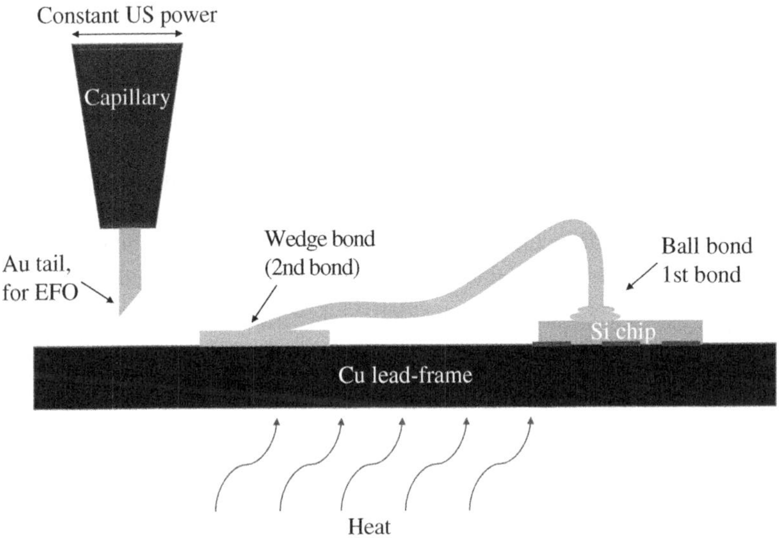


**Figure 2.3** First and second bonds in ultrasonic wire bonding.

process. Another consideration is cost. A common material used is aluminum wires with 1% silicon added to improve mechanical properties, or aluminum wires with 0.5–1% of magnesium to increase fatigue failure and avoid massive intermetallic compound formation. In the binary phase diagram of Au-Al system, there are five intermetallic compounds, which may form in the reaction between Au and Al, including $AuAl_2$, AuAl, $Au_2Al$, $Au_5Al_2$, and $Au_4Al$. Generally speaking, the intermetallic compounds are brittle in nature. Therefore, formation of too many intermetallic compounds would be harmful for the reliability issues of the joints.

Gold wires have the advantage of its antioxidation property, but the softness increases its difficulty for processing. Small amounts of impurities of beryllium, calcium, and copper added to toughen the gold for finer diameters. Copper, silver, and palladium wires are also used for its relative greater hardness than aluminum and gold. Due to the high cost of gold, Cu wires have been developed to replace the Au wires in certain devices.

The current trend in the industry, driven by the need for smaller bond pitches and greater number of interconnect count, gold wires have been slowly replaced by copper wires. Aluminum pads are still the most widely and commonly used, while other materials such as gold, silver, nickel, and copper are also used. All have certain downsides, such as aluminum pads have a gold passivation layer, an adhesion metal layer is needed for gold pads, silver oxidizes at 400 °C and thus cannot be used in high temperature processes, and nickel requires passive gas ambient environments to avoid oxidation. Cu has oxidation issues too, and Pd has been added to prevent Cu from oxidation.

The wire bonds can be fabricated only in the periphery of a Si chip, where no transistors or active devices can be located there. This is because the ultrasonic may damage the transistors or other devices. Therefore, the I/O (input/output) number per chip is limited. We will discuss this later.

## 2.3 Tape-Automated Bonding

Tape-automated bonding (TAB) is a process invented by Frances Hugle at General Electric. TAB places individual packaging circuit units one by one onto a strip-like flexible circuit board, see Figure 2.4. The process is completed by mounting a die onto a flexible packaging circuit unit, similar to a movie tape roll, designed for smoother and convenient assembly process. Holes are punched on the tape where dies will be positioned. The bonding sites are made of bumps of gold or silver, allowing direct connection of the die to the circuits during the mounting process.

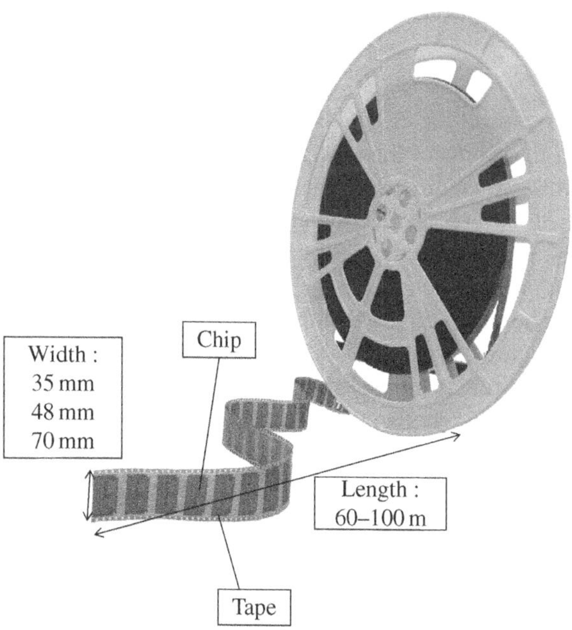


**Figure 2.4** General layout of tape circuit boards. Sprocket holes guide the tape for positioning, which can increase the die attachment rate.

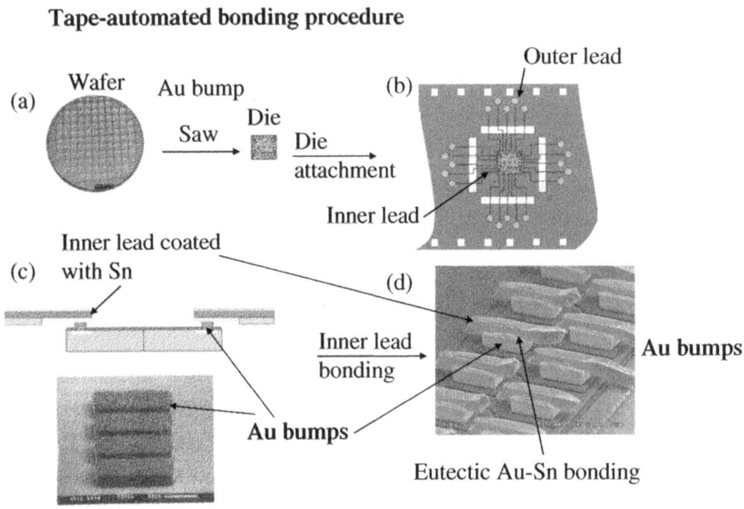


**Figure 2.5** Tape-automated bonding procedure. Inner lead bonds connect die to tape, and outer lead bonds connect tape to external substrates. (a) Electroplating Au bumps on a Si wafer. The wafer was cut into small dies afterwards. (b) A Si die is placed under the TAB tape and aligned with the inner leads. (c) Schematic cross-sectional view showing the Au bumps are aligned with the Cu leads coated with Sn. The bottom figure presents the image Au bumps obtained by scanning electron microscopy.

The TAB bonds that connect the die and tape are called inner lead bonds (ILB), and the bonds that connect the tape to external circuits are called outer lead bonds (OLB), see Figure 2.5 for the schematic drawing of procedure and structures. Au bumps are electroplated onto the Si wafers. After dicing of the Si wafers, Si dies were aligned with the tape, and then thermo-compression bonding or gang bonding is usually adopted for forming a bond between the gold bump on the die and the lead of a TAB circuit. Gang bonding, as shown in Figure 2.6, employs a specially designed bonding tool to apply force and temperature to all the leads and bumps at the same time. Without the use of ultrasonic energy, this type of bonding can be called thermo-compression bonding. Gang bonding offers a high-throughput rate.

After gang bonding, the bonded area along with part of the tape is covered with an encapsulant of epoxy or plastic, which provides protection to the circuit after curing. The die is then electrically tested, after which the useable part of the tape is punched from the tape for assembly into the final application.

TAB has the advantages of high automation due to its gang bonding process. This gives a lot of flexibility to the chip design, where the I/O density can be as high as 300 per chip, and the chip can be facing either up or down.

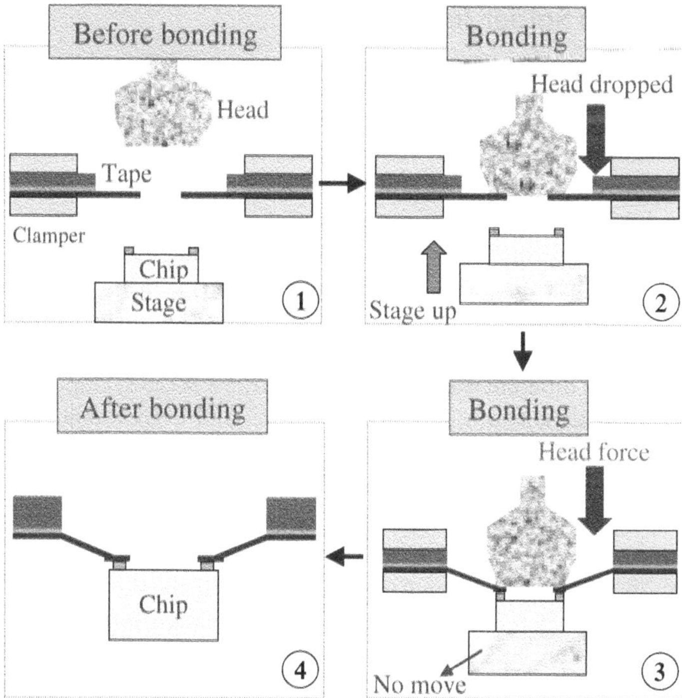


**Figure 2.6** Schematic diagram of gang bonding. A press applies force to multiple bumps at once, forming multiple bonds.

The low weight, volume, and flexible tape are all strong points. Therefore, it has been adopted for packaging of liquid crystal display (LCD) driver chips. However, the downside of the TAB processing limits the widespread and usage, because the process is not flexible due to the required equipment and testing tools. The development is relatively expensive, and it is difficult to find experts in this field.

Eutectic Au-Sn reactions serve as the bonding mechanism for the Si chip jointed to the TAB inner leads. The Cu inner leads are coated with a thin Sn layer by immersion. During the thermal compression process, interdiffusion of Au and Sn takes place, and eutectic melting occurs when the local composition reaches eutectic composition of $Au_{80}Sn_{20}$ (wt. %).

Similar to wire bonds, the TAB bonds can be fabricated only in the periphery of a Si chip. This is because the inner leads are free standing before bonding, they may entangle, and thus the circuit could become shorted if area array of leads is present. Therefore, the I/O (input/output) number per chip is also limited.

## 2.4 Flip-Chip Solder Joint Bonding

Flip-chip solder joint technology was first commercially introduced by International Business Machines Corporation (IBM) in the 1960s for individual transistors and diodes. [1–3] The controlled-collapse-chip-connection (C4) solder joints were designed to utilize solder bumps to attach chips to ceramic modules with a higher number of I/O connections. Since then it has become a common application in packaging technology and has seen many variations to be compatible with different packaging assemblies.

The typical manufacturing processes include fabrication of redistribution layers, under-bump-metallization (UBM), and solder bumps. Solders can be deposited in different ways on the entire surface of the Si chip, such as electroplating, stencil printing, pick-and-place, and evaporation. The chips are then flipped, hence the term "flip" chip and brought down for the solder joints to come into connection with the connectors on the substrate, see Figure 2.7. The solder is then remelted or reflowed to form electrical connections.

In Figure 2.8, a schematic diagram of the cross-section of a ceramic module joined to a large PCB for mainframe computers is shown. It is a two-level packaging scheme because there are two levels of solder joints, where the C4 solder joints are used to join the Si chip to ceramic substrate, which is then joined to PCB by ball-grid-array (BGA) solder balls. In manufacturing, the second-level solder joints (BGA) should have a lower melting point than the first-level solder joints (C4), which is to prevent the melting of the latter during the processing of the former.

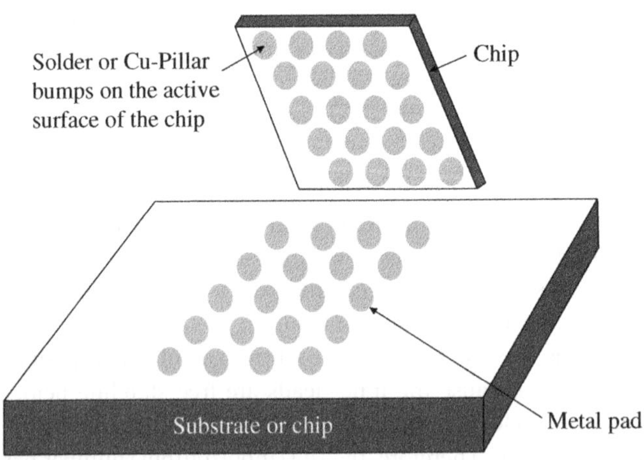


**Figure 2.7** Schematic diagram of flipping chip to form bonds at multiple solder joints.

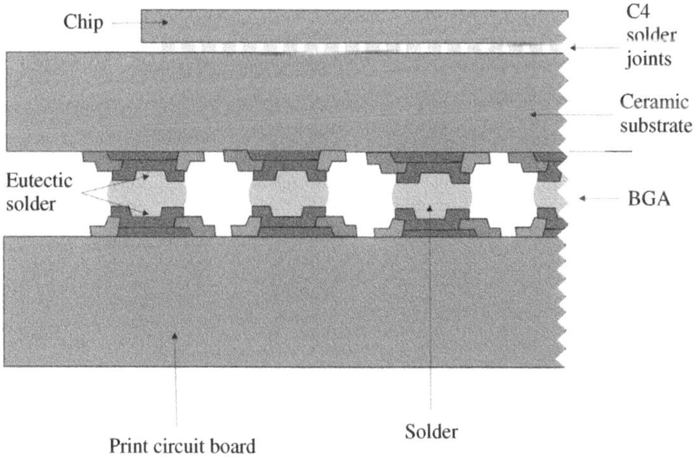


**Figure 2.8** Schematic diagram of the cross-section of a ceramic module joined to a large printed circuit board (PCB) for mainframe computers.

In Figure 2.9, a schematic diagram of the cross-section of a C4 flip-chip solder joint and its UBM is shown. On the multilayered Al interconnects, there is the tri-layer of Cr/Cu/Au, which serves as UBM, which allows the joining of the Al interconnect and the high-Pb Pb97Sn3 (wt%) solder joint. The Cr film of serves as adhesion layer; the Cu film will react with the solder to form interconnect; and the Au film protects the Cu layer from oxidation. It is a two-level packaging scheme because there are two levels of solder joints, where the C4 solder joints are used to join the Si chip to ceramic substrate, which is then joined to PCB by BGA solder balls. In manufacturing, the second-level solder joints (BGA) should have a lower melting point (183 °C for eutectic SnPb) than the first-level solder joints (C4), which is to prevent the melting of the latter during the processing of the former.

In Figure 2.10, the processing steps of C4 are depicted. First, the passivation opening was fabricated and ball-limiting metals (BLM) were evaporated, as shown in Figure 2.10a. The high-Pb solder, which has a melting point over 300 °C, is deposited on the UBM, see Figure 2.10b. The first reflow (around 350 °C) forms a ball-shaped bump on the UBM, as shown in Figure 2.10c. Because the surface of $SiO_2$ cannot be wetted by molten solder, the base of the molten solder bump is defined by the contact opening of the UBM, thus the molten solder bump balls up on the UBM contact. Therefore, the UBM controls the dimensions (height and diameter) of the solder ball when its volume is given by deposition. This is the "controlled -collapse -chip –connection," or "Controlled Collapse Chip Connection" Without the

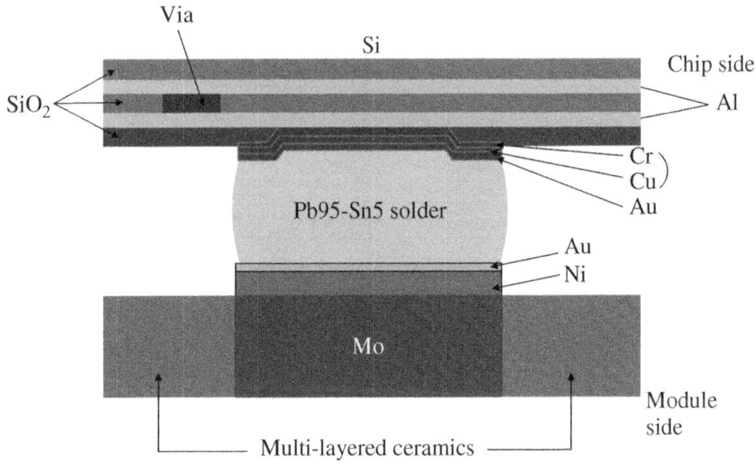


**Figure 2.9** Schematic diagram of the cross-section of a C4 flip-chip solder joint and its Cr/Cu/Au UBM.

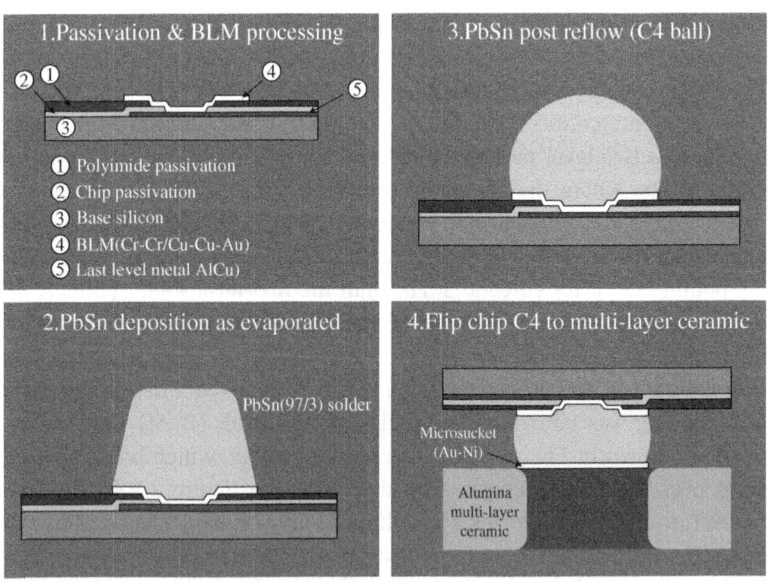


**Figure 2.10** Schematic diagram of the formation of a C-4 solder joint of solder joint on under-bump-metallization. (a) Cross-sectional view for the Si chip and the metallization layers on it. (b) After the evaporation of PbSn solder. (c) After the first reflow. (d) After alignment and the second reflow to bond with a ceramic substrate.

control, the solder ball will spread or collapse due to a low wetting angle, which may lead to linking of neighboring solder joints. The chips are then flipped, hence the term "flip" chip, and brought down for the solder joints to come into connection with the connectors on the circuit board, see Figure 2.10d. The solder is then remelted, or reflowed, to form an electrical connection.

One major advantage of flip-chip solder technology is self-alignment. To join the chip to the ceramic module as shown in Figure 2.11, a second reflow is needed to join the solder bump to the metallization on the ceramic substrate. During the second reflow, the surface energy of the molten solder bump provides self-aligning force to position the chip on the module automatically. Actually, when the solder melts, the chip will drop slightly and move slightly due to the reduction of surface energy of the molten solder ball, but it can achieve the exact alignment between the chip and the module, so it is a very forgiving technology.

As personal computers (PCs) became popular in 1980s, cost and weight became crucial considerations for flip-chip packaging. Therefore, the costly and heavy ceramic substrates were replaced by polymer ones. However, the change caused two problems. The first one was attributed to the low glass transition temperature of the polymer substrate, which was below 250 °C. Thus, the high-Pb solders cannot be used for flip-chip solders. The second one is the high coefficient of thermal expansion (CTE). The linear CTE is $2.6 \times 10^{-6}\ k^{-1}$ and $7.6 \times 10^{-6}\ k^{-1}$ for Si and alumina ($Al_2O_3$), respectively. The thermal stress is low when Si chips bonded to ceramic substrates. However, the CTE for polymer is typically large than $40 \times 10^{-6}\ k^{-1}$. The thermal stress increased significantly when the polymer substrates were used. For the former problem, eutectic solders or composite solders were adopted to reduce the reflow temperatures to 220 °C. IBM invented underfill to strengthen the package to provide the solution for the later problem. In 2003, due to environmental concern and the legislation of Restriction of

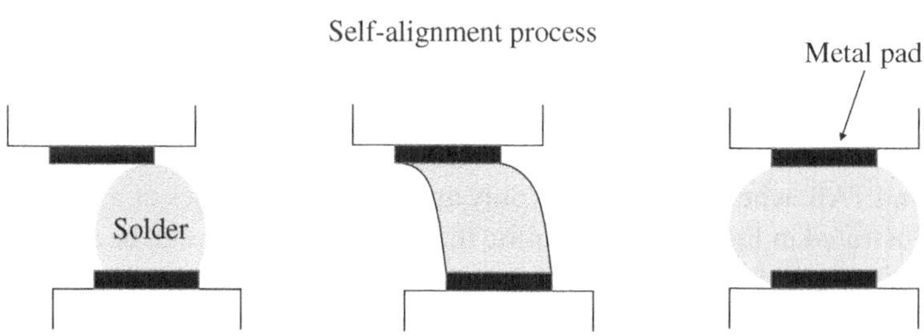


**Figure 2.11** Schematic diagram of automatic self-alignment process in the reflow of a C-4 solder joint.

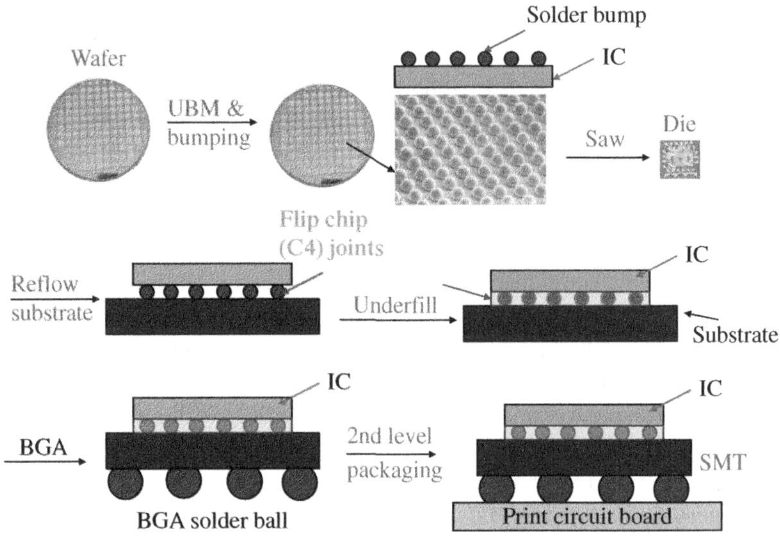


**Figure 2.12** Schematic diagrams of the procedure in flip-chip solder joint technology.

Hazardous Substances (RoHS) in 2003, packaging industry adopted Pb-free solders to replace eutectic SnPb alloys. SnAg solders are the most popular Pb-free solders.

Figure 2.12 shows the typical process and structure of flip-chip technology schematically. Electroplating of SnAg solders is performed on Si wafers after the fabrication processes of BEOL final metal routing, followed by dicing of Si wafers into individual dies. The chip is flipped over and aligned with a polymer substrate, and then reflows at 250–260 °C to form the flip-chip solder joints.

For flip-chip solder joint technology, the major advantage is that the solder joints can be fabricated on the entire area of a Si Chi, as shown in Figure 2.13, where the area array of solder joints is prepared on the surface of the Si chip, meaning that below the solder joints there are transistors. However, the melting and solidification of solder will not introduce a high mechanical stress on the transistors. In comparison to wire bonding and TAB, where the leads are only arranged on the edges of a Si chip, as illustrated in Figure 2.14, meaning that no wire bonding can be applied to the chip area having transistors because of the high vibrational mechanical stress. Thus, on a Si chip of size of 1 × 1 cm, we may have at the most 200 bond-pads on each edge with a pitch of 50 μm, so the number of bonds or the I/O leads is 800. But on the surface of the chip, the number of I/O solder joints can be as many as 100 × 100 = 10,000 with a

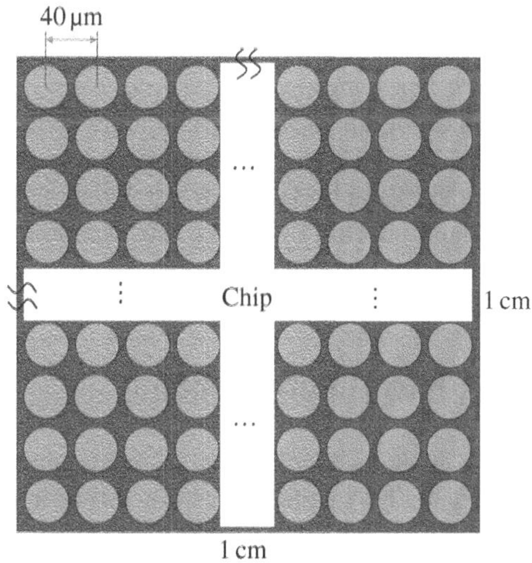


**Figure 2.13** Schematic diagram of the distribution of C-4 solder joint on the entire surface of a Si chip.

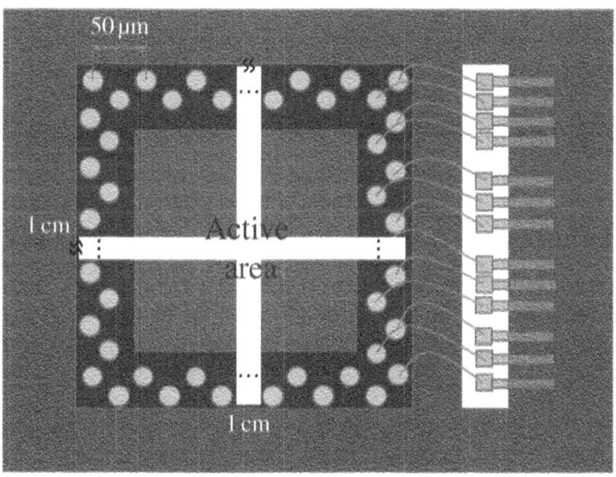


**Figure 2.14** Schematic diagram of the peripheral distribution I/O connections in wire bonding and tape-automated bonding.

pitch of 100 μm. With the advance of micro-electronic technology, the increase of functionality and memory requires more and more I/O. Thus, flip-chip solder joints have become the major packaging technology in the last several decades.

However, in the most advanced mobile cell phones, the pitch of neighboring solder joints becomes smaller and smaller and now it is limited to about 10 μm, for example in the packaging of lens on a high-end product. The limit is because during reflow, the melting of solder may lead to bridging of neighboring joints. For this reason, direct Cu-to-Cu bonding has been developed to replace solder joints, which will be discussed here later.

Early types of flip-chip substrates were manufactured using ceramic. If we make a comparison between Figure 2.8 here and Figure 1.6 in Chapter 1, we can recognize that the ceramic substrate is an early example of interposer. The ceramic served as a stable material that did not react with the UBM and has thin thickness. This complimented the high-Pb solder balls used, due to the higher reliability and high melting of high-Pb solders, which required high reflow temperatures of approximately 350 °C.

## 2.5 Micro-Bump Bonding

As we enter the big data era, 5G communication technology and artificial intelligence (AI) applications using mobile internets and movable devices are ubiquitous. It has affected our society, as well as every aspect of our life. The continuing demand for smaller size, more functionality, lower power consumption, and reduced cost in mobile devices is challenging, because Moore's law of miniaturization is ending. A paradigm change, from 2-dimensional integrated circuits (2D IC) to 3D IC, has occurred in microelectronics industry more than 10 years. Yet, the transition has not been successful due to high cost and low reliability. [4] Currently, the use of 5G technology in AI applications has provided the impetus to promote the 2.5D IC into mass production, instead of 3D IC. In Chapter 1, Figure 1.6 shows a schematic diagram of the cross-section of a 2.5D IC structure. Compared to 2D IC devices, the addition of a Si interposer in 2.5D IC devices requires one more redistribution layer (RDL) and one extra level of solder joints. This is because of the need of vertical interconnects by using through-Si-vias (TSVs) and micro-bumps.

In the past, when we study electromigration in Al or Cu lines, we may focus on stress-migration, but not thermomigration. This is because Soret effect occurs in an alloy. When the reliability of flip-chip C-4 solder joint or micro-bump became an issue, the study of thermomigration began. It was discovered that across a C-4 solder joint of 100 μm in diameter, if there is a temperature difference of 10 °C, the temperature gradient is 1000 °C/cm. This is even more severe in small structures; in a micro-bump of 10 μm in diameter, if there is a temperature difference of only 1 °C across it, the temperature gradient is 1000 °C/cm, and serious thermomigration can occur. In the dense 2.5D IC packaging, heat dissipation is very poor. To enhance heat dissipation, we need to have a high temperature gradient. Unfortunately, the gradient can lead to thermomigration. Furthermore, the use of Si interposer enhances the lateral heat transfer along the interposer, which has been found to have caused an unexpected thermomigration failure.

In 2.5D IC for mobile technology, the total device thickness is limited. In turn, the thickness of Si chips in stacking is thin, especially the interposer, same as the gap between them. Thus, Joule heating and thermal management become serious and affect all the reliability problems under consideration. Electromigration enhanced by Joule heating is the primary concern of failure, especially the early failure. When we include Joule heating into reliability study, we deal with multiple driving forces and kinetic responds. On driving forces, we have chemical potential gradient, stress potential gradient, temperature gradient, and electric field. On kinetic responds, we have heat flux, atomic flux, and electron flux. Furthermore, in atomic flux, we have lattice, grain boundary, and surface diffusion. It becomes harder and harder to perform a critical experiment in order to understand a specific reliability problem in the future microelectronic devices. In Chapter 9, we shall introduce entropy production to analyze these reliability problems.

At this moment, the design of 2.5D IC and the fabrication processes vary greatly among microelectronic companies. The state is fluid. Each company tends to select a particular integration and processing scheme, which is most suitable to their design of a specific device, so that they feel most comfortable based on their past experience in manufacturing, and which may be protected by their patents. For example, consider the processing of TSV, there are already the selections of via-first (before transistor formation), via-middle (after transistor formation but before metal interconnect formation), and via-last (after full device fabrication). Consequently, we do not expect Joule heating and electromigration-induced failure will be the same among all the 2.5D IC devices in production now, as well as in the near future. This is because each product will have its own unique "weak-link," which tends to fail early or to fail outside the statistical distribution.

Figure 2.15a and b show the schematic diagram and cross-sectional SEM image of a micro-bump, respectively. In the near future, the diameter of

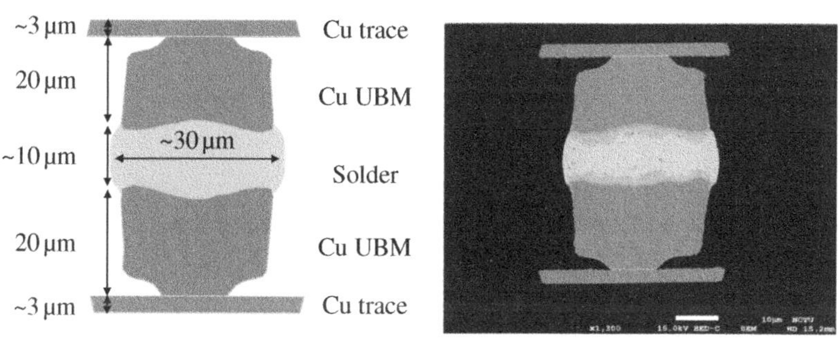


**Figure 2.15** Schematic diagram and SEM image of Cu/solder/Cu micro-bump.

solder joint may even go below 10 μm. However, when the diameter shrinks 10 times from that of C-4 solder joint, meaning from 100 to 10 μm, the volume shrinks 1000 times. The change in microstructure is dramatic. For example, if we assume a grain size of 10 μm, the μ-bump may have just one grain, yet the C-4 joint can have 1000 grains. The latter can be assumed to have isotropic properties, but not the former. This anisotropic issue is especially critical for β-Sn or white Sn because it has a body-centered tetragonal unit cell, where the c-axis is less than both a-axis and b-axis. It has been found that the diffusion of Cu is much faster along the c-axis than those along either a- or b-axis. Actually, it has been found that the reaction between Cu UBM and a Sn grain is indeed much faster if the c-axis is in contact with Cu, which means the interfacial reaction among a large number of micro-bump can have a wide distribution, so some of them may cause early failure in micro-bump bonding.

One of the biggest issues in scaling down solder joints is the side wetting phenomenon. It has been reported that reflow of Cu/Ni/solder/Ni/Cu micro-bumps at 260 °C for 20 minutes caused side wetting to occur. [5, 6] Sn diffused to the circumference of Ni and Cu UBMs to form $Ni_3Sn_4$ and $Cu_3Sn$. Because the amount of solder in a micro-bump is approximately 1% of a flip-chip solder joint, the side wetting may lead to necking or open failure of the micro-bump. Figure 2.16a and b show reflow of Cu/SnAg-solder/Cu micro-bumps at 260 °C for 20 and 40 minutes, respectively, resulted in side wall and porous $Cu_3Sn$ formation.

Another new issue about μ-bump is the interaction between the two reacting interfaces across the bump. Over the diameter of 100 μm in a C-4 solder joint, we seldom find chemical interaction between the two reactive

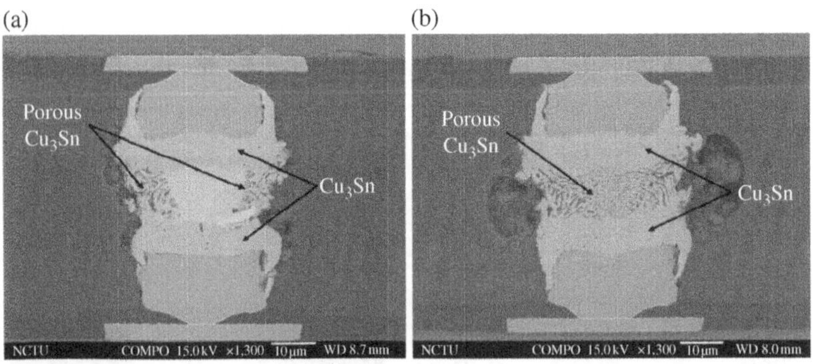


**Figure 2.16** SEM cross-sectional image of side-wall wetting and porous $Cu_3Sn$ formation in micro-bump. A Cu/SnAg-solder/Cu microbump was reflowed at 260 °C for (a) 20 minutes and (b) 40 minutes.

interfaces across a solder joint. But in a μ-bump, this has happened. Furthermore, if we consider a μ-bump of 1 μm in diameter, the volume reduction from that of a C-4 solder joint will be $10^6$ times. The entire μ-bump could possibly transform to intermetallic compound (IMC). Not only the mechanical and electrical properties of the joint will be greatly affected, but also the kinetic rate of IMC formation will change. The rate will also be severely affected by the side-wall wetting effect, which will induce voids and/or necking. This is because of the increase of surface/volume ratio, and we must consider surface diffusion, which will be reviewed in Chapter 5. In addition, the increase of surface/volume ratio has introduced new morphological changes in μ-bump technology: first is the side-wall reaction and second is porous $Cu_3Sn$ formation. Both of them will be covered in detail in Chapter 5.

Finally, Cu-to-Cu direct bonding has been a field of high interest in the semiconductor industry in recent years. It has surfaced as a promising technology to replace the μ-bumps, which have been widely regarded as the core packaging material, yet when it has been scaled down to smaller and smaller pitch sizes, the side wetting phenomenon and the porous $Cu_3Sn$ formation occurred. It has greatly limited its applications to a minimum pitch of 10 μm in order to ensure bonding reliability. The topic of Cu-to-Cu direct bonding and hybrid bonding will be covered below.

## 2.6 Cu-to-Cu Direct Bonding

Copper-to-copper direct bonding has been a field of high interest in the semiconductor industry in recent years due to the need of ultrahigh I/O and wider bandwidth for high performance computing and communication devices. [7–16] Figure 2.17 is a plot of number of I/O per chip for all types of bonding technology discussed so far. It has surfaced as a promising technology to replace the micro-bumps widely used as the core packaging material. The solder joint interconnects face process and reliability problems as electronic components shrink and scale down to smaller pitch sizes. There have also been many studies on the formation of IMCs after electromigration tests, and the IMCs are brittle for interconnects. [4–6]. Cracks were observed to propagate across a micro-bump, in which the entire solder had been transformed into IMCs after electromigration test under $4.6 \times 10^4$ A/cm$^2$ at 150 °C for 48 hours. These issues greatly limit the scaling down of the micro-bumps. Even though micro-bumps could be scaled down below 10 μm pitch, the high resistivity of Sn-based solders (11.5 μΩ-cm), $Cu_6Sn_5$ (17.5 μΩ-cm), and $Ni_3Sn_4$ (28.5 μΩ-cm) IMCs will cause RC delay of high-performance computing and communication devices.

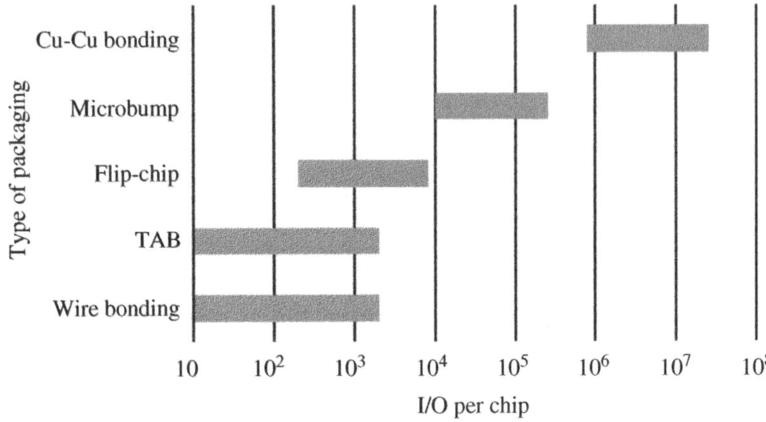


**Figure 2.17** A plot of the I/O number per chip vs. various types of bonding technology.

On the other hand, Cu possess excellent electrical conductivity (1.7 μΩ-cm), as well as thermal properties, and it is a promising material to replace solder micro-bumps in packaging technology. In comparison, Cu micro-bumps have better scalability, electrical, and thermal properties than solder micro-bumps. It has higher bandwidth and lower power consumption, so it could solve the side wetting issue in ultra-high-pitch interconnects. Cu-to-Cu direct bonding has been demonstrated to be feasible and to be scaled down to submicron size.

Two pieces of Cu can be bonded at high temperatures. However, it is challenging to bond two Cu films at low temperatures (<250 °C) because the melting point of Cu is 1085 °C, and so the lattice diffusivity of Cu at 250 °C is quite low. In addition, Cu is easily oxidized. In 1997, Prof. Suga adopted surface activated bonding to achieve Cu-to-Cu bonding. In 1999, Prof. Reif's group reported Cu-to-Cu bonding can be achieved at 450 °C for 30 minutes. Transmission electron microscope (TEM) images show grain growth during annealing and bonding, with a jagged interface representing the interdiffusion between the two Cu layers. TEM studies were performed on the formation and evolution of voids in the interface of the bonded Cu layers. It is revealed that large voids are present in the bonded layers at 300 and 400 °C.

### 2.6.1 Critical Factors for Cu-to-Cu Bonding

The procedures and results of Cu-to-Cu bonding depend greatly on the sample preparation and bonding process. Some of the deciding factors are surface roughness, surface contamination or activation, crystallographic orientation of surface grains (surface diffusivity), bonding temperature, bonding pressure, passivation surface, structure design, and grain size. The requirements of the bonding process are highly decided by the transistor components, with the ultimate goal to achieve the best bonding results while causing the least damage to the chip. For example, bonding pressure should be lower than 1 MPa for low K dielectrics, and bonding temperature should be below 220 °C for memory chips. To improving the bonding quality, it is possible to lower surface roughness, via surface planarization or polishing, or adopt surfaces with rapid surface diffusivity (such as the <111> planes for FCC) to improve the bonding quality.

Surface roughness plays a key role on the bonding temperature [17]. With low surface roughness, the Cu-to-Cu bonds possess small initial interfacial voids, and bonding can be achieved at a low temperature or short time. Figure 2.18 shows the comparisons of copper wafers bonded under different surface roughness [4], where Cu films deposited by different methods were bonded and its minimum bonding temperatures were plotted along with non-polished samples. The Cu films were planarized by chemical-mechanical-polishing (CMP), which was able to reach a surface roughness of 1.05–1.40 nm, and the minimum bonding temperature could reach as low as 175 °C for electrodeposited samples and room temperature for physical vapor deposition (PVD) samples. In contrast, samples without surface

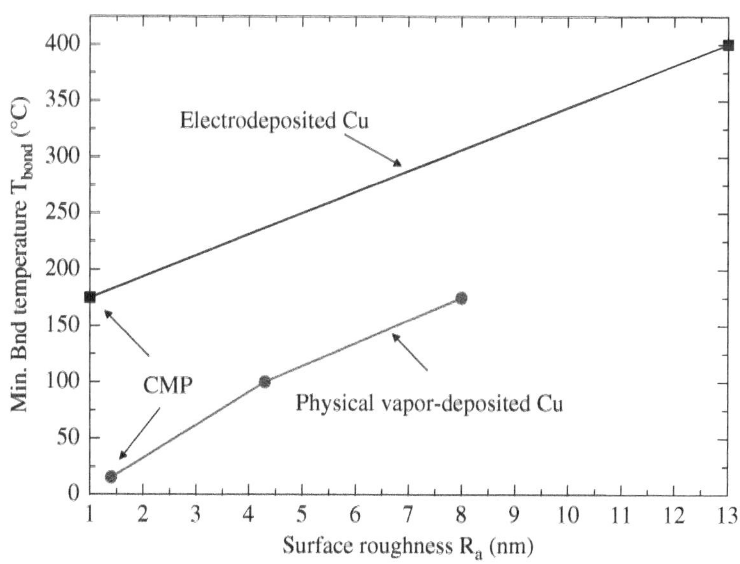


**Figure 2.18** Effect of surface roughness on minimum bonding temperature.

roughness polishing required at least 400 °C for electrodeposited samples and 150 °C for PVD samples. This shows the importance of low surface roughness for lowering the required bonding temperature.

Another method to lower surface roughness is by sputtering a passivation or capping layer on the Cu surface. This method could be implemented not only to lower surface roughness but to serve as an antioxidation barrier. Common capping materials are Au, Ti, and Pd. Au layer deposited by direct immersion lowered the surface roughness from 51 to 38 nm, and successful bonding was achieved at 350 °C for 15 minutes. Passivated titanium and palladium on Cu surfaces also aided bonding processes, due to lower activation energy at the surface. Cu had a tendency to diffuse toward surface, while the Ti/Pd diffused toward the silicon substrate. This allowed Cu to be bonded at 180 and 150 °C for capping of Ti and Pd, respectively [12, 18]. The introduction of passivated layers can greatly reduce the bonding temperature, becoming a viable method to apply in procedures where oxidation should be avoided.

Another method adopted to planarize the surface is by using a "fly-cut" method [13]. After fly cutting and accurately controlling Cu bump heights, bonding can be achieved. This method was demonstrated with a 40-μm bump pitch array, forming bump interconnects with no interface layer. The application of fly-cut to lower surface roughness provided a method to achieve bump consistency, which also shows potential to lower the bonding temperature and time to 300 °C/15 minutes.

To avoid surface contamination and improve bonding quality, surface activation via ion beam or plasma treatment serves as a method to clean the surface of absorbates. Surface activation was applied before Cu direct bonding, with successful results in film-to-film, bump-to-bump, and bump-to-film bonding experiments at 200 °C. Formic acid treatment to clean contaminants and create a nano-adhesion layer has also been introduced to aid bonding procedures. Surface roughness is controlled to 3 nm after CMP processes and formic acid treatment, and samples were successfully bonded at 200 °C for 10 minutes. The application of surface treatment shows potential and importance in lowering bonding temperature and time while improving overall bonding quality [19].

Bonding temperature plays a critical role on the quality of the Cu–Cu joints. During the thermal compression bonding, diffusion of Cu atoms occurs to fill the interfacial voids, and the diffusivity $D$ of Cu depends as follows:

$$D = D_0 exp\left(\frac{-\mathrm{E}_a}{\mathrm{kT}}\right) \tag{2.1}$$

where $D_0$ is prefactor, $E_a$ is activation for diffusion, $k$ is Boltzmann constant, and $T$ is absolute temperature. As the temperature increases, the diffusivity rises exponentially and the bonding time can be shortened significantly.

Crystallographic orientation of surface grains in Cu also has a profound influence on the bonding process. The crystal structure of Cu is face-centered cubic (FCC), and its (111) plane is the closest packing plane. Therefore, Cu atoms diffuse much faster on (111) plane than the other planes. Table 2.1 lists the calculated surface diffusivities on Cu (111), (100), and (110) planes at various temperatures using the parameters in the literature. The diffusivity on Cu (111) plane at 300 °C is $1.5 \times 10^{-5}$ cm$^2$/s, which is approximately three orders of magnitude higher than that on (100) plane. Therefore, the bonding temperature can be lowered from 300 to 150 °C when two bonding Cu surfaces are both (111)-oriented. We will introduce (111) unidirectionally oriented nanotwinned Cu in Chapter 3.

Grain size near the Cu surface also affects the bonding process. Several researchers reported that Cu films with small grain sizes with strains can be bonded at 150 °C because the fine grains can easily recrystallize at low temperatures. Furthermore, the bonding interface of the Cu-to-Cu joints disappeared at 200 °C for 30 minutes.

### 2.6.2 Analysis of Cu-to-Cu Bonding Mechanism

Because the Cu films/bumps remain in solid state during the bonding processes, the Cu-to-Cu bonding is classified as solid-state diffusion bonding [20–23]. Several researchers have performed theoretical modeling to investigate the effect of surface roughness and void shapes on the bonding time and fraction of contact area. Made et al. proposed an analytic model to

**Table 2.1** Cu surface diffusivity (cm$^2$/sec).

| | (111) | (100) | (110) |
|---|---|---|---|
| 300 °C | **1.51*10−5** | 1.48*10$^{-8}$ | 1.55*10$^{-9}$ |
| 250 °C | **1.22*10−5** | 4.71*10$^{-9}$ | 3.56*10$^{-10}$ |
| 200 °C | **9.42*10−6** | 1.19*10$^{-9}$ | 5.98*10$^{-11}$ |
| 180 °C | **8.36*10−6** | 6.27*10$^{-10}$ | 2.63*10$^{-11}$ |
| 150 °C | **6.85*10−6** | 2.15*10$^{-10}$ | 6.61*10$^{-12}$ |

*Surface Science* 515, 21–35 (2002).

fit experimental data quite well. They can predict the bonding strength as a function of bonding time at various temperatures [21].

A surface creep model has been proposed as the mechanism that takes place during the Cu-to-Cu bonding process. [8] Figure 2.19a shows a schematic drawing of the initial bonding interface with multiple contact points, and Figure 2.19b illustrates an drawing of the parameters at one of the bonding interfaces [39]. Considering that the surface of two copper films is rough and uneven in nanometer scale, the contacted area would be under a high compressive stress during a thermal compression bonding stage. The non-contacted areas might be in very low stress to form a stress gradient, which drives Cu atoms from the high-stress areas to the low-stress areas to "fill" in the gaps gradually. Because the bonding pressure usually does not exceed the yield stress of Cu, which is around 100 MPa, and the Cu films are under static stress at an elevated temperature, Cu atoms diffuse on the surfaces of Cu films. Therefore, the bonding mechanism belongs to surface creep. This surface creep phenomenon is a core theory to explain the increase in contact area and the elimination of voids during bonding.

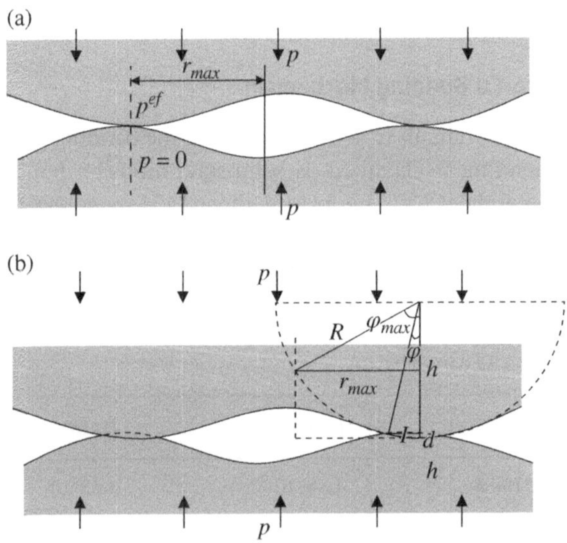


**Figure 2.19** Schematic diagram of the Cu-to-Cu interface. (a) The stress gradient between the contact region and the non-contact region is depicted. (b) Under thermal compression, the contact region becomes bigger by squeezing out atoms under compression. The parameters in the bonding model are indicated on the upper right corner.

Copper direct bonding can be divided roughly into four stages. The initial stage is mechanical contact of two Cu surfaces and a certain amount of plastic deformation may occur in small contact areas. The second stage is to form grain boundaries and isolated voids at the contact interface. The third stage is void ripening at the bonding interface. The fourth stage is to eliminate the bonding interface by grain growth across the interface. A model has been proposed to simulate the effect of bonding temperature, pressure, and surface roughness on the bonding time at various temperatures.

The main driving forces of irreversible bonding process are the lowering of Gibbs free energy of the system under fixed temperature and fixed external pressure and can be separated into several parts below.

1) After the first touching between the two "wavy" (not ideally even) surfaces, the residual free space between the two surfaces becomes topologically complicated empty (void) space in the system. The volume of the system can be shown as: $V_{total} = V_1 + V_2 + V$, where $V_{total}$ is the overall volume of the system, $V_1$ is the Cu volume at the top, $V_2$ is the Cu volume at the bottom, and $V$ is the empty intermediate space between $V_1$ and $V_2$. Thus, any decrease of this empty space during bonding corresponds to a decrease of the product $pV_{total}$ and therefore represents the first driving force of bonding. Here $p$ is the external pressure.
2) After the touching in the contact point, the atoms diffuse via surface diffusion to proper sites and form a grain boundary to replace the two free surfaces, for example a twist-type grain boundary forms between two (111) free surfaces in contact with a relative rotation. Usually, $\gamma_{GB}$ is smaller than $2\gamma_{surf}$, where $\gamma_{GB}$ is grain boundary energy and $\gamma_{surf}$ is surface energy. Thus, the change of interfacial energy during bonding $(\gamma_{GB} - 2\gamma_{surf}) \cdot \Delta A_{contact}$ is negative and favorable, where $\Delta A_{contact}$ is positive, meaning an increase of the contact interface, and $\Delta A_{contact}$ is the variation of contact area. Thus, $(\gamma_{GB} - 2\gamma_{surf}) \cdot \Delta A_{contact}$ represents the second driving force, which is often called capillary force.
3) In some cases, the newly formed contact interface may contain many defects (such as nanotwin boundaries, or stresses, or dislocation network). So, further recrystallization in the contact zone may represent the third driving force.

Here, only the first and second major driving forces of the irreversible direct bonding process will be discussed with simple geometric configurations, as shown in Figure 2.18b.

In pure (single element) system, the main driving forces for atomic flux are the stress gradient and Gibbs Thomson capillary effect. At the first stage, stress gradient is the dominated factor of Cu atomic diffusion, which starts from increasing the contact area by surface diffusion, and then after forming twist-type grain boundaries by grain boundary diffusion. Certainly, there is an ambiguous demarcation of surface and grain boundary diffusion, so there will be an effective driving force between them. In this study, bonding process at the first stage will be explained by a pressure-induced bonding model, see Figure 2.19b.

For simplicity, we assume periodic surface roughness, as depicted in Figure 2.19a, where two curved surfaces of a large curvature radius "$R$" are forming an interface of radius "$r$," and $r$ is much smaller than $R$. Bonding process stops (complete bonding) when $r$ reaches $r_{max}$. In this case, $r_{max}$ is the radius of a single contact. Critically, we note that bonding proceeds only due to the squeezing of atoms from the compressed contact interface of radius "$r$" to the free surfaces outside the contact interface. In Figure 2.19b, the distance $h$ to the curvature centers decreases and the contact radius $r$ increases. The (h – dh) is a change of distance from the contact spot to the curvature center. It means that the volume $2dh \cdot \pi r^2$ is being squeezed out from the contact interface via surface and grain boundary diffusion driven by the gradient of stress. The stress gradient is assumed to be zero outside the contact area at the free surface, but in the center of the spot, it is $p^{ef} = p\dfrac{\pi r_{max}^2}{\pi r^2}$, where $p^{ef}$ is the effective pressure at the contact area. Therefore, the conservation of mass gives:

$$-\pi r^2 \cdot 2dh \approx \delta \cdot 2\pi r \cdot \frac{D}{kT}\frac{\left(\frac{p r_{max}^2}{r^2}\Omega - 0\right)}{r}dt \tag{2.2}$$

where $\delta$ is the effective thickness of the atomic diffusion path, $D$ is the effective diffusivity of combining surface or grain boundary diffusion, $k$ is the Boltzmann constant, $T$ is the bonding temperature and $\Omega$ is Cu atomic volume, $\dfrac{D}{kT}$ is a mobility, $\dfrac{p r_{max}^2}{r^2}\Omega - 0$ is the energy difference between the atoms at the already formed contact and the atoms at the still free internal surfaces, $\dfrac{\left(\frac{p r_{max}^2}{r^2}\Omega - 0\right)}{r}$ is an average force of squeezing per atom, $\dfrac{D}{kT}\dfrac{\left(\frac{p r_{max}^2}{r^2}\Omega - 0\right)}{r}s$ is a rough estimate of squeezing outflux.

In Figure 2.17b, by taking $r = R\sin\varphi \approx R\varphi, h = R\cos\varphi \approx R\cdot\left(1-\frac{\varphi^2}{2}\right)$, $dh = -R\varphi d\varphi$, we obtain:

$$-\frac{r^4}{r_{max}^2}\cdot dh \approx \frac{\delta D}{kT}p\Omega dt$$

$$\frac{R^3}{\varphi_{\max}^2}\varphi^5 d\varphi \approx \frac{\delta D}{kT}p\Omega dt. \tag{2.3}$$

Integration of Eq. (2.3) from the start to the full bonding gives:

$$\frac{R^3}{\varphi_{max}^2}\frac{\varphi_{max}^6}{6} \approx \varphi_{max}\frac{r_{\max}^3}{6} \approx \frac{\delta D}{kT}p\Omega t_{bonding}.$$

Thus, the time of full bonding ($t_{bonding}$) starting from the single touching point due to pressure-induced atomic migration along contact interface is

$$t_{bonding} \approx \varphi_{max}\frac{r_{max}^3}{6\delta D}\frac{kT}{p\Omega} \tag{2.4}$$

In the above model, we assume that the process rate is controlled by grain-boundary diffusion along the radius of contact spot, because grain boundary diffusivity is much smaller than surface diffusivity along the unbonded free surfaces. At T = 300 ° C, the product $\delta D_{GB}$ for Cu should be:

$$\delta\cdot D_{GB}^{*} = 9.5\cdot10^{-15}\frac{m^3}{s}exp\left(-\frac{8.9RT_{melt}}{RT}\right) = 6.8\cdot10^{-24}\frac{m^3}{s}.$$

Figure 2.20 shows the surface morphology of a Cu surface analyzed by atomic force microscopy (AFM). The profile showed the wavy surface with a half wave length about 150 nm, and the hills and valleys have different amplitudes. Yet, we expect that very few nano-size hills will be under immense effective pressure and will be flattened until the effective pressure over all contact spots becomes less than the yield strength. To use Eq. (2.4), the wavy bonding surface can be cut into small units with single hill like Figure 2.18b. The $r_{max}$ can be defined by the half wavelength of local roughness. However, $r_{max}$ will not be equal to the half wavelength of local roughness under the real bonding condition, so the depths of valleys are different; and a certain number of voids must remain

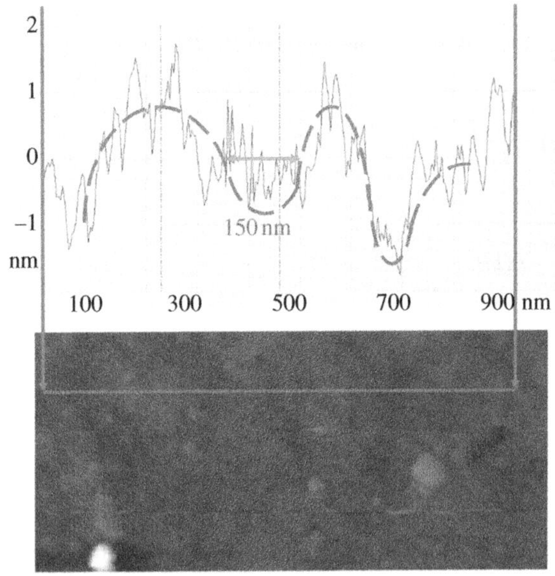


**Figure 2.20** The surface morphology of a Cu surface analyzed by atomic force microscopy (AFM).

after first-stage bonding process and followed by the second stage of ripening of voids.

Therefore, Eq. (2.4) is used again, but $r_{max}$ was replaced by 150 nm, which is the half wavelength of a wavy surface (see Figure 2.18). Now, $r_{max}$ is 100 times smaller than that of single starting contact and have more than 30000 contact points in the array of $\pi \cdot 100^2$. The effect of plastic deformation can be avoided. The bonding time becomes approximately 17 minutes, which is close to published data. Thus, Eq. (2.4) can be used to predict the bonding conditions with multiple contacts.

The theoretic results in Eq. (2.4) show that the surface roughness and temperature have huge influence on the bonding time. The bonding time may increase at least three orders of magnitude when the $r_{max}$ increases from 10 to 150 nm, indicating the surface roughness has significant influence on the bonding time. Figure 2.21a shows the effect of $r_{max}$ on the bonding time at 200 °C for various applied pressure. The applied pressure or the stress gradient has less effect on the bonding time. Furthermore, the temperature also has profound effect on the bonding time, as presented in Figure 2.21b. When the temperature increases from 150 to 300 °C, the bonding time was shortened by three orders of magnitude.

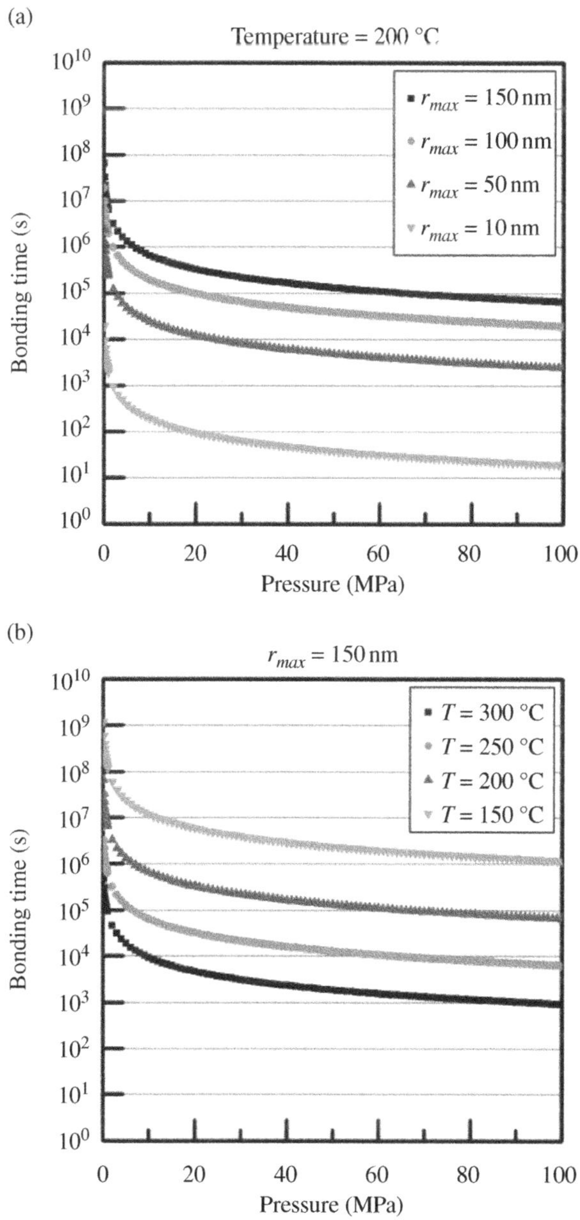


**Figure 2.21** (a) It shows the effect of surface roughness on the bonding time at 200 °C. (b) When the temperature increases from 150 to 300 °C, the bonding time was shortened by three orders of magnitude.

Because Cu direct bonding starts from two surfaces in contact to form a grain boundary, surface diffusion is a key factor at the beginning until a grain boundary is formed and then it became dominated by grain boundary diffusion. For instance, surface diffusivity on the (111) surface of Cu has at least four orders of magnitude larger than other surface orientation, so it have been reported that in Cu direct bonding, two (111) surfaces can bond together at 200 °C in 30 minutes without surface treatment like surface-activated bonding (SAB) or passivation layers. After the formation of a twist-type grain boundary, diffusivity along the twist-type grain boundary is much slower than that on the (111)-oriented surface. Furthermore, the contact area or the twist-type grain boundary increases continuously in the bonding process. Yet, there are always free surfaces surrounding the contact areas and keep the benefit of the highest surface diffusivity of highly (111)-oriented surfaces during the first stage of bonding.

If the grain boundary diffusivity in Eq. 2.4 is replaced by surface diffusivity, the bonding time will be reduced by about 2–6 orders depending on the surface orientation. For example, if the bonding pressure is 1 MPa, the bonding temperature is 200 °C and the $r_{max}$ is 150 nm, the bonding time of two (111) surface will be about two seconds. In fact, the total bonding time should include overcoming the blockage of oxides at the bonding interface or diffusing on not perfect (111) surface, so the real surface diffusivity may be only 1 or 0.1% of that on a perfect (111) surface. It leads to an underestimation of bonding time. For another example, if the bonding temperature is 300 °C, it is possible to bond two highly (111)-oriented surfaces together in 10 seconds [24]. On the contrary, SAB method can bond two ultrasmooth (roughness =1 nm) surfaces at room temperature. The bonding process will skip atomic diffusion for such planarized surfaces, and the metal bonds will form instantly when the high-energy surfaces come close to interatomic distance. It is noteworthy that the diffusion in the initial may be dominated by surface diffusion, as shown in the Figure 2.19a. However, when the gaps near the initial contact points are filled by Cu and form grain boundaries, the subsequent Cu atoms may have to diffuse through the grain boundaries to fill the remailing voids, as shown in Figure 2.19b. Therefore, the diffusion may be dominated by grain boundary at later stages of bonding.

### 2.6.3 Microstructures at the Cu-to-Cu Bonding Interface

After the bonding process, numerous nanoscale voids exist in the bonding interface, and they may not be eliminated completely even after high temperature annealing [25–28]. Figure 2.22a shows the cross-sectional TEM image for the Cu–Cu joint bonded at 200 °C for 10 minutes. There are many voids at the bonding interface. After the annealing at 200 °C for 30 minutes, as shown in Figure 2.22b, the voids seem to be much less than the sample in Figure 2.20a. A better way to observe the morphology of the interfacial voids is from plan-view TEM image, as shown in Figure 2.23a [40]. It can be seen that most of the voids locate at the junctions of grain boundaries, and their size is less than 100 nm. Figure 2.23b depicts the distribution of void size for the sample in Figure 2.23a. The void size ranges from 10 to 110 nm, with an average value of 53 ± 19 *nm*. In addition, the voids occupy 5.4% of the interfacial area, and thus they have very little effect on bump resistance increase. After annealing at a higher temperature or prolonged time, the nanoscale voids may ripen into large voids, but the number of voids will decrease [40]. Void distribution was also examined using 3D FIB image by Kagawa et al. Figure 2.24 presents their void distribution for the Cu-to-Cu bond after annealing at 150 °C. Most of the voids are less than 70 nm. How these voids affect the reliabilities is of interest. Further research is needed to provide a better understanding on the void evolution during annealing, electromigration, or temperature cycling tests.

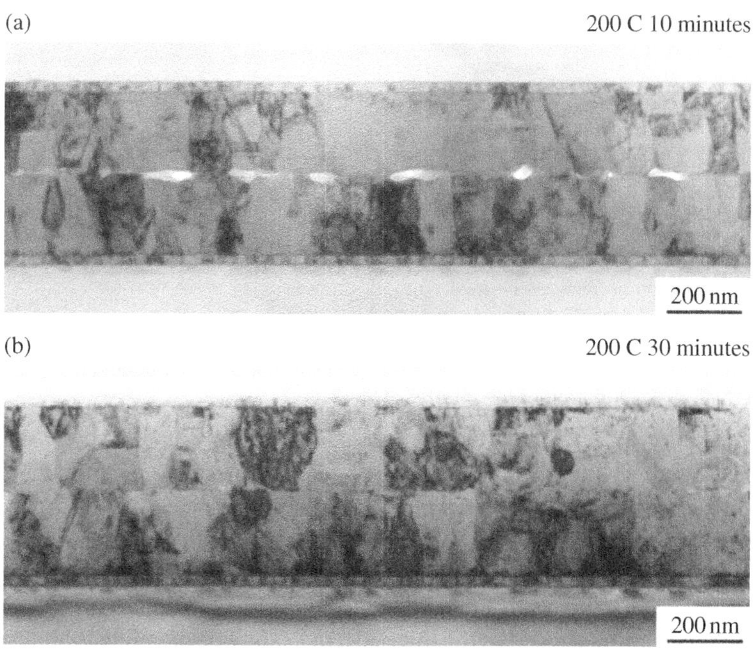


**Figure 2.22** Cross-sectional TEM images of voids in the Cu-to-Cu bonding interface after annealing at 200 C for (a) 10 minutes and (b) 30 minutes.

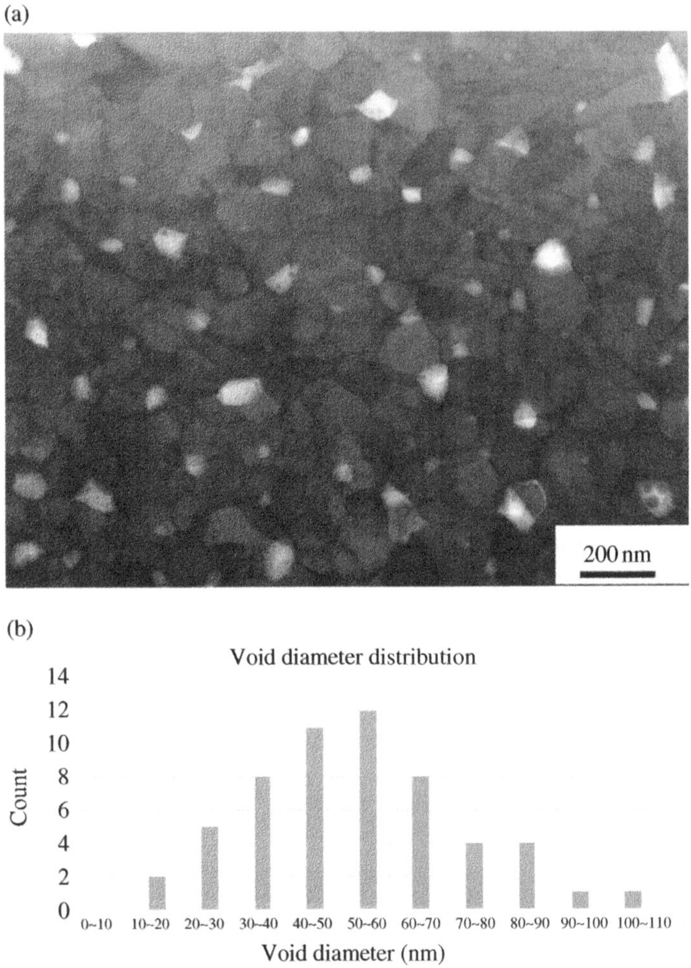


**Figure 2.23** (a) Plan-view of TEM image of voids in the Cu-to-Cu bonding interface. (b) Distribution of the interfacial voids.

The bonding interface may consist of regular low-angle or high-angle grain boundaries. Sometimes, twist-type boundaries may also form [8]. When two grains possess the same orientation, but with a rotation angle, twist-type grain boundaries may form. Figure 2.25a shows a high-resolution TEM image for a low-angle grain boundary at the bonding interface. Lattice atoms are continuous across the interface, indicating the bonding is almost perfect. Figure 2.25b shows a twist-type grain boundary with a misorientation of 3°, as indicated by the electron diffraction pattern in the upper right

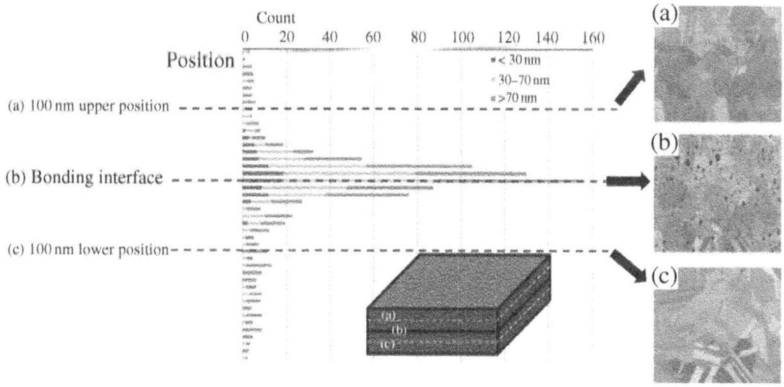


**Figure 2.24** Kagawa et al. presented their void distribution for the Cu-to-Cu bonding after annealing at 150 °C. Most of the voids are less than 70 nm.

corner. A hexagonal network can be observed in the TEM image and the interface consists of arrays of screw dislocations. More studies are needed in order to understand which boundaries can better resist electromigration and stress.

Elimination of bonding interface is another important issue in Cu-to-Cu bonding. Straight interface may indicate that no grain growth has taken

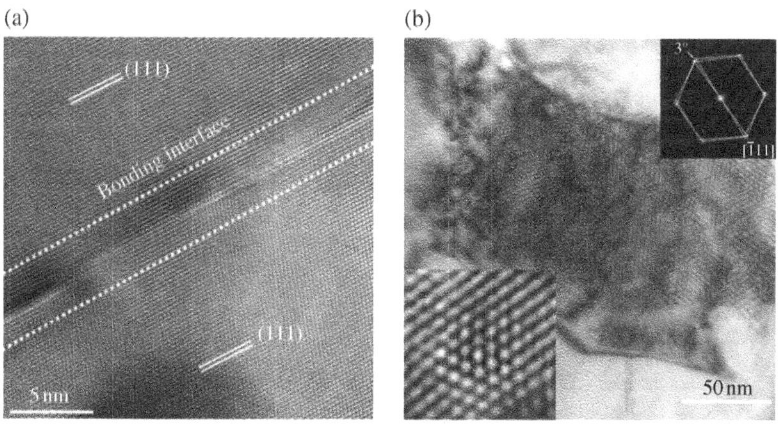


**Figure 2.25** (a) A high-resolution TEM image for a low-angle grain boundary at the Cu-to-Cu bonding interface. Lattice atoms are continuous across the interface, indicating the bonding is almost perfect. (b) A twist-type grain boundary with a misorientation of 3°, as indicated by the electron diffraction pattern in the upper right corner.

place across the interface, and the bonding strength is weak. Zig-zag interface represents some grain growth. The ideal condition is that grain growth can occur extensively to eliminate the bonding interface [29, 30]. Thus, the electric and mechanical properties of the Cu-to-Cu bonds is close to those of bulk Cu. Juang et al. performed shear tests on the Cu-to-Cu joints fabricated at 200, 250, 300, and 350 °C [31]. Figure 2.26a–h shows the microstructures of the four Cu-to-Cu joints, in which almost no grain growth occurred near the bonding interface for the samples bonded at 200 and 250 °C. The interfaces are still straight. The sample bonded at 300 °C had some amount of grain growth results in zig-zag bonding interface in part of the interface. For the sample joined at 350 °C, extensive grain growth occurred and a big grain grew across the interface. Thus, part of the interface was eliminated. These different interface morphologies possess different bonding strength. Figure 2.27 illustrates the measured shear strength for the four samples in Figure 2.24. The results indicate brittle fractures took place in the samples bonded at 200 and 250 °C, which had no grain growth, and their bonding strength is low. For the samples with grain growth, the failure switched to ductile fracture. The shear strength can reach 158 MPa for the samples with extensive grain growth. For regular electroplated and sputtered Cu, it is still unclear what kind of microstructures may trigger massive grain growth in order to eliminate the bonding interface. Chu et al. reported that the bonding interface can be eliminated completely by

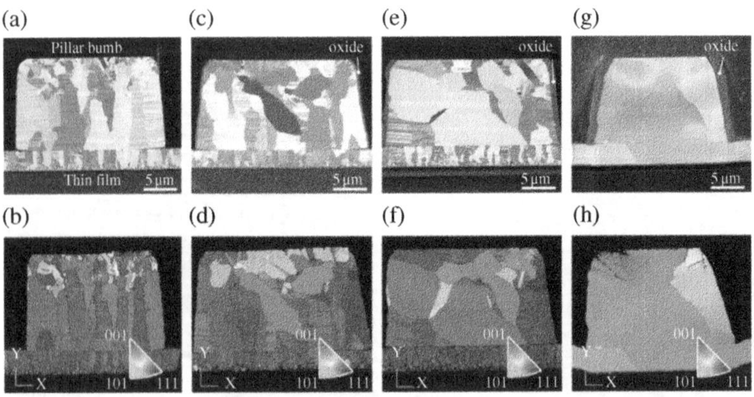


**Figure 2.26** Cross-sectional focus ion beam (FIB) and electron back-scatter diffraction (EBSD) images showing grain growth in the Cu points bonded at various temperatures. (a) FIB image, 200 °C. (b) EBSD image for the joint in (a). (c) FIB images, 200 °C. (d) EBSD image for the joint in (c). (e) FIB image, 300 °C. (f) EBSD images for the joint in (e). (g) FIG image, 350 °C, and (h) EBSD image for the joint in (g).

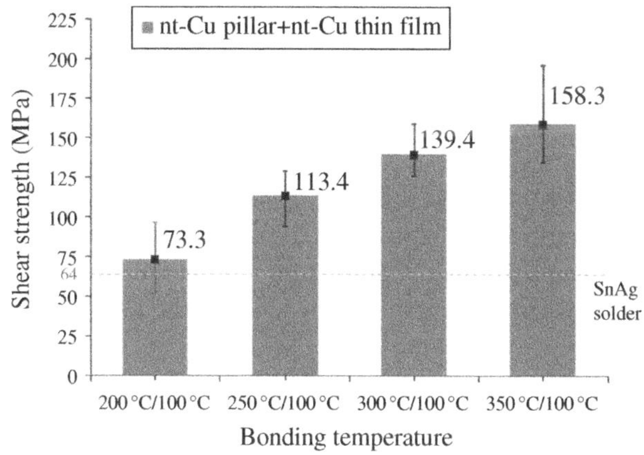


**Figure 2.27** A plot of shear strength against bonding temperature of Cu pillar on Cu thin fim.

anisotropic large grain growth in the highly (111)-oriented nanotwinned Cu. We will elaborate this in Chapter 3.

## 2.7 Hybrid Bonding

Hybrid bonding appears to be the most important technology to realize high-density Cu-to-Cu bonding for heterogeneous 3D integration. This technology was first developed by Ziptronix (Acquired by Tessera in 2015, now Xperi). Fine pitch of 1.6 μm has been achieved in Cu direct bonding. Patents US 7,387,944 B2 and US 7,807,549 B2 filed an etching or cleaning procedure that formed oxide bonds with a strength of at least 2000 mJ/m$^2$ [32, 33]. Figure 2.28a–c depicts the schematic drawing for the

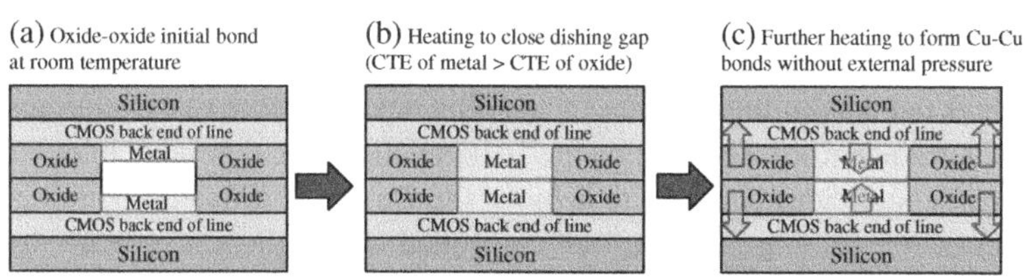


**Figure 2.28** Schematic diagrams depict the effect of Cu dishing on hybrid bonding. (a) Initial oxide-to-oxide bonding, and there is a gap between the upper and lower Cu. (b) Heating to close the gap between the two Cu. (c) Further heating can compress the two Cu together without external compression.

processes and structures of the hybrid bonding with $SiO_2$ dielectrics. Fabrication of Cu bumps embedded in oxide dielectrics can be done through (dual-) damascene approaches on Si wafers, which is compatible to semiconductor processes. The Cu bumps have to be slight dishing and the dishing has to be well controlled in nanometer scale, depending on the temperature for Cu-to-Cu bonding. Initially, oxide-to-oxide bonding is formed at room temperature by aligning two Si wafers and by thermal compression bonding, as shown in Figure 2.28a. Then the wafers are heated to an elevated temperature. Because the Cu has larger coefficient of temperature expansion (CTE) than the Si oxide, the Cu will expand more than the oxide to close the dishing gap between the Cu bumps, as seen in Figure 2.28b. Finally, the wafers are further heated to around 300 °C without external pressure. The expansion of Cu bumps causes compressive stress in the Cu–Cu bump interface, and provides the pressure needed for Cu–Cu bonding. On the other hand, the oxide layers are under tensile stress during the bonding process, as illustrated in Figure 2.28c.

This technology utilizes the differences in CTE between the Cu and Si oxide to provide the pressure needed to form Cu-to-Cu diffusion bonding. In general, the linear CTE difference can be expressed as

$$\Delta L = \Delta\alpha \, L_0 \Delta T \tag{2.5}$$

where $L_0$ is the initial length, $\Delta L$ is the length difference due to thermal expansion, $\Delta\alpha$ is the difference in CTE, and $\Delta T$ is temperature difference. Consider a Cu/oxide layer of 2 μm thick, we can estimate the Cu dishing required to achieve the hybrid bonding. The linear CTE is $5.6 \times 10^{-6}\,\mathrm{k}^{-1}$ and $16.6 \times 10^{-6}\,\mathrm{k}^{-1}$ for $SiO_2$ and Cu, respectively. If the bonding temperature is at 300 °C, the height that Cu would expand relative to $SiO_2$ layer is 6.4 nm, as shown below.

$$\Delta L = \Delta\alpha \, L_0 \, \Delta T = (16.6 - 5.6) \times 10^{-6} \times 2000\,\mathrm{nm} \times (300 - 25) = 6.4\,\mathrm{nm}$$

This is the maximum value for the dishing, meaning that two Cu bumps can touch each other. The dishing has to be smaller than this value in order to provide sufficient pressure to achieve the Cu-to-Cu diffusion bonding. However, the dishing must be well controlled across the entire wafer by CMP. When the Cu dishing is too large, the Cu bumps may not be able to touch each other and bond together. When the Cu dishing is too small, the Cu bumps will expand more than we need and causes delamination in the interface of oxide/oxide layers. As the bonding temperature decreases, the tolerance in dishing becomes even smaller. When the bonding temperature decreases to 150 °C, the difference in height is only approximate 3 nm, as shown below:

$$\Delta L = (16.6 - 5.6) \times 10^{-6} \times 2000\ \text{nm} \times (150 - 25) = 2.8\ \text{nm}$$

Therefore, low-temperature hybrid bonding becomes very challenging, and CMP plays the critical role on the hybrid bonding.

There are many advantages for the hybrid bonding, including the following:

1) The initial oxide–oxide bonding forms at room temperature. The wafers do not slide during the subsequent annealing process, and alignment can be preserved after the heat treatment.
2) No external pressure is needed in the final Cu-to-Cu annealing process. Therefore, the wafers can be annealed at a batch process, and the throughput can be increased.
3) The oxides seal the Cu-to-Cu bonds hermetically and prevent them from oxidation and corrosion. The oxide bonding provides bonding strength of approximately 2000 mJ/m$^2$, and it protects the Cu-to-Cu bonds mechanically.
4) The processes for hybrid Cu/oxide are similar to the dual Cu damascene technology in BEOL of semiconductor. The Cu bumps can be scaled down continuously below submicron.

Therefore, the Cu/oxide hybrid bonding is already implemented in many handheld devices. In 2016, Samsung has released a microstructure image of the CMOS image sensor (CIS) module in the Galaxy S7 [34]. Copper interconnects with a pitch of approximately 10 μm are bonded, which can be seen in a cross-sectional figure, as shown in Figure 2.29. The pitch has been demonstrated to be able to scaled down [35, 36]. Figure 2.30 shows

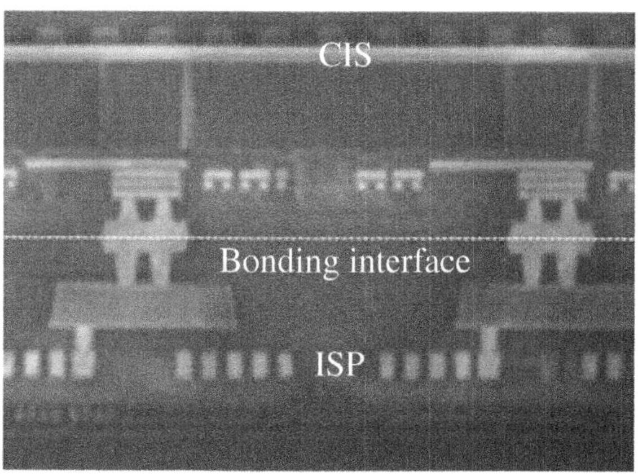


**Figure 2.29** Cross-sectional SEM image of Cu interconnects with a pitch of approximately 10 μm.

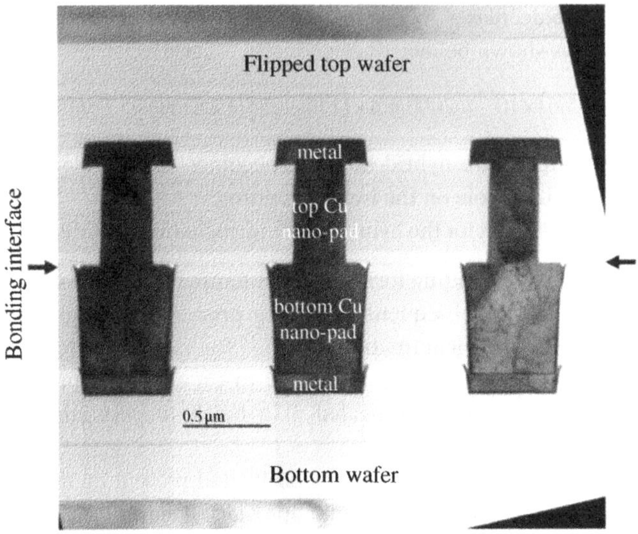


**Figure 2.30** It shows state-of-the-art of hybrid Cu-to-Cu bonds using the SiCN dielectrics. By stacking 270 nm Cu bump on 540 nm Cu bumps, the pitch is as small as 1.08 μm.

state-of-the-art hybrid Cu–Cu bonds using the SiCN dielectrics. By stacking 270 nm Cu bump on 540 nm Cu bumps, the pitch is as small as 1.08 μm. The TEM image shows that the three bonding interfaces, SiCN-to-SiCN, Cu-to-Cu, and SiCN-to-Cu, are free from voids. The voids at the bottom bump corners are attributed to galvanic corrosion during CMP process.

## 2.8 Reliability – Electromigration and Temperature Cycling Tests

Reliabilities of Cu-to-Cu bonds have not been well investigated so far, because most of the efforts have been made in process and yield. Several researchers have reported preliminary results of reliability tests of hybrid bonds, such as temperature cycling tests (TCT), high temperature storage (HTS), and electromigration (EM). The fundamentals of electromigration will be introduced in Chapter 10. It is well known that Cu possesses high electromigration lifetime than solders. However, the electromigration failure modes in Cu-to-Cu bonds are unclear so far. In 2016, Moreau et al. conducted electromigration tests in 3.6 × 3.6 μm Cu/oxide for CMS image sensors, and they reported that Cu-to-Cu bond itself did not fail after

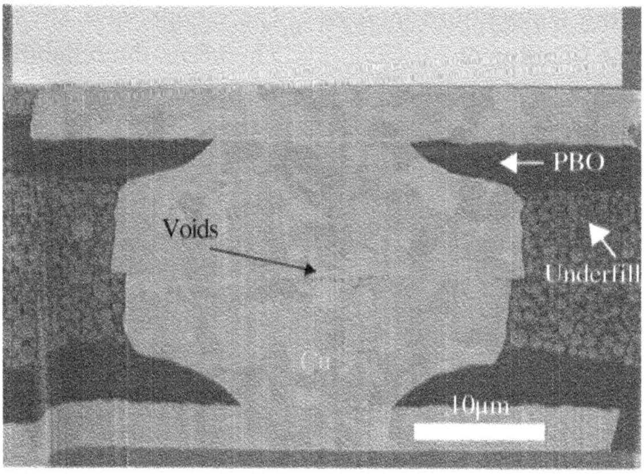


**Figure 2.31** It shows the microstructures for a Cu-to-Cu joint after 1000 cycles, and the resistance of the joint increased about 15% of its initial value.

electromigration tests by 20 mA at 350 °C [37]. Instead, the voids formed at the BEOL current feeding lines. It is noteworthy that the intrinsic voids at the bonding interface were not affected by the current stressing. Therefore, the weakest link is the Cu lines in BEOL. Shie et al. performed electromigration tests in Cu-to-Cu bonds with 30 μm in diameter. After current stressing under 1.5 A ($2.12 \times 10^5 A/cm^2$) at 150 °C for approximately 1600 hours, the resistance of the joint increased 20% of its initial value. Voids formed in the interface between the Cu bump and the BEOL Cu line. The morphology of the intrinsic voids at the original bonding interface did not change much. In some of the joints, the electromigration test enhances the grain growth of Cu, and thus the bonding interface was eliminated.

Preliminary results of TCT were also reported for Cu-to-Cu bonds with 30 μm in diameter with polymer underfill [38]. The temperature range is from −55 to 125 °C; the soak time is five minutes; the ramp rate is 18 °C/min. Due to the difference in CTE, the Cu-to-Cu joints were subjected to tensile and compressive stress during the temperature cycling. Figure 2.31 shows microstructures for the Cu-to-Cu joint after 1000 cycles, and the resistance of the joint increased about 15% of its initial value. The cracks formed at the middle of the bonding interface. It indicates that the weakest link is at the original bonding interface with intrinsic voids. More studies need to be performed to provide a better understanding of the failure mechanism. For the hybrid Cu/oxide bonds for CMOS image sensors, silicon oxides are adopted as dielectric materials. Because the CTE of the silicon oxide is smaller than that of Cu, the Cu-Cu joints would be under compressive stress as the temperature increases. The compressive stress may further improve the bonding of Cu-Cu joints. Therefore, the reliabilities of the Cu-Cu/oxide hybrid joints would be much better than the Cu-Cu joints with polymer dielectric materials.

## Problems

**1.1** What are the main functions of microelectronic packaging?

**1.2** In 1960s, IBM's flip-chip packaging consisted of ceramic substrates and high-Pb solders. IBM adopted thin-film under-bump-metallization (UBM).

**1.3** (a) Why IBM adopted ceramic substrates? (b) Why IBM chose high-Pb (Pb97Sn3) solders? (c) Draw the cross-section diagram of the thin-film UBM and explain the functions of each layer.

**1.4** High-Pb solders and ceramic substrates have been used in packaging of microelectronic devices in 1970s due to their excellent mechanical properties. When personal computers came out, cost was an important consideration. Therefore, polymer substrates and eutectic SnPb solders were adopted to replace the high-Pb solders and polymer substrates in 1990s. (a) What are the two main problems arisen due to this change? (b) What are the solutions to these problems? In 2000s, Pb-free solders were adopted to replace eutectic SnPb solders due to environmental concerns. (c) What are the two main problems arisen due to this change? (d) What are the solutions to these problems?

**1.5** Flip-chip solder joints have been adopted for high-end chip packaging due to their excellent performance. Compared with wire bonding, what are the advantages of flip-chip in electric and thermal performances.

**1.6** For **tape-automated bonding (TAB)** technology, (a) draw the schematic picture for the inner lead and bumped chip before bonding. Label the material in the picture you draw. (b) Usually the bonding can be achieved by thermal compression at 350 °C for 0.5 seconds without ultrasonic. Explain why the bonding can be formed in such a short time. (c) Explain why TAB technique can NOT be used to fabricate area-array bonding?

**1.7** (a) Describe the process flow for **thermosonic wire bonding** technology. (b) Explain why the first bond is in ball shape and the second one is in crescent shape? Why we can achieve the bonding in such a short time and temperature below 200 °C?

**1.8** The C4 solder joints can achieve about 10 times higher than the wire bond in terms of connection per chip. (a) Explain why; (b) for a chip with 1 cm × 1 cm and with a pitch 50 μm, estimate the connection per chip for both C4 and wire bond.

**1.9** Flip-chip solder joints have been adopted for high-end chip packaging due to their excellent performance. (a) One of their advantages is self-alignment. Describe how self-alignment works. Explain what is the driving force for the self-alignment. (b) Compared with wire bonding, what are the advantages of flip-chip in electric and thermal performances. (5+5 points)

**1.10** Copper wires have been developed successfully to replace some of the Au wires. (a) What are the main advantages of Cu wires. (b) The main problem for Cu wire is that Cu can be oxidized easily. How to solve the problem? (4+4 points)

**1.11** Cu-to-Cu direct bonding has been adopted for ultrafine packaging method. As shown below, Cu-to-Cu direct bonding of 10 μm pitch has been realized in CMOS image sensor. List two advantages for this technology.

## References


1 K. Puttlitz and P. Totta, "Area Array Technology Handbook for Micro-Electronic Packaging", Kluwer Academic, Norwell, MA (2001).

2 J. H. Lau, Flip Chip Technology, McGraw-Hill, New York (1996).

3 K. N. Tu, Solder Joint Technology; Materials Properties, and Reliability Springer, New York, (2007).

4 C. Chen, D. Yu, and K. N. Chen, "Vertical interconnects of microbumps in 3D integration". *MRS Bulletin*, 40(3), 257–263, (2015).

5 Y. C. Liang, C. Chen, and K. N. Tu, Side wall wetting induced void formation due to small solder volume in microbumps of Ni/SnAg/Ni upon reflow *ECS Solid State Letters* 1, 4, 60–62 (2012)

**6** Jie-An Lin, Chung-Kuang Lin, Chen-Min Liu, Yi-Sa Huang, Chih Chen, David T. Chu and K. N. Tu, Formation mechanism of porous Cu3Sn intermetallic compounds by high current stressing at high temperatures in low-bump-height solder joints, *Crystals*, 6, 12 (2016)

**7** K. N. Chen, A. Fan, C. S. Tan, R. Reif, & C. Y. Wen, «Microstructure evolution and abnormal grain growth during copper wafer bonding», *Applied Physics Letters*, 81(20), 3774–3776 (2002)

**8** C. M. Liu, H. W. Lin, Y. C. Chu, C. Chen, D. R. Lyu, & K.N. Chen, and K. N. Tu, Low-temperature direct copper-to-copper bonding enabled by creep on highly (111)-oriented Cu surfaces, *Scripta Materialia* 78–79 (2014) 65–68.

**9** C. M. Liu, H. W. Lin, Y. S. Huang, Y. C. Chu, C. Chen, D. R. Lyu, K.N. Chen and K. N. Tu, "Low-temperature direct copper-to-copper bonding enabled by creep on (111) surfaces of nanotwinned Cu", *Scientific Reports*, 5, 9734 (2015)

**10** B. Rebhan, and K. Hingerl, "Physical mechanisms of copper-copper wafer bonding", *Journal of Applied Physics*, 118, 135301 (2015).

**11** H. Noma, T. Kamibayashi, H. Kuwae, N. Suzuki, T. Nonaka, S. Shoji, & J. Mizuno, "Compensation of surface roughness using an au intermediate layer in a Cu direct bonding process", *Journal of Electronic Materials*, 47(9), 5403–5409 (2018)

**12** Y. P. Huang, Y. S. Chien, R. N. Tzeng, M. S. Shy, T. H. Lin, K. H. Chen, H. M. Tong & K.N. Chen, "Novel Cu-to-Cu bonding with Ti passivation at 180 °C in 3-D integration", *IEEE Electron Device Letters*, 34(12), 1551–1553 (2013)

**13** Hu, Y.H., Liu, C.S., Lii, M.J. et al. (2012). 3D stacking using Cu–Cu direct bonding for 40um pitch and beyond. 4th Electronic System-Integration Technology Conference (pp. 1–5). IEEE.

**14** Riko I Made, Chee Lip Gan, Liling Yan, Katherine Hwee Boon Kor, Hong Ling Chia, Kin Leong Pay, And Carl V. Thompson, "Experimental characterization and modeling of the mechanical properties of Cu-to-Cu thermocompression bonds for three-dimensional integrated circuits", *Acta Materialia*, 60, 578–587 (2012).

**15** W. Yang, M. Akaike, & T. Suga, "Effect of formic acid vapor in situ treatment process on Cu low-temperature bonding", *IEEE Transactions on Components, Packaging and Manufacturing Technology*, 4(6), 951–956 (2014).

**16** Y. Kagawa, N. Fujii, K. Aoyagi, Y. Kobayashi, S. Nishi, N. Todaka, & K. Tatani, "Novel stacked CMOS image sensor with advanced Cu2Cu hybrid bonding" *IEEE International Electron Devices Meeting (IEDM)* (2016), 208–211.

17 H. Takagi, R. Maeda, T. R. Chung, N. Hosoda, and T. Suga, "Effect of surface roughness on room-temperature wafer bonding by Ar beam surface activation", *Japanese Journal of Applied Physics*, 37, 4197–4203 (1998).

18 Y.P. Huang, Y.S. Chien, R.N. Tzeng, and K.N. Chen, Demonstration and electrical performance of Cu–Cu bonding at 150 °C With Pd passivation, *IEEE Transactions on Electron Devices*, 62(8), 2587–2592 (2015)

19 Suga, T., Yuuki, F., and Hosoda, N. (1997). A new approach to Cu-Cu direct bump bonding. IEEE 1st Joint International Electronic Manufacturing Symposium and the International Microelectronics Conference (IEMT/IMC) Symposium.

20 R. A. Nichting, D. L. Olson, and G. R. Edwards, "Low-temperature solid-state bonding of copper", *Journal of Materials Engineering and Performance*, 1, 35–44 (1992).

21 Made RI, Gan CL, Yan L, Kor KHB, Chia HL, Pey KL, et al. Experimental characterization and modeling of the mechanical properties of Cu–Cu thermocompression bonds for three-dimensional integrated circuits. *Acta Materialia* 2012;60(2);578–587.

22 Derby B, Wallach ER. Diffusion bonds in copper. *Journal of Materials Science* 1984;19(10);3140–3148.

23 Ashby MF. A first report on deformation-mechanism maps. *Acta Metallurgica* 1972;20(7);887–897.

24 K. C. Shie, J. Y. Juang, and C. Chen, "Instant Cu-to-Cu direct bonding enabled by ⟨111⟩-oriented nanotwinned Cu bumps", *Japanese Journal of Applied Physics*, 59, SBBA03 (2020).

25 C. S. Tan, R. Reif, N. D. Theodore, and S. Pozder, "Observation of interfacial void formation in bonded copper layers", *Applied Physics Letters*, 87, 201909 (2005).

26 C.M. Liu, H.W. Lin, Y.C. Chu, C. Chen, D.R. Lyu, K.N. Chen and K.N. Tu, Low-temperature direct copper-to-copper bonding enabled by creep on highly (111)-oriented Cu surfaces, *Scripta Materialia* 78–79 (2014) 65–68.

27 A. Fan, A. Rahman, and R. Reif, "Copper wafer bonding", *Electrochemical and Solid-State Letters*, 2, 534–536 (1999).

28 Mudrick, J.P., Sierra-Suarez, J.A., Jordan, M.B. et al. (2019). Sub-10μm pitch hybrid direct bond interconnect development for die-to-die hybridization. IEEE Electronic Components and Technology Conference (ECTC).

29 Y. C. Chu, and C. Chen, "Anisotropic grain growth to eliminate bonding interfaces in direct copper-to-copper joints using< 111>-oriented nano-twinned copper films", *Thin Solid Films*, 667, 55–58 (2018).

**30** S.Y. Chang, Y.C. Chu, K.N. Tu, and C. Chen, Effect of anisotropic grain growth on improving the bonding strength of <111>-oriented nanotwinned copper films, *Materials Science and Engineering A* 804 (2021) 140754.

**31** J.Y. Juang, C.L. Lu, Y.J. Li, K. N. Tu and C. Chen, Correlation between the microstructures of bonding interfaces and the shear strength of Cu-to-Cu joints using (111)-oriented and nanotwinned Cu, *Materials*, 11, 2368 (2018)

**32** Three dimensional device integration method and integrated device. US patent (2006), US 6984571 B1, Ziptronix, Inc.

**33** Method for low temperature bonding and bonded structure. US Patent (2000), US 6902987, Ziptronix, Inc.

**34** Chipworks (2016). Samsung Galaxy S7 Edge Teardown Report. https://www.techinsights.com/zh-tw/node/16563

**35** Beyne, E., Kim, S.W., Peng, L. et al. (2017). Scalable, sub 2μm pitch, Cu/SiCN to Cu/SiCN hybrid wafer-to-wafer bonding technology. IEEE International Electron Devices Meeting (IEDM).

**36** Kim, S.W., Fodor, F., Heylen, N. et al. (2020). Novel Cu/SiCN surface topography control for 1 μm pitch hybrid wafer-to-wafer bonding. IEEE Electronic Components and Technology Conference (ECTC).

**37** S. Moreau, A. Fraczkiewicz, D. Bouchu, P. Bleuet, P. Cloetens, J. C. Da Silva, H. Manzanarez, F. Lorut, and S. Lhostis, "Correlation between electromigration-related void volumes and time-to-failure by high resolution x-ray tomography and modeling", *IEEE Electron Device Letters*, 40, 1808–1811 (2019).

**38** Shie, K.C., Hsu, P.N., Li, Y.J. et al. (2021). Electromigration and temperature cycling tests of Cu-Cu joints fabricated by instant copper direct bonding. IEEE Electronic Components and Technology Conference (ECTC), 995–1000.

**39** K.C. Shie, A.M. Gusak, K.N. Tu C. Chen, "A kineticmodel of copper-to-copper direct bonding under thermal compression", *Journal of Materials Research and Technology*, 15, 2322 (2021).

**40** https://doi.org/10.1016/j.matchar.2021.111459

# 3

# Randomly-Oriented and (111) Uni-directionally-Oriented Nanotwin Copper

## 3.1 Introduction

Why nanotwinned Cu (nt-Cu) is important? Technically, Cu is the most widely used conductor in microelectronic technology, including both on-chip and in packaging. Scientifically, nt-Cu is the fifth mechanism in strengthening metals. This is because in K. Lu's Science paper in 2004 [1], it is reported that the mechanical yield strength can increase 10 times from about 100 MPa for ordinary Cu to 1000 MPa for nt-Cu. At the same time, electrical resistivity does not change much. In classical metallurgy, one of the most important challenges was how to strengthen a metal? Typically, there are already four mechanisms. First is by work hardening due to dislocation entanglement. Second is solution hardening. For example, pure gold is too soft, so it cannot hold diamond. Therefore, we need to add some Ag or Cu to make 18 or 14 k gold to make diamond ring. Third is precipitation hardening as in the case of GP zone formation in Al(Cu) alloy, where the precipitation of $Al_2Cu$ has to go through θ", θ', and θ phases. Fourth is grain size reduction, as given by the Hall-Pitch relation. Now, nanotwin formation is the fifth one. [2–16]

Furthermore, nt-Cu lines possess higher electromigration lifetimes than regular Cu lines. Chen et al. adopted in situ high-resolution transmission electron microscopy (HRTEM) to observe electromigration of Cu lines, and they observed that the migration rate of Cu atoms was retarded at the twin boundaries, as shown schematically in Figure 3.1. The time to remove the Cu at the twin boundary is approximately 10 times longer than that in regular lattice. [2] Thus, nt-Cu lines serve as a potential material for redistribution layer (RDL) for three-dimensional integrated circuits (3D IC) integration.

*Electronic Packaging Science and Technology*, First Edition. King-Ning Tu, Chih Chen, and Hung-Ming Chen.

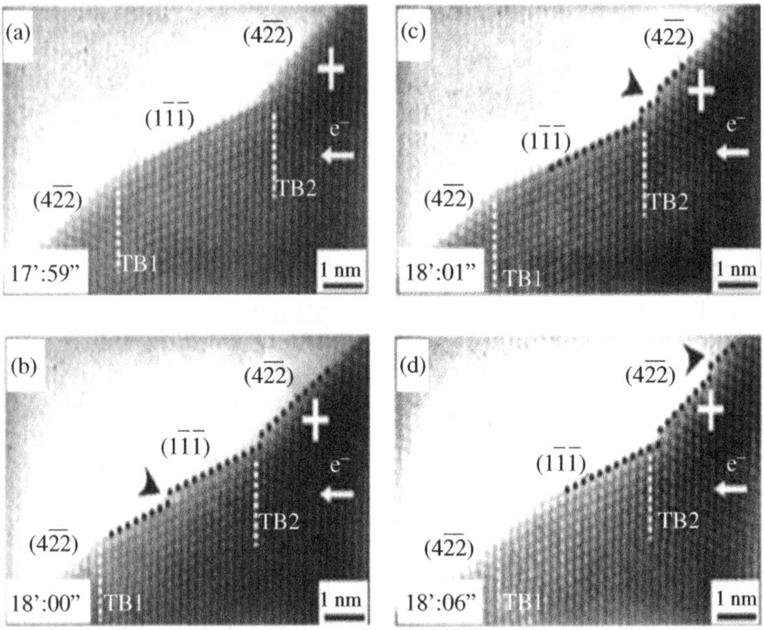


**Figure 3.1** High resolution TEM images showing the atom migration under electromigration at $2.5 \times 10^6$ A/cm$^2$ as a function of time at electrons drifted from right to left. The lower-left corner shows the time of the image capture in minutes and seconds. TB stands for twin boundary. (a) At 17":59"; (b) At 18":00"; (c) At 18":01"; (d) At 18":06".

Currently, the technical trend of 3D IC requires new packaging technology, such as copper direct bonding and fine-line fan-out technology. Integrated fan-out wafer-level packaging (InFOWLP) technology has made groundbreaking advances and is at the moment the mainstream packaging technology for consumer electronic products. Compared to traditional flip-chip wafer-level packaging (FCWLP) technology, InFOWLP possesses advantages like higher pin count, lower power consumption, higher electrical performance, better cooling, and overall lower cost. However, InFOWLP used an epoxy molding compound (EMC) to integrate with Si, causing thermal stress to become a reliability issue because of different thermal expansion coefficients. Thus, the high strength nt-Cu could be a solution.

In this chapter, we shall explain the mechanism of nanotwin formation. More importantly, we shall emphasize the formation of unidirectionally oriented (111) nt-Cu for unique fan-out interconnect and direct bonding applications in 3D IC devices. [1–8]

## 3.2 Formation Mechanism of Nanotwin Cu

The crystal structure of a face-centered cubic (FCC) metal can be represented by the stacking of (111) plane in a repeating sequence of ABCABC. When an error of stacking occurs and changes the stacking to ABCACBA, in which the middle plane of A can be regarded as a mirror of the atomic planes on both sides, a twin is formed.

The stacking error increases the internal energy of the crystal only slightly because each of the atom in the twin plane still has 12 nearest neighbors. Thus, twin is considered to be a coherent plane defect of low energy, which is of the same order of magnitude of a stacking fault. For metals, such as Cu and Ag, which has a low stacking fault energy, they tend to have micro-twins, meaning the twin plane spacing is of the order of microns. On the other hand, Al and Ni have high stacking fault energy, so they seldom show twins. The coherent twin boundary energy is only approximately 3% of that of a grain boundary.

The formation energy of twin plane has been studied by the first principle total energy calculation using the Vienne ab initio simulation package (VASP) with the local density approximation (LDA). Periodic nanoscale twinning structure has been constructed, and strain has been introduced to estimate the evolution of the total energy under increasing strain.

In the calculation, a biaxial stress state was used. It assumes the structure of a Cu thin film on a substrate so that it is stress-free along the normal direction to the film surface. Figure 3.2 shows the periodic nanotwinned structure by repetition of the chosen supercell. Each supercell contains 12 (111) atomic planes in a stacking sequence of . . .ABCAB[C]BACBA[C]ABCAB[C]BACBA[C]. . ., so that in the 24 atomic planes shown in Figure 3.2, there are two twin planes in each of

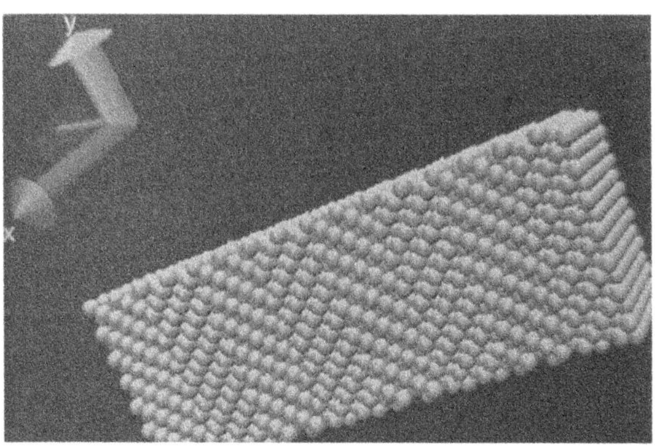


**Figure 3.2** Schematic diagram of the periodic nanotwin structure by repetition of the chosen super cell.

the two supercell, as indicated by the four [C] planes. There are 6 (111) planes in between 2 twin planes, and these planes can be increased to 9 or 12 planes or more in order to increase the twin spacing or to decrease the twin density.

The twin fault energy can be calculated by using the equation below,

$$\gamma_{twin} = \frac{E - NE_{fcc}}{2A}$$

where $E'$ is the total energy of the super cell, $N$ is the number of atoms in the non-twinned supercell, $E_{fcc}$ is the energy per atom in an ideal FCC Cu, $A$ is the surface area of each twin plane in the supercell, and the factor of 2 accounts for the presence of two twin planes in a supercell.

To study the influence of stress/strain on twin formation in Cu thin films, isotropic biaxial stress was introduced and the supercell with strain in the 3D structure were created using the Poisson's ratio of 0.34 of bulk Cu. Two types of biaxial stresses were studied as depicted in Figure 3.3. They are biaxial stress in the {111} plane with x-axis //[110] and y-axis //[112], and biaxial stress in the {112} plane with x-axis//[110] and z-axis//[111].

The calculated twin fault energy is very small, only about two-thousandth of 1 eV/atom, which is close to other theoretical and experimental studies. We recall that interatomic pair potential energy is typically 0.5 eV/atom.

Figure 3.4 shows the evolution of the total energies for non-twinned FCC Cu (the lower curve) and nt-Cu (the upper curve) when they respond

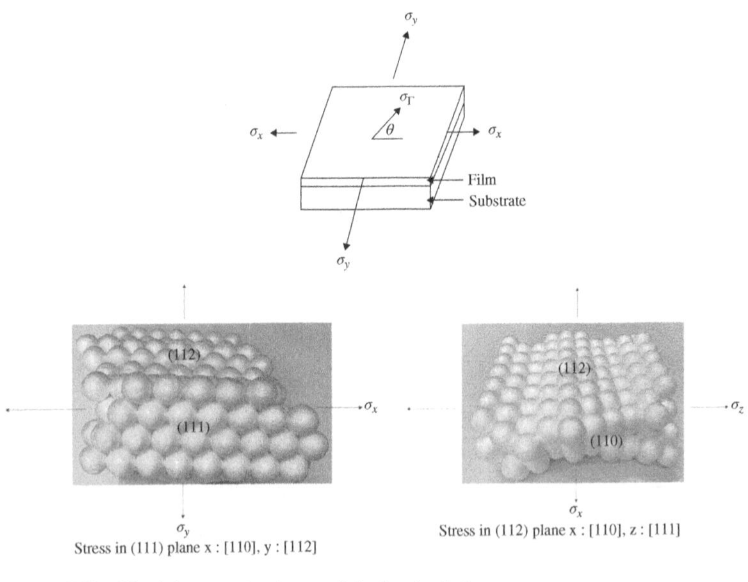


**Figure 3.3** Biaxial stress in the model of calculation.

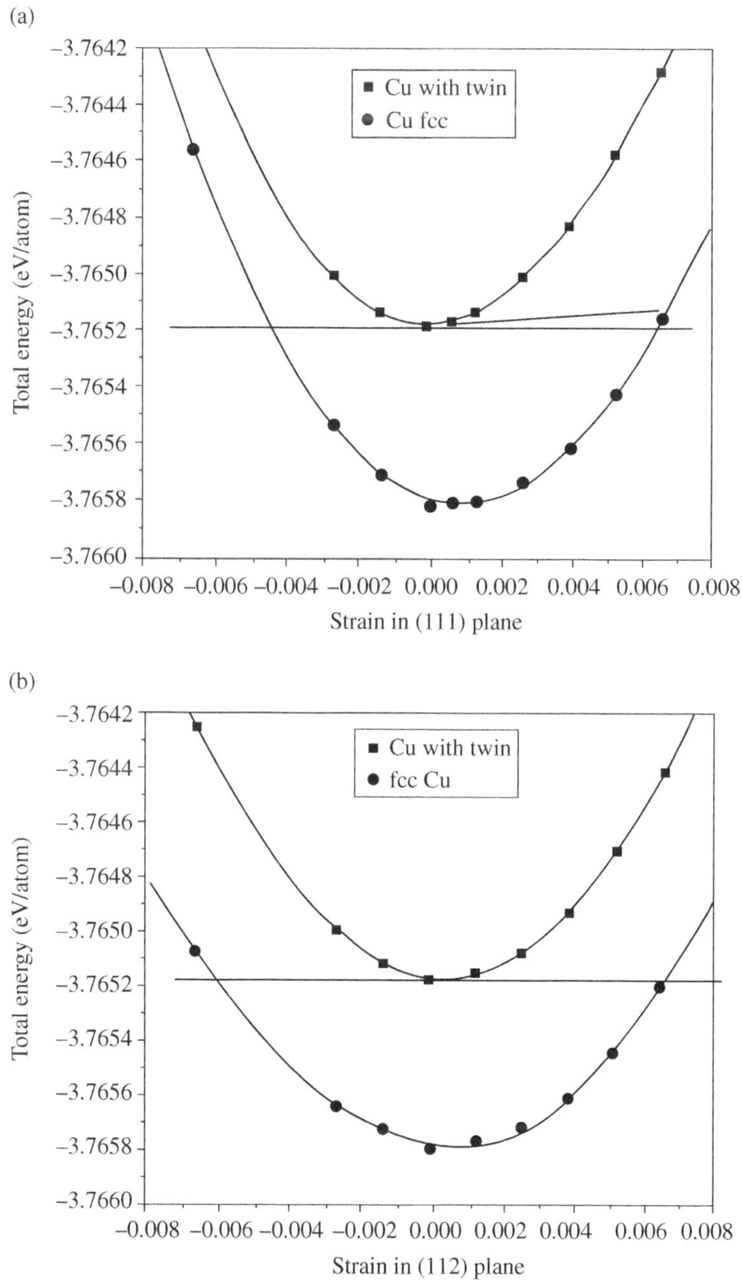


**Figure 3.4** (a) Calculated curves showing the evolution of the total energies for FCC Cu (the lowercase) and the nt-Cu (the upper curve) when they response to the strain in (111) plane. (b) Calculated curves of the results when the biaxial stresses are in the (112) plane.

to biaxial stress in the {111} plane. The pair of curves indicated that when the strain in the non-twinned Cu thin film is larger than a certain value, the total energy of the strained FCC Cu becomes larger than the total energy of the strain-free nt-Cu. To see it clearly, we draw a horizontal straight line tangential to the bottom of the upper curve in Figure 3.5. The horizontal line will cut the lower curve at 0.008 on the left-hand side (the compressive strain side) and at 0.011 at the right-hand side (the tensile strain side). This means that if we strain the non-twinned Cu by a compressive strain of more than 0.8%, it will be higher than the energy of the strain-free Cu having the nanotwins, so the strain energy can be traded to form nanotwins of twin spacing of 6 {111} atomic planes to reduce energy. Or if we strain the non-twinned Cu by a tensile strain more than 1.1%, it can be traded to form nanotwins of twin spacing of 6 {111} atomic planes to reduce energy. We can conclude that a highly strain nt-Cu is energetically less stable than a strain-free nt-Cu. It is obvious that the nt-Cu may not be completely strain-free. This is because under a given strain, the rate of nucleation and growth of nanotwins may be insufficient to compensate the entire strain.

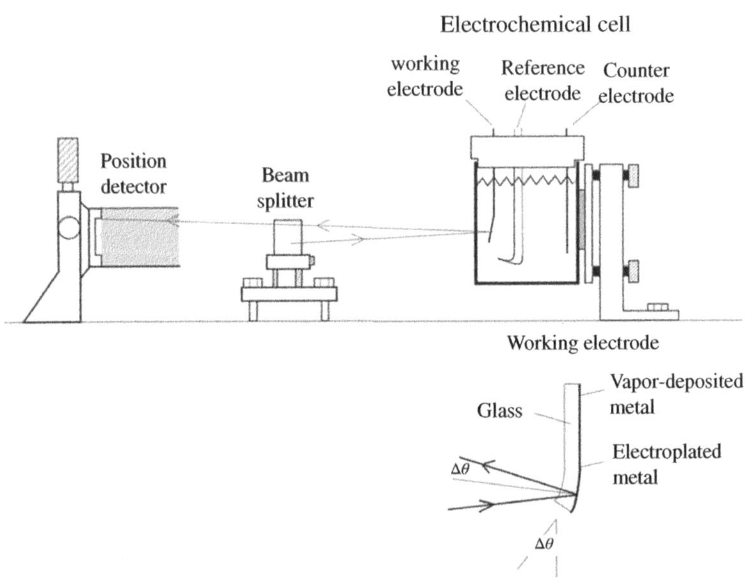


**Figure 3.5** Schematic diagram depicts the in situ stress measurement system. It consists of an electrochemical cell. An optical system that can detect with high sensitivity the beaming of the cantilever beam on which the nt-Cu film is being deposited, and a signal processing unit.

## 3.3 In Situ Measurement of Stress Evolution During Nanotwin Deposition

In situ stress measurement by bending beam method was performed during high-frequency pulse electroplating of nanotwinned Cu thin films. It shows repeating cyclic events of stress build-up and relaxation in the pulse plating. Figure 3.5 shows the in situ stress measurement system. It consists of an electrochemical plating cell, an optical system that can detect with high sensitivity the bending of the cantilever beam on which the nt-Cu film is being plated, and a signal processing unit. The laser reflection response was recorded every 1 ms, which is fast enough to record the bending during every pulse deposition. When the bending or the radius of the cantilever beam was measured, the stress was calculated on the basis of Stoney's equation as shown below [17, 18].

$$\sigma_f = \left(\frac{Y}{1-\nu}\right)\frac{t_S^2}{6rt_f}$$

where $\sigma_f$ is the biaxial stress in the film, $Y$ and $\nu$ are the Young's modulus and Poisson's ratio of the cantilever beam substrate, $t_s$ is the substrate thickness, $t_f$ is the deposited film thickness, and $r$ is the radius of the cantilever beam. The equation shows that it requires knowing the film thickness in order to calculate the biaxial stress. Typically, the film thickness was measured after deposition. However, in the pulse electroplating, we cannot measure in situ the deposited film thickness during each pulse. In Stoney's equation, we can move the film thickness of $t_f$ from the right-hand side to the left-hand side, so it is the product of ($\sigma_f \times t_f$) that will be measured during each pulse deposition when the radius is obtained.

Figure 3.6 shows the evolution of the product of stress and film thickness as a function of pulse deposition time. The product evolves with a periodicity that is consistent with the pulse cycle. During the pulse-on time period, the stress-thickness product moves in the tensile stress direction, whereas during the pulse-off period, approximately 60% of this tensile stress is relaxed. The twin structure was examined by X-ray diffraction and scanning and transmission electron microscopies. The twin spacing was about 20 nm, and the twins are randomly oriented.

To tune the stress in Cu thin films, we have examined the effect of various bath temperatures on residual stress of both the random-oriented Cu films and the highly (111)-oriented nt-Cu films by synchrotron radiation X-ray diffraction measurements. The bath temperature varied from 15 to 40 °C. The results indicate that the typical residual stress in the highly (111)-oriented

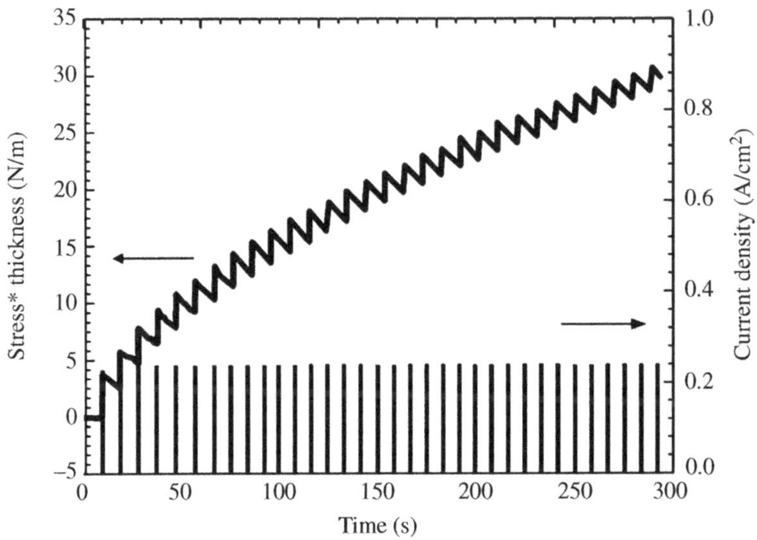


**Figure 3.6** The evolution of the product of stress and film thickness as a function of pulse deposition time.

nt-Cu films is above that in the randomly oriented Cu films. However, the strain within the highly (111)-oriented nt-Cu decreases with increasing bath temperature. The typical residual stress is often reduced from 253 MPa electroplated at 15 °C to 95 MPa under a lower temperature of 35 °C. We can successfully tune and measure the residual stress of the Cu thin films. The films with low residual stress prevent warpage from occurring on the substrate and lower the processing failure in Cu-to-Cu direct bonding and other processes that require a good alignment.

After performing thermal cycling tests, the Cu interconnects within the RDL across the interface between EMC and Si show serious fracture failure. To overcome the high-stress issues, nt-Cu has been proposed as a promising replacement material for interconnects, thanks to its higher mechanical strength than bulk Cu while maintaining an equivalent resistivity. Furthermore, nt-Cu increases electromigration lifetime of the interconnects while also possessing high thermal stability. The characteristic of a low oxidation rate of the (111)-oriented nt-Cu could significantly decrease the cleansing time in IC processing.

We can decrease the residual stress of Cu thin films by increasing electroplating temperatures. At higher bath temperatures of 35 °C, the highly (111)-oriented nt-Cu thin film has rock bottom residual stress of 95 MPa. The microstructure, grain orientation, grain size, and residual stress of

Cu thin films were suffering from different bath temperatures within the electroplating process. The grain size increases with increasing bath temperatures. However, there is still much effort to form highly (111)-oriented nt-Cu thin films suitable for future applications, like understanding their thermal behavior and thermal stress in further research. The residual stress of the as-deposited Cu thin films can help to better understand the film quality and life prediction in future reliability researches.

## 3.4 Electrodeposition of Randomly Oriented Nanotwinned Copper

Randomly oriented nt-Cu films have been fabricated by pulsed electrodeposition in 2004. [1] Figure 3.7 shows the electroplated nt-Cu film with densely packed nanotwins and the distribution of twin lamella. The average grain size is 400 nm. As described in 3.3, during the pulse-off process, formation of twins relaxed the stress of the Cu films.

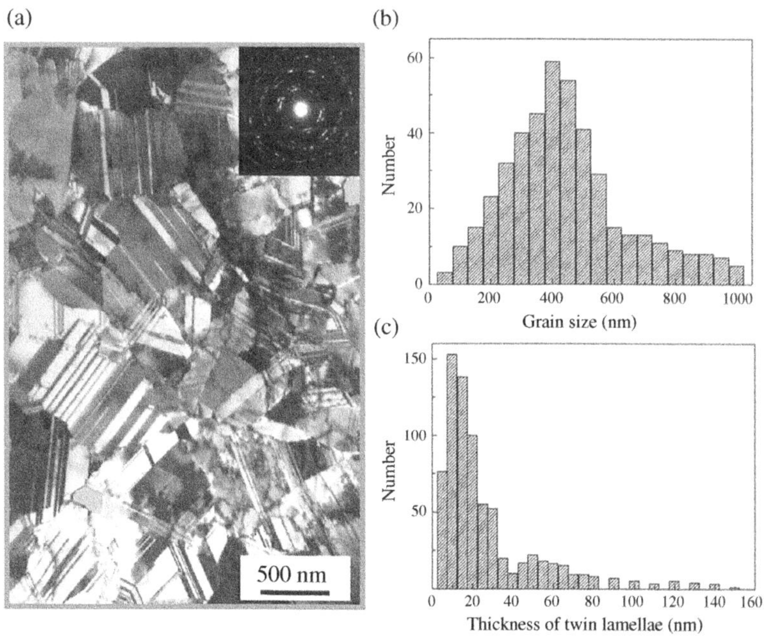


**Figure 3.7** Microstructures of pulsed electrodeposited randomly oriented nt-Cu. (a) TEM image; (b) distribution of grain size; (c) distribution of twin lamellae.

It is well known that grain boundary can strengthen crystalline metals. Equation (3.1) depicts the Hall–Petch equation for the effect of grain size on yield strength.

$$\sigma_y = \sigma_0 + \frac{k_y}{\sqrt{d}} \tag{3.1}$$

where $\sigma_0$ is the friction stress; $k_y$ is strengthening coefficient, and $d$ is average grain diameter.

In addition, twin planes are parallel to each other in a grain, so the confined layer slip (CLS) model can be adopted to explain the twin strengthening mechanism. Substituting the limiting thickness with twin spacing in the CLS model can render Eq. (3.2),

$$\sigma_{cls} = M\beta\frac{\mu b}{\lambda}\ln\frac{\alpha\lambda}{b} \tag{3.2}$$

where $M$ is the Taylor factor, $\mu$ is the shear modulus, $b$ is the Burger's vector component, $\lambda$ is the twin spacing, $\alpha$ is the material constant related to the dislocation core, and $\beta$ is Poisson's ratio. Taylor's law could explain deformation in multi-slipping systems. Figure 3.8 shows the possible slip systems of dislocations between two coherent twin boundaries. For FCC metals, the slip systems are $\{111\}\langle 1\bar{1}0\rangle$, and there are 12 slip systems. The slip systems between the two twin boundaries could be divided into three types. In type 1, the slip plane and slip direction of the dislocation are parallel to the twin boundaries. In type 2, the slip plane and slip direction of the dislocation are both unparallel to the twin boundaries. In type 3, the slip plane is unparallel to the twin boundaries, but the slip direction is parallel to the twin boundary.

It is reported that the tensile strength of nt-Cu increases as the twin spacing decreases, and it reaches a maximum value of approximately 1050 MPa when the twin spacing is 15 nm. Then it decreases to 700 MPa as the twin spacing further deduces to 4 nm.

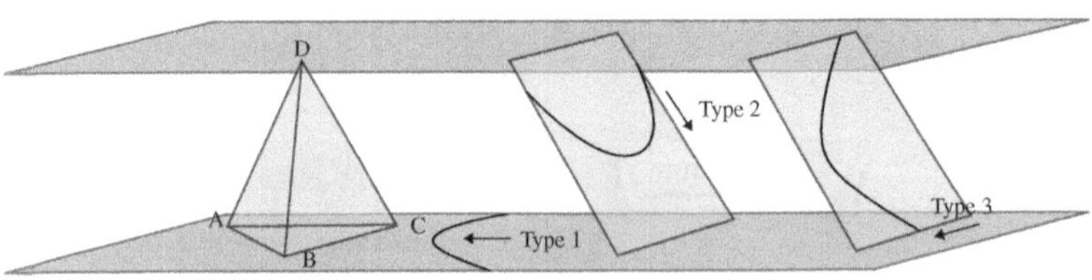


**Figure 3.8** The schematic drawing for dislocation slip between two parallel twin planes.

## 3.5 Formation of Unidirectionally (111)-oriented Nanotwin Copper

An evolution from 2D to 3D IC is underway within the microelectronics industry because the limits of very-large-scale integration in chip technology are approached. In essence, 3D IC is meant to bring packaging technology and chip technology together. Concerning the scaling of density of solder bumps on a chip surface, the diameter of a flip-chip solder bump is about 100 μm today, and it should be possible to be scaled down to 1 μm. This improvement will increase the density of bumps per unit area by 4 orders of magnitude, yet the solder volume is going to be reduced by 6 orders of magnitude. However, the freezing point of the solder remains unchanged, meaning that every micro-bump may contain only a couple of grains after processing. Variation within the grain orientation may cause a wide distribution of the orientation-dependent properties of μ-bumps, which can lead to early failure and low reliability. This is because certain grain orientations of β-Sn allow ultrafast interstitial diffusion and electromigration. [7] Because there is a large number of μ-bump on a stack of chips in a 3D IC device, how to achieve a consistent microstructure in several thousands of μ-bumps on each chip may become a critical issue?

Maintaining control over the microstructure of the μ-bumps is nontrivial due to reflow. Reflow means the solder bump is melted in order to realize the joining of chip-to-chip or chip-to-substrate. It is very difficult to regulate the solid-state microstructure as a molten solder bump solidifies; typically, the grains are randomly oriented.

Furthermore, during the solid-state aging of the μ-bumps, the solder will transform into IMCs. These compounds will dominate the properties of the μ-bumps because the solder thickness decreases to about 10 μm or less in each μ-bump in order to attach the through-Si-vias (TSVs).

Because the twin plane is the (111) plane of Cu, and if we can deposit this plane to be parallel to the substrate surface during electroplating, we have a unidirectional growth of (111)-oriented nt-Cu film, or we have a very highly textured Cu film. The reason to do so is that if we can control the microstructure of Cu thin films, in turn we can affect the interfacial reaction between the Cu and solder or with itself in Cu-to-Cu direct bonding.

In this section, we discuss the deposition of the highly oriented [111] nt-Cu grains with densely packed nanotwins by direct-current electroplating. [8] It has been reported that nt-Cu with a [111] preferred orientation can be deposited via magnetron sputtering. However, in interconnect and packaging technology, Cu interconnects are deposited by electroplating because electroplating can have better capability to fill narrow trenches and vias.

During DC electroplating, surfactants are added to the electrolyte, and the stirring speed of the electrolyte, starting from 600 to 1200 revolutions per min (rpm), was found to be critical to the electroplating result. Figure 3.9 shows the cross-sectional focused ionic beam (FIB) image of the nt-Cu electroplated with a current density of 80 $mA/cm^2$ at 1200 rpm for 30 minutes. The thickness of the Cu film is larger than 20 μm, and the columnar Cu grains with unidirectionally oriented nanotwins are obtained. All the [111]-oriented nanotwins are aligned nearly vertical to the film growth direction. X-ray diffraction indicates that these Cu grains possess a really strong [111] preferred orientation. In Figure 3.9, we note that a lot of cones appear on the film surface. Images from twins were observed in the cross-sectional TEM, and the [111] orientation was verified by corresponding electron diffraction patterns. The spacing of the nanotwins ranges from 10 to 100 nm. Figure 3.9a shows a FIB image, which has both the top view and cross-sectional view of an oriented nt-Cu. On the cross-section, columnar Cu grains are shown and in each grain there are layered parallel twin planes. Figure 3.9b shows X-ray diffraction spectrum of the sample, where only a very strong <111> and <222> reflections of Cu are shown. After electro-polishing, the sample in Figure 3.9a was analyzed by electron backscattered diffraction (EBSD), and the results are presented in Figure 3.10. Almost all the surface grains in the nt-Cu films are [111] oriented. Thus, the nt-Cu film possesses a very high [111] preferred orientation.

The growth mechanism for [111] preferred nt-Cu may be different from randomly oriented nt-Cu film. In general, surfactants play an important role during the DC electrodeposition. The surfactants may have higher adsorption ability on non-[111]-oriented grains, thus exposing [111] grains to the electrolyte. However, it takes time for [111] columnar grains to prevail. Figure 3.11 shows the cross-sectional FIB image for the electroplated nt-Cu film. On top of the Cu seed layer, there exists a nanoscale randomly oriented layer, which is denoted as transition layer. Then, the large-grained columnar [111] Cu grain grew after the transition layer. This layer is typically 0.2 μm to few microns, depending on the electroplating conditions. The transition layer can be reduced by periodic reverse waveforms. It is because the non [111] grains can be dissolved during the reverse waveforms.

**Figure 3.9** (a) FIB images of the both the top view and cross-sectional view of an oriented and nanotwinned Cu film. (b) X-ray diffraction spectrum of the oriented and nanotwinned Cu film; only a very strong (111) and (222) reflections of Cu were detected.

**Figure 3.10** Plan-view EBSD showing the orientation of surface grains of the nt-Cu film.

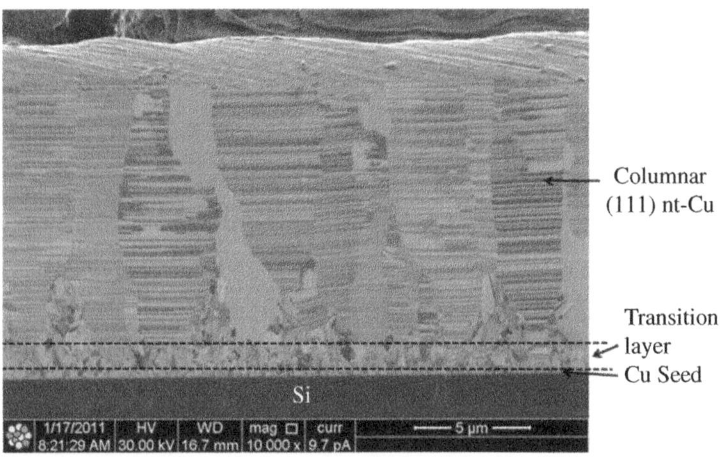


**Figure 3.11** Cross-sectional FIB image of an electroplated [111]-oriented nanotwinned Cu film.

## 3.6 Grain Growth in [111]-Oriented nt-Cu

In this section, we introduce the interesting grain growth behaviors in electroplated [111]-oriented nt-Cu [19–23]. The transition layer is thermally unstable because it consists of nanoscale grains. For magnetron sputtered [111] nt-Cu films, there is almost no transition layers. Thus, the sputtered nt-Cu grains are stable up to 800 °C. However, for electroplated [111] nt-Cu films, grain growth takes place upon annealing and the grains will transform to other orientations. At low annealing temperatures, the large [111] columnar grain consume the small grains in the transition, resulting even higher [111] preferred orientation in the nt-Cu film. Figure 3.12 shows cross-sectional FIB images for the same specimen before and after the annealing at 250 °C for one hour. To observe the grain growth in the same grains, FIB was used to cut a marker in the specimen. The grains A, B, and C grew downward, not only to consume the small grains in the transition layer but also to form densely packed annealing twins. Then, the electroplated columnar grains may be stable up to 500 °C, depending on the thickness of the transition layer. Thus, the thermal stability can be tailored by controlling the thickness of the transition layer.

However, the electroplated [111] nt-Cu films undergo a unique anisotropic large grain growth when it is annealed at 400 °C. Figure 3.13 shows a sequence of the transformation of [111] nt-Cu to (100)-oriented large grain Cu without twins. In Figure 3.13a, the EBSD cross-section image of the as-deposited [111]-oriented nt-Cu is shown. Upon annealing at 400 °C for 20 minutes, Figure 3.13b shows that transformation to an extremely large grain of (100) orientation has occurred from the bottom side (transition layer) of the nt-Cu film. The size of the (100) grain has a diameter about 290 μm but a thickness about 7 μm. Figure 3.13c shows the EBSD image of

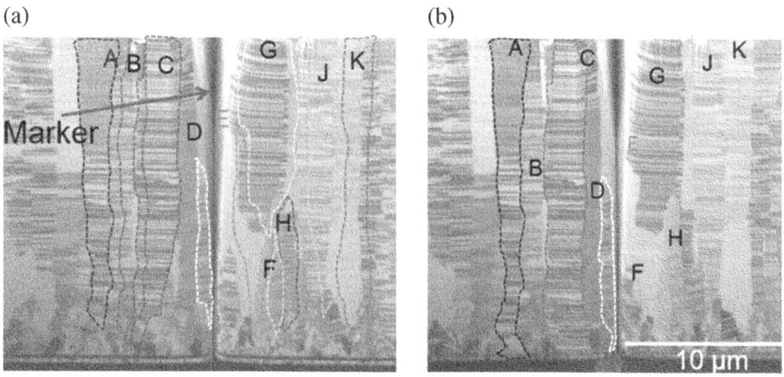


**Figure 3.12** Grain growth in [111] nt-Cu film. Cross-sectional FIB images for the same specimen. (a) as-plated; (b) after 250 °C-one hour annealing.

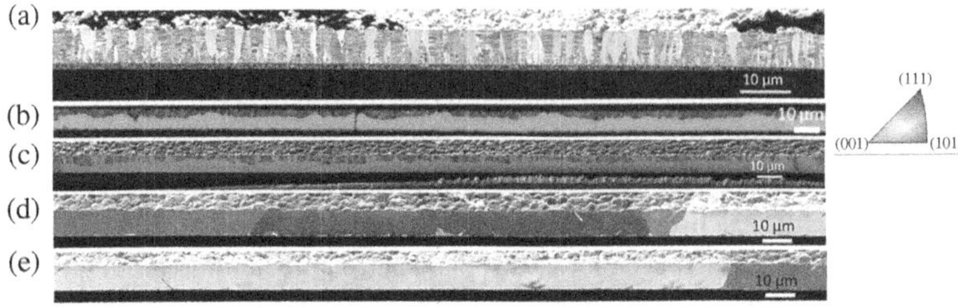


**Figure 3.13 Transformation of the oriented (111) nt-Cu to large grain (100) Cu.** (a) Cross-sectional FIB image of the as-electroplated nt-Cu film. (b) Cross-sectional orientation image map of the large <100> crystal with a lateral crystal size of 290 μm. (c) Cross-sectional FIB image of the sample in b. (d) and (e) Cross-sectional FIB images for the sample with all of the <111> grains transformed into large <100> crystals.

the cross-sectional view after annealing at 400 °C for 40 minutes. After the annealing at 400 °C for 60 minutes, all the columnar nt-Cu grains were transformed into large (100) grains, as shown in Figure 3.13d–e. The growth of the (100) grain is highly anisotropic; it has a very slow growth in the vertical direction, but the growth in the lateral direction is about 50 times faster.

Figure 3.14a and c show the top view of the microstructure of the (100) grains after one hour annealing at 400 and 450 °C, respectively. Figure 3.14b

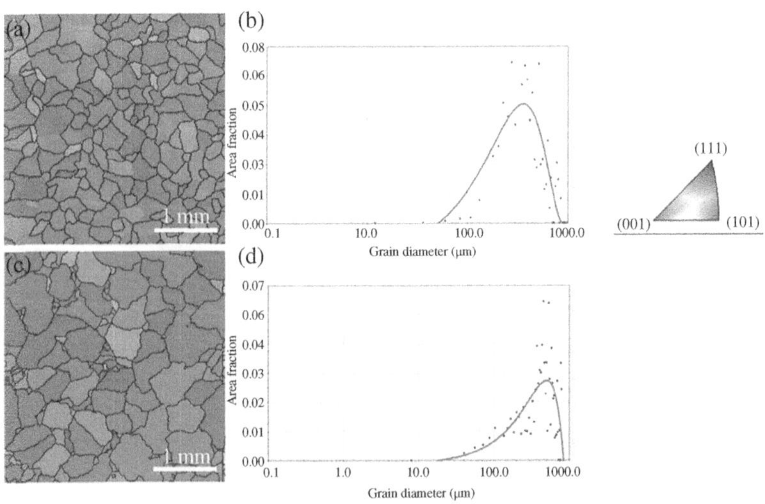


**Figure 3.14** The growth of large (100) grains from the oriented (111) nt-Cu. (a) Plan-view EBSD OIM image for a 7 μm nt-Cu film annealed at 400 °C for one hour. (b) Grain size distribution for the (100) grains in (a). (c) Plan-view EBSD OIM image for a 7 μm nt-Cu film annealed at 450 °C for one hour. (d) Distribution of the (100) grains in (c).

and d show the grain size distribution, respectively. The average grain size in Figure 3.14c is about 400 μm. The transformation temperature can be altered from 150 to 500 °C by varying the thickness of transition layer and microstructures of the columnar nt-Cu grains. In some application, such as Cu-to-Cu bonding, we prefer to trigger the anisotropic grain growth at low temperatures, so that the bonding interface can be eliminated. We will discuss it later in this chapter.

## 3.7 Uni-directional Growth of $\eta$-$Cu_6Sn_5$ in Microbumps on (111) Oriented nt-Cu

In Section 3.2, we reviewed why a high density of nanotwin can form and what is the energy of formation of nt-Cu. Then, in Section 3.5, we discuss how to control the growth direction of nanotwins so that they have an oriented growth in [111] direction, so that all the twin planes are parallel to the substrate surface. It means we can control the growth of the nanoscale microstructure. With the formation of the oriented nanotwins, we can have two interesting transformations. First, we will have a unidirectional growth of $Cu_6Sn_5$ IMC during the wetting reactions of molten solder on the [111]-oriented nt-Cu. Second, we can transform the [111]-oriented nt-Cu to a really large single crystal of (100) Cu having no twins, due to a particularly anisotropic abnormal grain growth of Cu, as discussed in Section 3.6. This transformation will be significantly important in Cu-to-Cu direct bonding.

Figure 3.15 shows the EBSD OIM image for the orientation relationships between the [111] nt-Cu under-bump-metallization (UBM) and the $Cu_6Sn_5$ intermetallic compounds (IMCs) grown on the nt-Cu. The nt-Cu remained highly [111] oriented after the reflow process at 260 °C for three minutes. Figure 3.15b depicts the $Cu_6Sn_5$ grains only and orientation analysis shows that the $Cu_6Sn_5$ grains possess a preferred orientation of $[2\bar{1}\bar{1}3]$. The orientation relationship is mainly due to small lattice mismatch between the [111] Cu and $[2\bar{1}\bar{1}3]$ $Cu_6Sn_5$ [24].

In addition, it has been reported that severe Kirkendall voids form in the $Cu_3Sn$/Cu interface after solid-state aging in solder joints and greatly weaken the mechanical properties of the joint. The cross-sectional FIB image in Figure 3.16a of a μ-bump after solid-state aging at 150 °C for 500 hours shows that the solder has been transformed completely into $Cu_6Sn_5$ and $Cu_3Sn$. The $Cu_3Sn$ layer is 2.14 μm thick, yet only a couple of voids are observed within the $Cu_3Sn$/Cu interface. After aging at 150 °C for 1000 hours, see Figure 3.16b, the $Cu_3Sn$ layer thickness grew to 3 μm, and no obvious Kirkendall voids were found. The cross-sectional TEM image of the

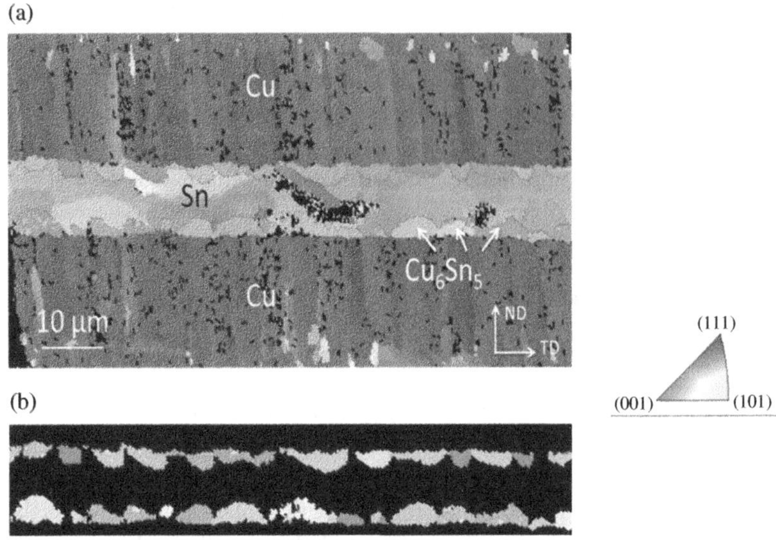


**Figure 3.15** (a) Cross-sectional EBSD orientation image map for a SnAg microbump with [111]-oriented nt-Cu metallization. The inverse pole figure represents the direction along the normal direction (ND). The solder joints were reflowed at 260 °C for three minutes, and (b) EBSD image showing the orientation of $Cu_6Sn_5$ grains only. The $Cu_6Sn_5$ grains possess a preferred orientation of $[2\bar{1}\bar{1}3]$.

interface of $Cu_3Sn$/nt-Cu showed that some nanotwins end in $Cu_3Sn$ grains, see Figure 3.16c. Although the twin boundaries are coherent, there are steps and kinks, which will function as vacancy sinks. The high density of nanotwins will provide a high density of vacancy sinks. Therefore, no supersaturation of vacancies or nucleation of Kirkendall voids occurs [25].

## 3.8 Low Thermal-Budget Cu-to-Cu Bonding Using [111]-Oriented nt-Cu

Low thermal-budget direct Cu-to-Cu bonding enabled by surface creep on (111) surfaces of nt-Cu is of wide interest because it has the potential to replace solder joints in high-end packaging technology, including the packaging of mainframe computers. [26–38] The process of Cu direct bonding can be divided roughly into four stages. The first stage is mechanical contact of two Cu surfaces and a certain amount of plastic deformation may occur. Under thermal compression, the second stage is to form twist-type grain boundaries at the contact interface by surface creep, which achieves the

**Figure 3.16** Microstructures after solid-state aging of nt-Cu/solder/nt-Cu at 150 °C. (a) Cross-sectional FIB image after 500 hours; (b) Cross-sectional FIB image after 1000 hours; (c) TEM image showing the interface of nt-Cu and $Cu_3Sn$.

initial bonding. To verify the formation of a twist-type grain boundary with a hexagonal network of dislocations, plan-view TEM was performed, see Figure 2.24. The third stage is void ripening at the bonding interface. The fourth stage is (100) grain growth across the bonding interface to eliminate the bonding interface.

Under ultrahigh vacuum (UHV) conditions, direct Cu-to-Cu bonding is often achieved at room temperature; however, the Cu surface must be cleaned before bonding by employing a surface-activated approach, and therefore, the process is rather time-consuming. In ordinary vacuum conditions on the order of $10^{-3}$ to $10^{-4}$ torr, direct Cu-to-Cu bonding is often obtained by thermal compression at temperatures exceeding 300 °C for over 30 minutes under a compression of roughly 100 psi. Nevertheless, the temperature is above the melting temperature of eutectic SnPb and Pb-free SnAgCu solders. The standard reflow temperature of these solders is below 250 °C. Thus, the key challenge in direct Cu-to-Cu bonding is to find the way to lower the bonding temperature to 250 °C in an ordinary vacuum.

The essential process of direct bonding is found to be surface diffusion, where atoms may jump across the bonding interface or along the interface. The surface diffusion coefficient on (111) surface of Cu is 3–4 orders of magnitude faster than that on other surfaces, see Table 2.1, indicating that the bonding temperature can be lowered on (111) surfaces [26, 39]. However, it is unrealistic to expect the utilization of (111)-oriented single crystals of Cu in cubic inch size in applications. To overcome this challenge, the surface of a large (111)-oriented single crystal of Cu is replaced by a highly [111]-oriented nt-Cu polycrystalline thin film. Furthermore, when an oriented nt-Cu is deposited with the (111) twin plane parallel to the free surface, direct bonding down to 150 °C are often achieved. To realize low temperature direct Cu-to-Cu bonding, the critical requirement is that one of the Cu pieces to be bonded must have a (111) surface. The bonding can occur at a quicker pace if both Cu pieces are (111)-oriented nt-Cu. Figure 3.17 shows the microstructures of a [111] nt-Cu film bonded to a [111] nt-Cu film at 150 °C for one hour at 0.78 MPa in $10^{-3}$ torr. Both the [111] nt-Cu films were electroplated by DC waveforms. Due to the high surface diffusivity and low oxidation rate on (111) surface [40], good bonding

**Figure 3.17** Cross-sectional FIB image showing the microstructures of nt-Cu film bonded to nt film. The bonding conditions were at 150 °C for one hour at 0.78 MPa in $10^{-3}$ torr.

can be achieved at 150 °C. Few tiny voids remained at the bonding interface. The percentage of the residual voids is approximately 5%.

For tensile tests of the bonding, the wafers were cut into 2 × 2 $cm^2$ squares. The tensile tests were performed using an auto inserting Pulling Force Tester 1220 S (SE Test Systems Co., Taiwan). Before tensile testing, each sample was adhered to a screw by applying a thermosetting composition adhesive at 150 °C for two hours. After curing, the sample was placed within the auto-inserting pulling force tester. The strain rate is 10 μm/min. The pull test showed that the bonded interface is stronger than the interface between Cu and its substrate.

In Figure 3.18, a schematic diagram depicts a local region of the initial contact between two (111) surfaces. Because the two surfaces are not ideally flat, there are many gaps or voids between them. To consider the bonding interface, a distribution of cavities within the interface is assumed. A cavity of radius *r* is taken into account here. The matrix surrounding the cavity is assumed to be under a compressive stress of σ. The strain applied in this study was within the elastic regime. The free surface of the cavity has no normal stress. Thus, the strain potential drops between the matrix and center of the cavity, as depicted in Figure 3.18. This stress potential drop must be a gradual phenomenon, assuming that there is no stress center within the compressive region.

The bonding time can be shortened to 30 minutes by increasing the bonding temperature to 200 °C. It yielded an interface that is almost free from residual voids. Bonding of (111)-oriented nt-Cu to ordinary Cu also can be achieved at this condition. The bonding time is further reduced to less than 10 minutes by increasing the bonding temperature to 250 °C. Almost no voids are present within the interface; therefore, the nanotwinned columnar grains remain after the bonding process. The bonding of (111)-oriented nt-Cu to ordinary Cu at 250 °C yielded the result, in which only a few voids

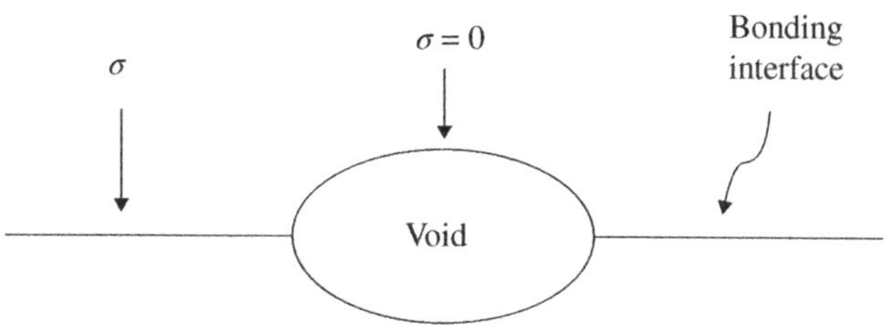


**Figure 3.18** Schematic diagram of a void in the bonding interface. Under thermal compression, there is a stress potential gradient between the bonded area and the unbonded void surface, which is stress-free. Creep will drive atoms to move from the stressed region to the void region.

were found. No grain growth was observed after annealing at 250 °C for one hour. However, grain growth happened after 250 °C for two hours; this might be attributed to the thermal instability of the transition layer.

It is worth mentioning that in the columnar grains of unidirectionally oriented nt-Cu, all the grain boundaries are tilt-type grain boundaries, with a distribution of the tilt angles. When two pieces of Si chip having the (111)-oriented columnar grains are bonded face-to-face, the bonding interface becomes twist-type grain boundaries. Kinetically, surface creep occurs. This is shown in Figure 3.18, where the regions in contact is under compression, yet the void surface is stress-free. Thus, there is a stress potential gradient to drive atoms to move from the stressed region to the unstressed region. It is a creep process. We recall that the creep model of Nabarro-Herring assumes lattice diffusion, and Coble creep model assumes grain boundary diffusion. Here, we consider creep by surface diffusion. At this stage, there are voids in the bonding interface, and bonding interface still exists, so the mechanical shear strength of the bonding interface is weak. However, big improvement can occur if we can eliminate the bonding interface.

Elimination of bonding interface is also critical to electric and mechanical properties of the Cu-to-Cu joints. For regular Cu films, it is hard to eliminate the bonding interface below 300 °C. However, by using electroplated [111]-oriented nt-Cu films, the bonding interface can be eliminated below 250 °C by anisotropic large grain growth shown in Figure 3.13. Figure 3.19a–c show the grain growth for two 1.5 μm thick Cu films after annealing at 250 °C for two hours. The bonding interface was completely eliminated and very large grain grew across the entire films [41–44]. Extremely large grains grew to consume all the small nt-Cu grains. As discussed in Chapter 2, the electric and mechanical properties of the joint will be close to bulk Cu. The research on void ripening in the Cu-to-Cu bonds is still active at this moment, and more studies are needed in order to provide a better understanding of this issue.

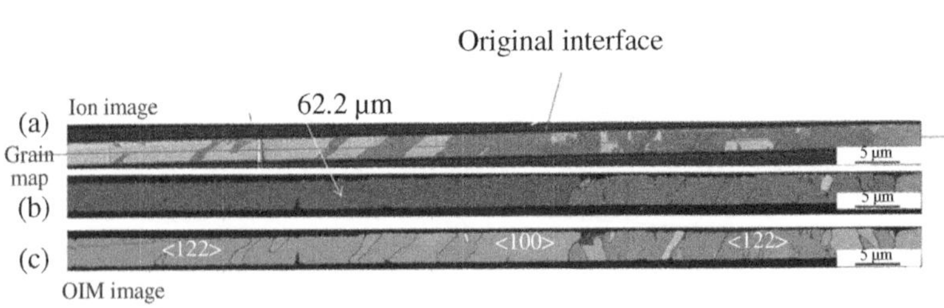


**Figure 3.19** Microstructures analysis of the bonding interface for a [111] nt-Cu film bonded to a [111] nt-Cu film at 250 °C for two hour. (a) FIB ion image, (b) EBSD grain map, and (c) EBSD OIM image.

## 3.9 Nanotwin Cu RDL for Fanout Package and 3D IC Integration

For high-performance computing and communication devices, the power continues to increase, so the current density in the RDL continues to rise. Thus, electromigration has become a reliability issue in RDL in 3D IC integration and fanout packages. In addition, the AI chip and package size continue to increase due to the requirement of performance. The chips are embedded in epoxy mold compound, as illustrated in Figure 3.20. Also, the thermal stress in the fanout package continues to increase as the package size increases. The toughness of the Cu RDLs become very important. If the toughness of the Cu RDL is low, they would fail by forming cracks in the RDL during fabrication process or under reliability tests of thermal cycling. Furthermore, the Cu RDLs have to be capped by polyimide (PI) to prevent oxidation and corrosion, and the curing temperature for the low-temperature PI is around 230 °C. There may be up to three layers of Cu RDLs in a fanout package. The Cu RDLs should possess high thermal stability to be stable at 230 °C for three hours. Therefore, industry is looking for Cu lines with high thermal stability, high electromigration lifetime, and high toughness.

Nanotwinned Cu possesses high toughness and high thermal stability up to 300 °C [45, 46]. Figure 3.21a presents the tensile stress-strain curve for the as-electroplated 20 μm thick nt-Cu and the regular Cu foils. The tensile strength is 800 and 500 MPa for nt-Cu and regular foil, respectively. The elongation is less than 4% for both Cu foils. After annealing at 250 °C for three hours, the tensile strength decreases to 656 and 280 MPa for nt-Cu and regular Cu foil, respectively. Yet the elongation increases to approximately 12% for both foils. The area under the stress-strain curve is denoted as toughness, which

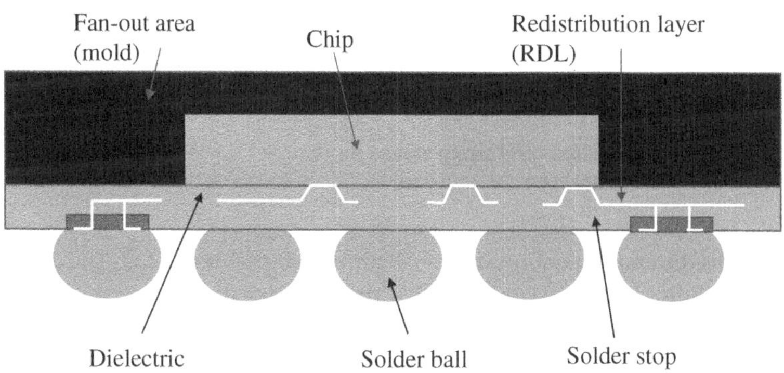


**Figure 3.20** Schematic drawing for a fanout package. A chip is embedded in epoxy mold compound. Cu RDLs are electroplated to connect the chip and solder joints.

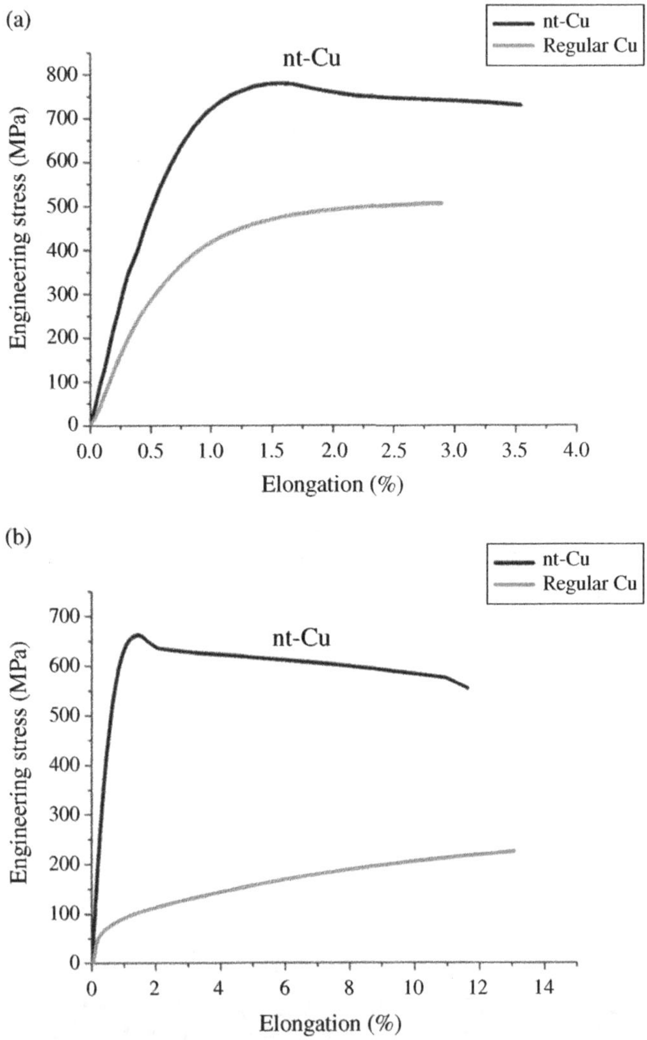


**Figure 3.21** Tensile stress and strain curves for regular Cu and nt-Cu films of 20 μm thick. (a) As-deposited; (b) annealed at 250 °C for three hours.

represents the energy per unit volume required to break the Cu foil. The calculated toughness is $65\times10^6$ J/m$^3$ for nt-Cu after the annealing, which is approximately twice of that of the annealed regular Cu foil ($30\times10^6$ J/m$^3$). This difference is mainly attributed to the extensive grain growth occurring in the regular Cu during the annealing. Nevertheless, the columnar grain structure with nanotwins still remained in the nt-Cu foil after the annealing. The twin

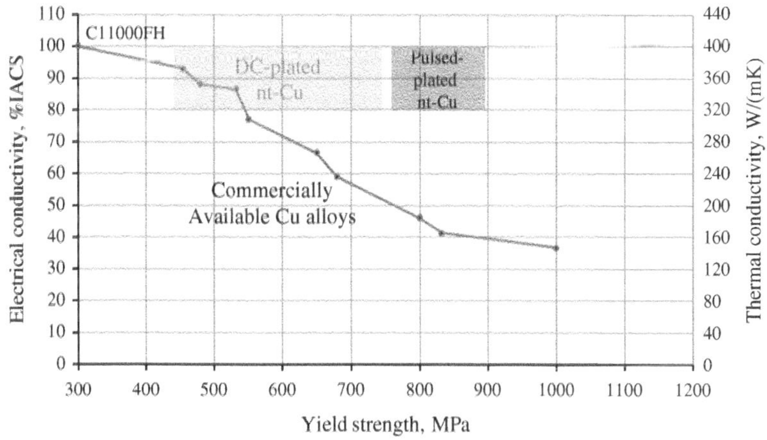


**Figure 3.22** Plot of yield strength against electrical and thermal conductivities for commercially available Cu alloys, DC-electrodeposited, and pulsed-electrodeposited nt-Cu films.

spacing slightly increased and results in the slight decrease in the tensile strength. Figure 3.22 summarizes the mechanical-electrical-thermal properties of commercially available Cu alloys, DC electroplated nt-Cu and pulsed-electroplated nt-Cu films. For most of the high-strength Cu alloys, the trade-off is the loss of electrical and thermal conductivities. The nt-Cu films possess high mechanical strength with excellent electrical and thermal conductivities. Therefore, they have high potential to be adopted for next-generation RDL and connectors for microelectronics devices.

Electromigration failure mechanism is investigated in the nt-Cu and regular Cu RDLs capped with polyimide dielectrics. Figure 3.23a and b show the

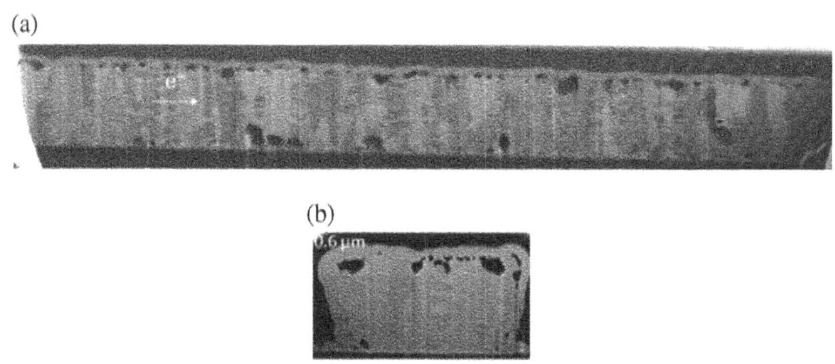


**Figure 3.23** Voids and oxidation formation in the nt-Cu line after current stressing by $1 \times 10^6$ A/cm$^2$ at 200 °C. (a) Along long axis; (b) along short axis.

typical microstructures along the long and short axis, respectively. The width of the Cu line is 10 μm. The nt-Cu line was stressed by $1 \times 10^6$ A/cm$^2$ at 200 °C, and the current was terminated when the resistance increase of the line reached 20% of its initial value. Results indicate that the electromigration lifetime of nt-Cu RDLs is approximately four times longer than that of the regular Cu RDLs. Two important failure mechanisms contribute to the lifetime increase. First is void formation at the interface of Cu and polyimide, as well as at the transition layer. Because there were many nanoscale grains at the transition layer, grain boundaries are continuous there. Thus, grain boundary diffusion can cause fast migration of Cu atom, resulting in void formation there. Second, oxidation occurred at the Cu/polyimide interface. Oxygen atoms diffused through the polyimide layer, and thus Cu was oxidized during electromigration. As shown in Figure 3.23, the oxide thickness is approximately 0.4 μm. The oxidation causes 10% resistance increase in the $10 \times 5$ μm Cu line [46]. The oxidation issue will become more critical to the electromigration of the Cu RDLs as the dimension continues to scale down, because the surface-to-volume ratio for narrow RDLs is higher than the wider Cu RDLs. For example, with the same oxide thickness of 0.4 μm, the resistance would increase 35% in a $2 \times 2$ μm Cu line. How to prevent oxidation for fine Cu RDLs becomes a key issue to prolong the electromigration lifetime of the Cu RDLs.

To summarize, electroplated [111] nt-Cu has many advantages, including high strength, high thermal and electric conductivity, high surface diffusivity, low surface oxidation rate, and high electromigration lifetime. Therefore, it may be adopted as the material for Cu-to-Cu bonding and RDLs for 3D IC integration.

## Problems

**3.1** (a) Explain what twin boundaries are.

**3.2** (b) Growth twins become more and more important because twin spacing can be controlled by electroplating. What are the growth mechanism for nt-Cu fabricated by pulsed electroplating.

**3.3** Twin boundaries have become fifth strengthening mechanism for metals. Explain how twin boundaries strengthen metals.

**3.4** Describe what integrated fanout wafer-level package technology is. What are its key reliability issues?

**3.5** Usually, electroplated Cu exhibits low thermal stability, and grain growth may take place at room temperature, which is called self-annealing. However, (111)-oriented nanotwinned Cu films are very stable. Explain what the reason is.

**3.6** Explain why Cu {111} surfaces have lowest oxidation rate among all the planes in Cu crystals.

**3.7** Compared the mechanical strength of randomly- oriented and (111)-oriented nt-Cu films, which one may have higher ultimate tensile strength? Why?

**3.8** Upon annealing at 250 °C, why the transition layer with nanocrystalline grains are converted by the large [111]-oriented grains?

**3.9** Controlling orientation of intermetallic compounds become more and more important in solder micro-bumps. For Cu/SnAg-solders/Cu micro-bumps, give an example how to control the orientation of $Cu_6Sn_5$ intermetallic compounds.

**3.10** What are the key parameters for Cu-Cu bonding? List three factors that may reduce the bonding temperatures or time.

**3.11** There exists a bonding interface between two Cu films or bumps. How do you eliminate it?

## References


1 L. Lu, Y. Shen, X. Chen, L. Qian, K. Lu, Ultrahigh strength and high electrical conductivity in copper, *Science*, 304, pp. 422–426 (2004)

2 K.C. Chen, W.W. Wu, C.N. Liao, L.J. Chen, K. N. Tu, "Observation of atomic diffusion at twin-modified grain boundaries in copper," *Science*, Vol. 231, pp. 1066–1069 (2008).

3 K. Lu, L. Lu, S. Suresh, *Science* 324, 349 (2009)

4 L. Lu, X. Chen, X. Huang, K. Lu, *Science* 323, 607 (2009).

5 Xiaoyan Li, Yujie Wei, Lei Lu, Ke Lu, Huajian Gao, *Nature* 464, 877 (2010).

6 D. Xu, W. L. Kwan, K. Chen. X. Zhang, V. Ozolins, K. N. Tu, «Nanotwin formation in Cu thin films by strss/strain relaxation in pulse electrodeposition», *Applied Physics Letters*, 91, 254105 (2007).

7 L. Minhua, D-Y. Shih, P. Lauro, P. Goldsmith, and D. W. Henderson, «Effect of Sn rain orientation on electromigration degradation mechanism in high-Sn based Pb-free sodlers,» *Applied Physics Letters*, 92, 211909 (2008).

8 H. Y. Hsiao, C. M. Liu, H. W. Lin, T. C. Liu, C. L. Lu, Y. S. Huang, C. Chen, K. N. Tu, "Unidirectional growth of microbumps on (111) oriented nano-twin copper," *Science*, Vol. 336, 1007 (2012).

9 B.Z. Cui, K. Han, Y. Xin, D.R. Waryob, and A.L. Mbaruku, "Highly textured and twinned Cu films fabricated by pulsed electrodeposition", *Acta Materialia*, 55, 4429–4438 (2007).

10 I. J. Beyerlein, X. Zhang, and A. Misra, "Growth twins and deformation twins in metals", *Annual Review of Materials Research*, 44, 329–363 (2014).

11 S. Lia, Q. Zhu, B. Zheng, J. Yuan, and X. Wang, "Nano-scale twinned Cu with ultrahigh strength prepared by direct current electrodeposition", *Materials Science and Engineering A*, 758, 1–6 (2019).

12 Julia R. Greer, "Nanotwinned metals: it's all about imperfections", *Nature Materials*, 12, 689–690 (2013).

13 Q. Pan, H. Zhou, Q. Lu, H. Gao, and L. Lu, "History-independent cyclic response of nanotwinned metals", *Nature*, 551, 214–217 (2017).

14 H. P. Chen, C. W. Huang, C. W. Wang, W. W. Wu, C. N. Liao, L. J. Chen and K. N. Tu, "Optimization of the nanotwin-induced zigzag surface of copper by electromigration", *Nanoscale*, 8, 2584–2588 (2016).

15 O. Anderoglu, A. Misra, H. Wang, and X. Zhang, "Thermal stability of sputtered Cu films with nanoscale growth twins", *Journal of Applied Physics*, 103, 094322 (2008).

16 X. H. Chen, L. Lu, and K. Lu, "Electrical resistivity of ultrafine-grained copper with nanoscale growth twins", *Journal of Applied Physics*, 102, 083708 (2007).

17 K. N. Tu and A. M. Gusak, in Chapter 10 on "formation and transformation of nanotwins in copper" "Kinetics in Nanoscale Materials, K. N. Tu and A. M. Gusak, Wiley (2014).

18 D. Xu, V. Sriram, V. Ozolins, J.M. Yang, K. N. Tu, G. R. Stafford, and C. Beauchamp, In situ measurements of stress evolution for nanotwin formation during pulse electrodeposition of copper, *Journal of Applied Physics* 105, 023521 (2009)

19 Y. S. Huang, C. M. Liu, W. L. Chiu, and C. Chen, "Grain growth in electroplated (111)-oriented nanotwinned Cu", *Scripta Materialia*, 89, 5–8 (2014).

20 T. C. Liu, C. M. Liu, H. Y. Hsiao, J. L. Lu, Y. S. Huang, and C. Chen, "Fabrication and characterization of (111)-oriented and nanotwinned Cu by DC electrodeposition", *Crystal Growth and Design*, 12, 5012–5016 (2012).

21 C. H. Tseng, and C. Chen, "Growth of highly (111)-oriented nanotwinned Cu with the addition of sulfuric acid in $CuSO_4$ based electrolyte", *Crystal Growth & Design*, 19, 81—89 (2019).

**22** T. L. Lu, Y. A. Shen, J. A. Wu, and C. Chen, "Anisotropic grain growth in (111) nanotwinned Cu films by DC electrodeposition", *Materials*, 13, 134 (2020).

**23** C. L. Lu, H. W. Lin, C. M. Liu, Y. S. Huang, T. L. Lu, T. C. Liu, H. Y. Hsiao, C. Chen, J. C. Kuo, and K. N. Tu, "Extremely anisotropic single crystal growth in nano-twinned copper", *NPG Asia Materials*, 6, e135 (2014).

**24** H. W. Lin, J. L. Lu, C. M. Liu, C. Chen, D. Chen, J. C. Kuo, and K. N. Tu, "Microstructure control of uni-directional growth of $\eta$-Cu6Sn5 in microbumps on <111> oriented and nanotwinned Cu", *Acta Materialia*, 61, 4910–4919 (2013).

**25** T. C. Liu, C. M. Liu, Y. S. Huang, C. Chen, and K. N. Tu, "Eliminate Kirkendall voids in solder reactions on nanotwinned copper", *Scripta Materialia*, 68, 241–244 (2013).

**26** C. M. Liu, H. W. Lin, Y. S. Huang, Y. C. Chu, C. Chen, D. R. Lyu, K. N. Chen, and K. N. Tu, Low-temperature direct copper-to-copper bonding enabled by creep on (111) surfaces of nanotwinned Cu, *Scientific Reports*, Vol. 5, 9734 (2015).

**27** Juang, J.Y., Shie, K.C., Hsu, P.N. et al. (2019). Low-resistance and high-strength Cu direct bonding in no-vacuum ambient using highly (111)-oriented nano-twinned Cu. *Proceedings of IEEE 69th Electronic Components and Technology Conference (ECTC)*, 642–647.

**28** Yu-Jin Li, K. N. Tu, and Chih Chen, "Tensile properties and thermal stability of uni-directionally <111>-oriented nanotwinned and <110>-oriented micro-twinned copper," *Materials*, Vol. 13, 1211 (2020).

**29** J. Wang, C. S. Ku, T. N. Lam, E. W. Huang, K. N. Tu, and Chih Chen, "Tuning stress in Cu thin films by developing highly (111)-oriented nanotwinned structure," *Journal of Electronic Materials*, Vol. 49, pp. 109–115 (2020).

**30** Yu-Jin Li, K. N. Tu and Chih Chen, «Tensile properties and thermal stability of unidirectionally <111>-oriented nanotwinned and <110>-oriented microtwinned copper», *Materials*, Vol. 13, 1211, doi:10.3390/ma13051211(2020).

**31** Suga, T., Yuuki, F., and Hosoda, N. (1997). A new approach to Cu-Cu direct bump bonding. 1st Joint International Electronic Manufacturing Symposium and the International Microelectronics Conference, pp. 146–151.

**32** Kagawa, Y., Fujii, N., Aoyagi, K. et al. (2016). Novel stacked CMOS image sensor with advanced Cu2Cu hybrid bonding. IEEE International Electron Devices Meeting (IEDM), pp. 8.4.1–8.4.4.

**33** K. N. Chen, A. Fan, C. S. Tan, R. Reif, and C. Y. Wen, "Microstructure evolution and abnormal grain growth during copper wafer bonding", *Applied Physics Letters*, 81, 3774–3776, (2002).

34 B. Rebhan, and K. Hingerl, "Physical mechanisms of copper-copper wafer bonding", *Journal of Applied Physics*, 118, 135301 (2015)

35 H. Noma, T. Kamibayashi, H. Kuwae, N. Suzuki, T. Nonaka, S. Shoji, and J. Mizuno, "Compensation of surface roughness using an Au intermediate layer in a Cu direct bonding process", *Journal of Electronic Materials*, 47(9), 5403–5409 (2018)

36 Y. P. Huang, Y. S. Chien, R. N. Tzeng, M. S. Shy, T. H. Lin, K. H. Chen, and H. M. Tong, "Novel Cu-to-Cu bonding with Ti passivation at 180 °C in 3-D integration", *IEEE Electron Device Letters*, 34, 1551–1553 (2013).

37 Y. P. Huang, Y. S. Chien, R. N. Tzeng, and K. N. Chen, "Demonstration and electrical performance of Cu–Cu bonding at 150 °C with Pd passivation", *IEEE Transactions on Electron Devices*, 62, 2587–2592 (2015).

38 W. Yang, M. Akaike, and T. Suga, "Effect of formic acid vapor in situ treatment process on Cu low-temperature bonding", *IEEE Transactions on Components, Packaging and Manufacturing Technology*, 4, 951–956 (2014)

39 C. M. Liu, H. W. Lin, Y. C. Chu, C. Chen, D. R. Lyu, K. N. Chen, and K. N. Tu, "Low-temperature direct copper-to-copper bonding enabled by creep on highly (111)-oriented Cu surfaces", *Scripta Materialia*, 78–79, 65–68 (2014).

40 C. H. Tseng, K. N. Tu, and Chih Chen, "Comparison of oxidation in uni-directionally and randomly oriented Cu films for low temperature Cu-to-Cu direct bonding", *Scientific Reports*, 8, 1–7 (2018).

41 Y. C. Chu, and C. Chen, "Anisotropic grain growth to eliminate bonding interfaces in direct copper-to-copper joints using< 111>-oriented nanotwinned copper films", *Thin Solid Films*, 667, 55–58 (2018).

42 J. Y. Juang, C. L. Lu, K. J. Chen, C. C. A. Chen, P. N. Hsu, C. Chen, and K. N. Tu, "Copper-to-copper direct bonding on highly (111)-oriented nanotwinned copper in no-vacuum ambient", *Scientific Reports*, 8, 1–11 (2018).

43 J. Y. Juang, C. L. Lu, Y. J. Li, K. N. Tu, and C. Chen, "Correlation between the microstructures of bonding interfaces and the shear strength of Cu-to-Cu joints using (111)-oriented and nanotwinned Cu", *Materials*, 11, 2368 (2018).

44 C. H. Tsen, I. H. Tseng, Y. P. Huang, Y. T. Hsu, J. Leu, K. N. Tu, and Chih Chen, "Kinetic study of grain growth in highly (111)-preferred nanotwinned copper films", *Materials Characterization*, 168, 110545 (2020).

45 Li, Y.J., Tseng, C.H., Tseng, I.H. et al. (2019). Highly (111)-oriented nanotwinned Cu for high fatigue resistance in fan-out wafer-level packaging. 2019 IEEE 69th Electronic Components and Technology Conference (ECTC), pp 758–762.

46 Tseng, I.H., Li, Y.J., Lin, B. et al. (2019). High electromigration lifetimes of nanotwinned Cu redistribution lines, 2019 IEEE 69th Electronic Components and Technology Conference (ECTC), pp 1328–1332.

# 4

# Solid–Liquid Interfacial Diffusion Reaction (SLID) Between Copper and Solder

## 4.1 Introduction

Solder has been used to join copper wires to form electrical circuitry connections in every electronic product, as well as to join copper pipes for plumbing in every modern house. The basic process in solder joint bonding is the chemical reaction between molten solder (typically the eutectic SnPb alloy or Pb-free SnAg alloy) and copper to form intermetallic compounds (IMC) of Cu-Sn, which have strong metallic bonds. [1, 2] The most important feature in solder joint bonding is the use of flux, which is a chemical reagent to remove the oxide on the Cu surface so that the molten solder can wet the clean Cu surface. The wetting angle, the smaller the better, is a direct indication of the success of a solder joint.

Often, people wonder that in today's big data era, why do we still study Cu-Sn, which is a bronze-age material and has been used for a long time? This is because first we cannot find a better replacement. Nevertheless, there have been very active research on Cu-to-Cu direct bonding to replace solder joints as discussed in Chapters 2 and 3. Second, solder joint has become extremely small in size in microelectronic packaging technology for consumer electronic products. Now it is less than 10 μm in diameter, so many unexpected problems in solder joint yield and reliability keep coming up due to the fact that the operation temperature of electronic products is about 100 °C, and Cu-Sn interfacial reaction occurs, which can be enhanced by surface diffusion. Third, its microstructure becomes unstable when the atomic diffusion is enhanced by electric current or Joule heating. Fourth, in electronic packaging technology, typically on a Si chip surface, there is a large number of solder joints, which can be several thousands of them and they have to be formed simultaneously in one reflow. Finally, in

*Electronic Packaging Science and Technology*, First Edition. King-Ning Tu, Chih Chen, and Hung-Ming Chen.

micro-bump technology, the solder joint microstructure is anisotropic, then if just one of them failed early, the device is broken.

To begin the study of solder joint reaction, we took a shallow glass breaker, which contains flux and placed it on a hot-plate at a temperature about 20–30 °C above the melting point of the solder. In the flux, we place a piece of Cu plate. Then we dropped a small solder ball, about 1 mm in diameter, on the Cu plate. The ball melted and formed a circular mount on the Cu. We removed the glass breaker from the hot-plate to cool it down, so that we could take out the solder mount on the Cu plate for examination. The procedure is simple, and thus to begin a research project on solder joint is easy.

Figure 4.1a shows the top view of a circular mount of solder on Cu. We made a cross-section of the mount and examined the interface between the

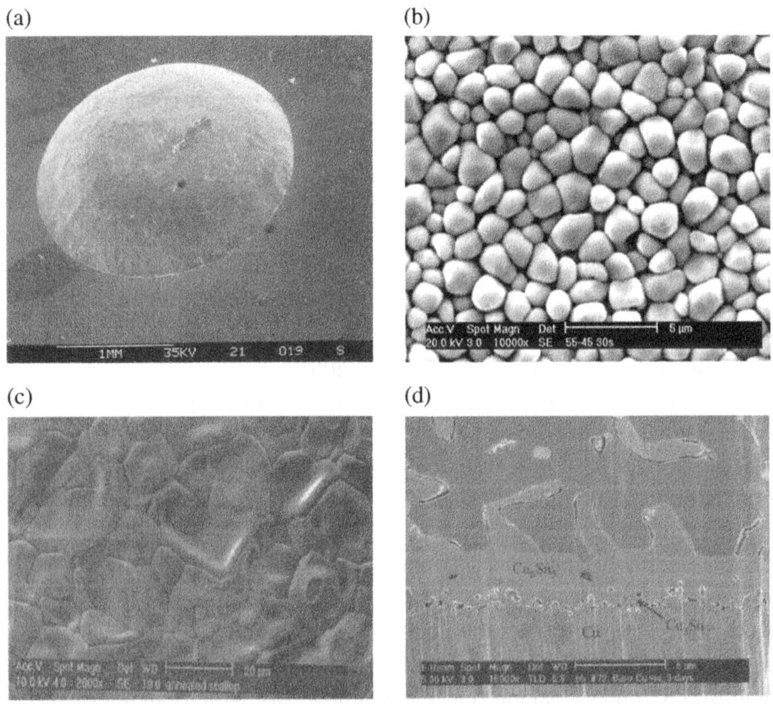


**Figure 4.1** (a) A circular solder mount or cap on a copper substrate. (b) The top-view SEM image of the surface of $Cu_6Sn_5$ scallops right after SLID and followed by removing the unreacted solder in etching. (c) The top-view SEM image of the surface of $Cu_6Sn_5$ grains after solid-state aging at 150 °C for 24 hours and followed by removing the unreacted solder. The solid-state aging has transformed the scallops into polycrystalline grains with a grain boundary between two of them. (d) In a long SLID reaction, the growth of $Cu_3Sn$, between $Cu_6Sn_5$ and Cu, tends to be accompanied by the formation of Kirkendall voids.

mount and the Cu. We can measure the wetting angle, and we can examine the IMC formation in the interface. The IMC ($Cu_6Sn_5$) grains has the shape of scallops. [3, 4] The scallop-type IMC is unique that its growth does not become a diffusion barrier, unlike the growth of a layer-type IMC. The kinetics of scallop-type growth will be presented later.

Figure 4.1b is the top-view of SEM image of the surface of $Cu_6Sn_5$ scallops right after SLID and followed by etching the unreacted solder away. Figure 4.1c is the top-view of SEM image of the surface of $Cu_6Sn_5$ grains after solid-state aging at 150 °C for 24 hours and followed by removing the unreacted solder. The solid-state aging has transformed the scallop-type IMC into a layer-type IMC having polycrystalline grains with a grain boundary between two of them. [5] The image of the deep trenches in Figure 4.1b is very different from that of the grain boundaries in Figure 4.1c.

A big advantage of scallop-type $Cu_6Sn_5$ is that it is not a diffusion barrier, so it can be dissolved by molten solder quickly. Therefore, the solder joint can go through several reflows without problem, which is needed in device manufacturing. However, a longtime reflow will enable $Cu_3Sn$ to form, which may contain Kirkendall voids.

In the binary phase diagram of Cu-Sn, there are only two IMCs, which are $Cu_6Sn_5$ and $Cu_3Sn$. Typically, the two do not form together in SLID until the former has grown to a thickness of nearly 5–10 μm. However, the growth of a thick $Cu_3Sn$ tends to be accompanied by the formation of Kirkendall voids, see Figure 4.1d, which has been one of the very harmful phenomena of solder joints, to be discussed later.

## 4.2 Kinetics of Scallop-Type IMC Growth in SLID

The growth of the scallops obeys the kinetics of ripening under the constraint of a constant contact area or a constant surface area. [4] This is because the contact area between the solder mount and Cu is constant, due to either the design of the contact opening in under-bump-metallization (UBM) or the fixed wetting angle between molten solder and Cu. The scallop-type ripening is different from the classic ripening model according to LSW theory [6, 7], which assumes a constant volume of all the particles during ripening. But the growth of scallops has the total volume of scallops increasing with time, yet the total surface area of scallops is constant. The volume increase comes from the rapid diffusion of Cu via the channels between scallops, and the channels are assumed to have liquid solder for fast Cu diffusion.

On the kinetics of growth of scallops, morphology controls kinetics. The unique morphology of scallops on a constant contact area requires that the kinetics of ripening growth of the scallops is supply-controlled, rather than diffusion-controlled or interfacial-reaction-controlled. When the wetting time is extended, the diameter of the solder mount (or cap) does not increase, thus the interfacial area between the cap and Cu reminds unchanged. However, within the cap, the SLID reaction continues and the scallops grow bigger, so the average diameter of scallops increases, yet at the same time, the number of scallops decreases. Morphologically, the scallops are nearly touching each other; they are separated by a channel, thus the growth of anyone of them is at the expense of its nearest neighbors. So the neighboring scallops shrink, and it is a ripening process.

Consequently, on a fixed interfacial area, while the average size of scallops has increased with time, the total number of scallops decreases with time. The ripening process is nonconservative, meaning the total volume of the scallop increases with time. The volume increase is due to the continuing Cu-Sn reaction, which enables the growth of the $Cu_6Sn_5$ scallops by the diffusion of Cu through the channels into the molten solder.

A very unique kinetic behavior of the scallop-type morphology is that the growth of scallops does not become a diffusion barrier to its own growth, unlike the growth of a layer-type IMC. In a layer-type growth, it obeys a diffusion-controlled growth; typically, the thicker the layer, the slower the growth, so the layer thickness should have a square-root dependence on time. In scallop-type growth, the diameter of the scallop was measured to have a cubic-root dependence on time, which is quite similar to the ripening kinetics described by the LSW theory. However, in the classic ripening process, it is diffusion-controlled, yet in the scallop-type ripening, it is supply-controlled and flux-driven.

While the total volume of all the scallops increase with time, the interfacial area (or the contact area) between the scallops and the Cu is fixed. This is the first constraint of the reaction. Furthermore, if the scallops are assumed to be hemispheres, it follows that the total surface area of all the hemispherical scallops is also fixed, to be just twice of the contact area, independent of the size distribution of the hemispheres. Therefore, we have a ripening process, which has a constant surface area, rather than a constant volume.

In the classical ripening process analyzed by the LSW theory, the driving force is the reduction of the total surface area. But in the scallop-type ripening, the driving force is the increase of the total volume of scallops or the gain in the formation energy of IMC.

In the scallop-type ripening, all the Cu atoms diffusing through the channels into the molten solder is consumed by scallop growth. This is the reason for supply-controlled reaction or flux-driven ripening because the reaction rate depends on the supply of the Cu flux. In the following, we shall present a kinetic analysis of the distribution and growth of scallops. First, a simple model of the growth of mono-size scallops is presented, followed by a general model of the distributional growth of different-sized scallops.

## 4.3 A Simple Model for the Growth of Mono-Size Hemispheres

Figure 4.2 is a schematic diagram of the cross-section on an array of hemispherical $Cu_6Sn_5$ scallops grown on Cu, represented by the solid curves. [4] In the simple model, we assume that all scallops have the same diameter. The broken curve represents the growth of a single hemisphere on the Cu of radius R. The surface area of the single hemisphere is $S^{total} = ½(4\pi R^2) = 2\pi R^2$, which is twice the contact area of $\pi R^2$. Now, the sum of the contact area of the array of small hemispheres is also $\pi R^2$. Then the total surface area of the array of hemispheres is $2\pi R^2$, which is the same as that of the single hemisphere. Thus, we conclude that the ripening of the scallops is under the constraint of a constant surface area.

To consider the SLID reaction by the diffusion of Cu atoms through the channels, we recall the first Fick's law of diffusion that $J = - D(dC/dx)$,

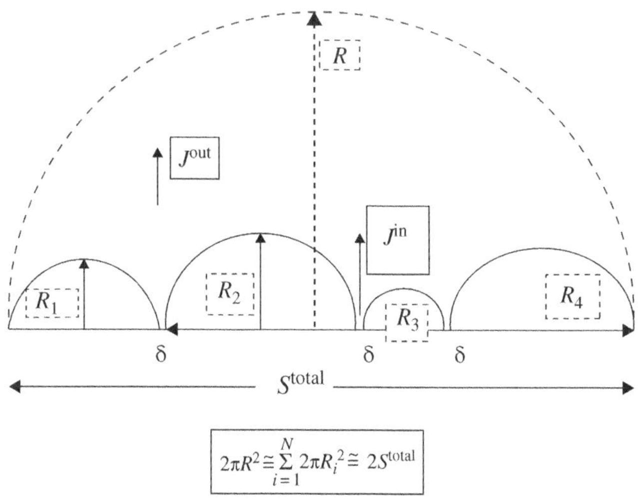


**Figure 4.2** A schematic diagram of the cross-section on an array of hemispherical $Cu_6Sn_5$ scallops grown on Cu, represented by the solid curves.

where $J$ is defined as atomic flux = number of atoms per unit area per unit time, $D$ is diffusivity, and $dC/dx$ is concentration gradient. Due to the definition of $J$, we have $JAt$ = number of atoms, where $A$ is the cross-sectional area of the diffusion flux, and $t$ is time. Then, if we consider the diffusion of Cu atoms through the scallop channels, we have the second constraint due to the conservation of mass or the number of atoms,

$$J_{in}At = n_i C_i V_i \tag{4.1}$$

where $J_{in}$ is the atomic flux of Cu diffusing through the scallop channels into the molten solder, $A$ is the total channel area, which is equal to $S^{free} = N2\pi R(\delta/2)$, where $N$ is the number of the mono-size scallops in the solder mount, $R$ is the radius of scallops, and $\delta$ is the effective width of the channel. Then, $n_i$ is atomic density in IMC or number of atoms per unit volume in IMC, $C_i$ is atomic fraction of Cu in IMC, which is 6/11 in $Cu_6Sn_5$, and $V_i$ is the volume of the reaction product of IMC scallops, and

$$V_i = N\frac{2\pi}{3}R^3 = \frac{2}{3}S^{total}R$$

To proceed, we take the in-flux of Cu

$$J_{in} \approx nD\frac{(C^b - C^e)}{R} \tag{4.2}$$

where $n$ is the atomic density or number of atoms per unit volume in the molten solder. The meaning of $C^b$ and $C^e$ are, respectively, the equilibrium concentration of Cu in the melt with the $Cu_6Sn_5$ phase and with the Cu surface, as shown in Figure 4.3. By substituting all the above into Eq. (4.1), we have

$$n_i C_i \frac{2}{3}S^{total}\frac{dR}{dt} = nD\frac{C^b - C^e}{R}\left(\frac{\delta}{R}S^{total}\right)$$

which gives

$$R^3 = kt$$

where $k = \dfrac{9}{2}\dfrac{n}{n_i}\dfrac{D(C^b - C^e)\delta}{C_i}$

Note that the surface tension is absent in the expression for the rate constant despite the ripening-like time law.

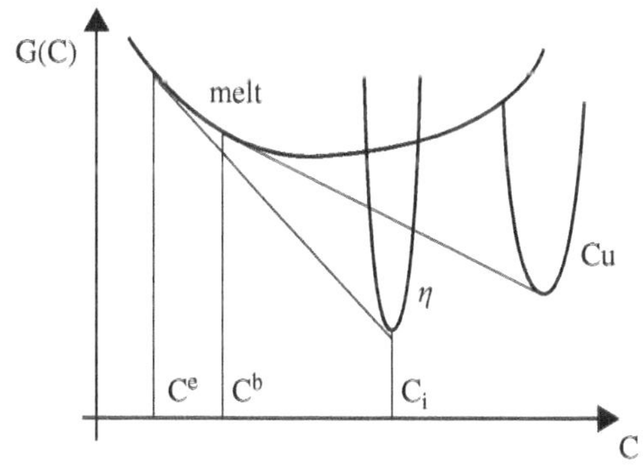


**Figure 4.3** The physical meaning of $C^b$ and $C^e$ are, respectively, the equilibrium concentration of Cu in the melt with the $Cu_6Sn_5$ phase and with the Cu surface.

If we take $n/n_i \approx 1$, $C_i = 6/11$, $D \approx 10^{-5}$ cm$^2$/sec, $\delta \approx 5 \times 10^{-6}$ cm, $C^b - C^e \approx 0.001$, where the concentration $C^b$ is taken for equilibrium of melt with the $Cu_6Sn_5$ phase, the rate constant k $\approx 4 \times 10^{-13}$ cm$^3$/sec. For example, after an annealing time of 300 seconds, it gives $R \approx 5$ μm, which agrees very well with experimental data.

## 4.4 Theory of Flux-Driven Ripening

For scallops having a distribution of size, we take f(t, R) to be the size distribution function of scallops, so that the total number of scallops is equal to [4]

$$N(t) = \int_0^\infty f(t, R)dR \tag{4.3}$$

And the average value is

$$\langle R^m \rangle = \frac{1}{N}\int_0^\infty R^m f(t, R)dR \tag{4.4}$$

In mean-field theory of ripening, the physical meaning of <*R*> is that those scallops with radius bigger than <*R*> with grow, and those scallops with radius smaller than <*R*> will shrink.

The first constraint of constant interface area takes the form of

$$\int_0^\infty \pi R^2 f(t, R)dR = S^{total} - S^{free} \cong S^{total} = const. \tag{4.5}$$

where the cross-sectional area of all channels for Cu supply is

$$S^{free} = \int_0^{\infty} \frac{\delta}{2} 2\pi R f(t, R) dR \tag{4.6}$$

The total volume of the scallops is,

$$V_i = \int_0^{\infty} \frac{2}{3} \pi R^3 f(t, R) dR \tag{4.7}$$

The second constraint is the conservation of mass, and we have

$$n_i C_i \frac{dV_i}{dt} = J_{in} S^{free} \tag{4.8}$$

where $n_i$ is the atomic density in the IMC, i.e. number of atoms per unit volume, and $C_i$ is the atomic fraction of Cu in IMC, which is 6/11 in $Cu_6Sn_5$. The in-flux, $J_{in}$, has been given by Eq. (4.2).

Because scallops must grow and shrink atom by atom, the distribution function should satisfy the usual continuity equation in the size space,

$$\frac{\partial f}{\partial t} = -\frac{\partial}{\partial R}(f \cdot u_R) \tag{4.9}$$

where $u_R$ is the velocity in the size space and it is simply the growth rate of scallop with radius $R$, and it is determined by the flux density, $j(R)$, on each individual scallop. The expression for $j(R)$ is usually found as a quasi-stationary solution of diffusion problem in infinite space around a spherical grain with fixed super-saturation of $<c> - C_e$ at infinity.

$$u_R = \frac{dR}{dt} = -\frac{j(R)}{n_i C_i}$$

where $j(R)$ can be obtained by assuming that the flux on (or out of) each individual scallop should be proportional to the difference between the average chemical potential of Cu in the reaction zone (we take it to be the same everywhere due to mean-field approximation) and the chemical potential at the curved scallop/melt interface.

$$-j(R) = L\left(\mu - \mu_\infty - \frac{\beta}{R}\right)$$

$$\frac{dR}{dt} = \frac{L}{n_i C_i}\left(\mu - \mu_\infty - \frac{\beta}{R}\right)$$

where $\beta = 2\gamma\Omega$. The parameters of L, $\beta$, $\mu - \mu_\infty$ are determined self-consistently from the above-mentioned two constrains of constant surface and mass conservation. We obtain

$$u_R = \frac{dR}{dt} = \frac{k}{9}\frac{1}{\langle R^2\rangle - \langle R\rangle^2}\left(1 - \frac{\langle R\rangle}{R}\right)$$

and $k = \frac{9}{2}\frac{n}{n_i C_i} D(C^b - C^e)\delta$

During the nonconservative flux-driven ripening, the rate of growth/shrinkage of each scallop is determined not only by diffusivity and the average size of all scallop, <$R$>, but also by the capacity of channel to supply Cu for the reaction. Thus, in the mean-field approximation, the basic equation for the distribution function has the following form;

$$\frac{\partial f}{\partial t} = -\frac{k}{9}\frac{\langle R\rangle}{\langle R^2\rangle - \langle R\rangle^2}\frac{\partial}{\partial R}\left(f\left(\frac{1}{\langle R\rangle} - \frac{1}{R}\right)\right)$$

where the rate coefficient $k$ is determined by the incoming flux conditions, which in turn are determined by the channels.

The formal solution of the distribution function is

$$f(t,R) = \frac{B}{bt}\frac{R}{(bt)^{1/3}}\exp\{\int_0^\eta \frac{3 - 4\xi}{\xi^2 - 3\xi + 9/4} d\xi = \frac{B}{\tau}\phi(\eta)$$

where $\tau = bt$, $\eta = R/(bt)^{1/3}$, and $B = \dfrac{S^{total}}{\pi\int_0^\infty \xi^2 \varphi(\eta) d\eta}$

The parameter $b$ should be found self-consistently. A standard integration gives,

$$\phi(\eta) = 0, when\, \eta > (3/2)$$

$$\phi(\eta) = \frac{\eta}{\left[(3/2) - \eta\right]^4}\exp\left(-\frac{3}{(3/2) - \eta}\right), when\ 0 < \eta < (3/2)$$

A plot of $\varphi(\eta)$ versus $\eta$ is shown in Figure 4.4. To check the above derivations, we take $n/n_i \approx 1$. $C_i = 6/11$, $D \approx 10^{-5}$ cm$^2$/sec, $\delta = 5 \times 10^{-6}$ cm, $C^b - C^e \approx 0.001$, where the concentration $C^b$ is taken for equilibrium of melt with the $Cu_3Sn$ phase, the rate constant $k \approx 4 \times 10^{-13}$ cm$^3$/sec. For example, for annealing time $t = 300$ seconds, it gives $R \approx 5$ μm, which agrees very well with experimental data.

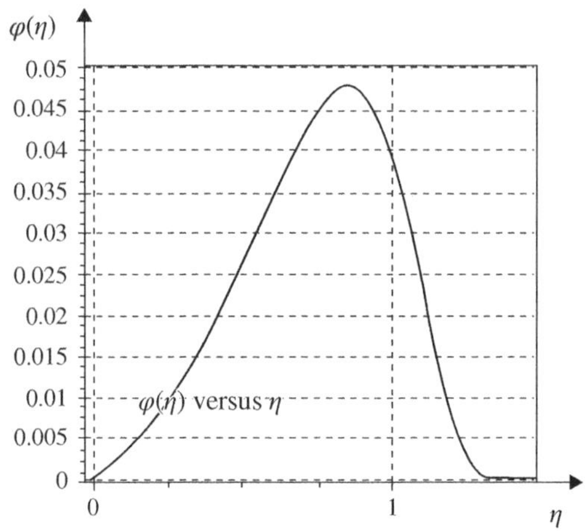


Figure 4.4 A plot of φ(η) versus η.

## 4.5 Measurement of the Nano-channel Width Between Two Scallops

In the above kinetic analysis, the liquid channel is an important parameter. There were attempts to measure experimentally the width of the channel, but without success. This is because when the channels were examined by TEM, they appeared to look like grain boundaries. This is because in TEM study near room temperature in solid state, the channels closed up. We must examine the channels at the temperature of SLID, which has not been done so far. However, we have solved it by using an extremely thin solder on Cu and keep the SLID reaction until it has consumed all the solder. Thus, when the sample is cooled down to room temperature, there is no more unreacted solder to close the channel. Figure 4.5 shows the cross-sectional view of one of the channel and we can measure the width of the channel to be about several nm.

## 4.6 Extremely Rapid Grain Growth in Scallop-Type Cu6Sn5 in SLID

In 3D IC technology, when the size of micro-bump solder joint is reduced to 10 μm, the scallop-type $Cu_6Sn_5$ grains on the opposite sides of a joint can interact by direct touching of each other during the SLID reaction. Figure 4.6 shows the opposite rows of IMC scallops in the micro-bump solder joint of eutectic SnAg solder about 10 μm in thickness, after reflow at 260 °C for three minutes. With one more minute of reflow, Figure 4.7a shows the

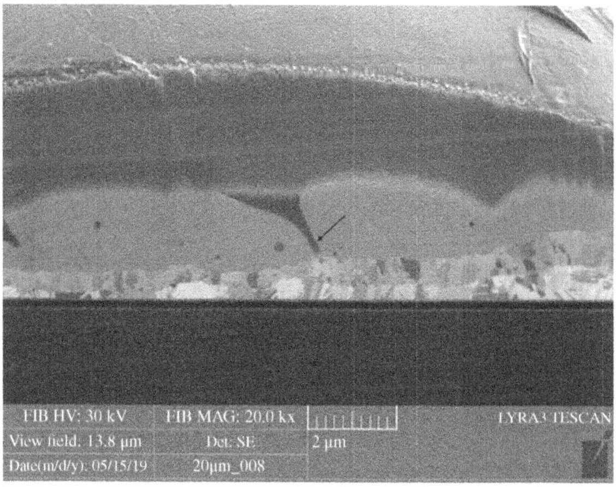


**Figure 4.5** In a sample having a very thin layer of Sn, when the Sn was completely consumed by SLID reaction, we can observe the channel between two scallops. The cross-sectional view of one channel, and we can measure the width of the channel to be about a few nm.

EBSD image where the scallops on both sides have joined and formed a large grain across the entire joint. The results indicate that the grain growth of scallop-type $Cu_6Sn_5$ is very fast in molten solder. Roughly, the growth rate is about 5 μm in one minute.

The extremely rapid grain growth of the scallops occurs and merges into a single large grain across the entire joint has been analyzed by assuming that the liquid solder wets the grain boundary between scallops and

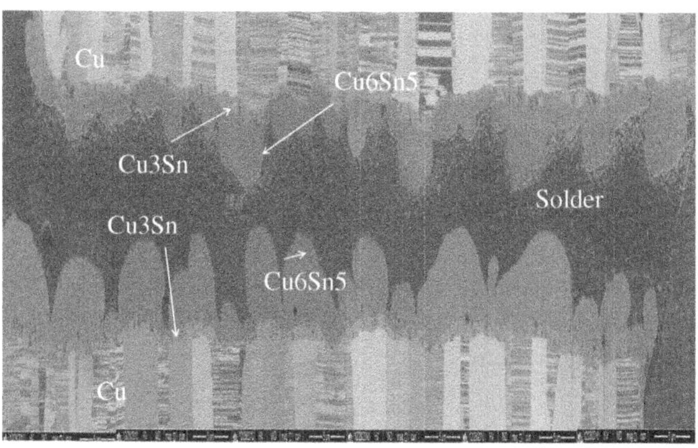


**Figure 4.6** SEM image of the opposite rows of IMC in a micro-bump solder joint after reflow at 260 C for three minutes. The SnAg solder has a thickness of 10 μm.

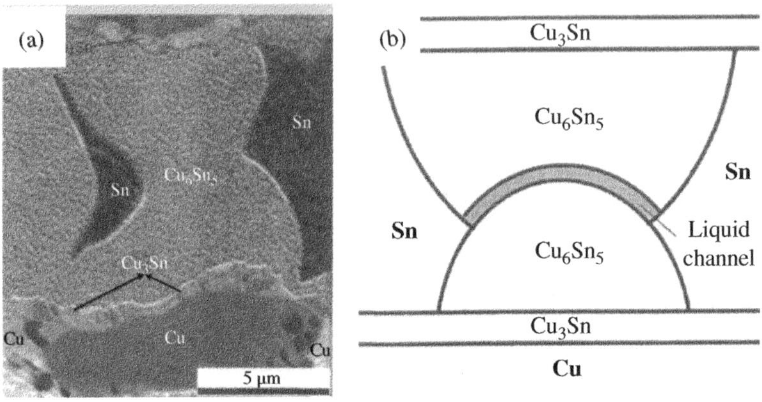


**Figure 4.7** (a) After reflow at 260 C for four minutes of the same SnAg solder joint, the two rows of scallops joined and form large grains of Cu6Sn5 across the entire join. (b) Schematic diagram of two scallops in contact to form a liquid boundary between them. The upper scallop can grow rapidly to consume the lower grain.

provides an extremely fast kinetic path of liquid boundaries for grain growth, as shown in Figure 4.7b. The kinetic path can provide the rapid reaction rate, because the liquid boundary enables a rapid atomic rearrangement for grain growth so that the upper grain can consume the lower grain very rapidly because the diffusivity in the liquid boundary is very faster. The width of the liquid boundary has been estimated to be about 15 nm. [8]

On the other hand, in a C-4 solder joint of 100 μm in diameter, the $Cu_6Sn_5$ scallops on both sides of the joint are far apart so that they do not make direct contact to show the rapid grain growth.

## Problems

**4.1** Why do we tend to use eutectic alloy for solder? What are the unique properties of a eutectic alloy?

**4.2** Why do we need to use under-bump-metallization (UBM) in C-4 solder joint? Typical UBM consists of Au/Cu/Cr or Au/Cu/Ti. What is the function of each of them?

**4.3** To apply flux is very important in solder joint reaction, can we have a flux-less solder joint?

**4.4** To measure the wetting angle, typically we need to make a cross-sectional cut of the solder cap. Can we determine the wetting angle of a solder cap without cut it?

**4.5** Make a comparison between the classic LSW ripening and the new flux-driven ripening. What are the key differences between them? In flux-driven ripening, what is the meaning of the flux?

## References


**1** J. H. Lau, "*Flip Chip Technology*," McGraw-Hill, New York (1996).

**2** K. N. Tu, "*Solder Joint Technology: Materials, Properties, and Reliability*," Springer, (2007), 1–368.

**3** H. K. Kim and K. N. Tu, "Kinetic analysis of soldering reaction between eutectic SnPb alloy and Cu accompanied by ripening, "*Physical Review B* 53 16027–16034 (1996).

**4** A. M. Gusak and K. N. Tu, "Kinetic theory of flux-driven ripening," *Physical Review B*, 66, 115403 (2002).

**5** K. N. Tu, F. Ku, and T. Y. Lee, "Morphology stability of solder reaction products in flip-chip technology," *Journal of Electronic Materials*, 30, 1129–1132 (2001).

**6** I. M. Lifshiz and V. V. Slezov, "The kinetics of precipitation from super-saturated solid solutions," *Journal of Physics and Chemistry of Solids*, 19 (1/2) 35–50 (1961).

**7** C. Wagner, Theorie der Alterung von Niederschlagen durch Umlosen (Ostwald-Reifung), *Zeitschrift fur Elektrochemie*, 65 581–591 (1961).

**8** A M. Gusak, K. N. Tu, and Chih Chen, "Extremely rapid grain growth in scallop-type $Cu_6Sn_5$ during solid-liquid interdiffusion reactions in micro-bump solder joints," *Scripta Materialia*, 179, 45–48 (2020).

# 5

# Solid-State Reactions Between Copper and Solder

## 5.1 Introduction

The solder joint after SLID reaction will be annealed around 150 °C up to 1000 hours for reliability consideration. Thus, we need to study solid-state interfacial reaction between Cu and solder. In the binary Cu-Sn system, there are two IMCs: Sn/$Cu_6Sn_5$/$Cu_3Sn$/Cu. Thus, $Cu_6Sn_5$ is unstable on Cu, and the reaction between them will form $Cu_3Sn$. For comparison, in SLID, we have molten solder/$Cu_6Sn_5$/Cu. If the reflow time is over several minutes, we will have a thin layer of $Cu_3Sn$ between the $Cu_6Sn_5$ and Cu. In solid-state reactions, $Cu_3Sn$ forms quickly, and often it is accompanied by Kirkendall voids. Furthermore, Sn whiskers can form on the surface of the solder. This is because the diffusion of Cu into Sn will generate compressive stress in Sn to grow whiskers. A detailed discussion on Sn whisker growth will be presented in Chapter 12 on stress-migration. We shall discuss Kirkendall void formation in a later section in this chapter.

It is worth noting that in wetting reaction, the time is about 1 min, which is much shorter than the time of 1000 hours in solid-state reaction. However, it has been found that the amount of IMC formation in these two reactions is about the same, 5–10 μm thick. We recall that in wetting reaction, the IMC is scallop-type and the diffusion of Cu is via the channels between scallops. The channel has molten solder and the diffusivity of Cu through the molten solder in the channel is about $10^{-5}$ cm$^2$/sec, as presented in the last chapter. But in solid-state reaction, the Cu diffusion has to go through the layer-type IMC, the diffusivity is around $10^{-13}$ cm$^2$/sec at 150 °C, about eight orders of magnitude slower than that in molten solder. If we take the simple relationship of $x^2 = Dt$, we can see that indeed the solid-state reaction will

*Electronic Packaging Science and Technology*, First Edition. King-Ning Tu, Chih Chen, and Hung-Ming Chen.

be eight orders of magnitude slower, if $x$ is the same! The difference supports the statement that morphology controls kinetics.

While the kinetics of growth of a layer-type IMC has been widely studied, it has a build-in difficulty. [1–3] This is because if we assume the IMC to be a stoichiometric compound, there is no concentration gradient across the IMC, which means we will not be able to use Fick's first law to describe the diffusional flux in the IMC. It has been a long-standing problem! In this chapter, we shall introduce Wagner's diffusivity to overcome it. [4]

Another important consideration in solid-state reaction occurs in μ-bump in 3D IC technology when we reduce the size of solder joint to 10 μm and below, where surface diffusion becomes important, and it has caused side-wall reactions and the formation of porous $Cu_3Sn$.

## 5.2 Layer-Type Growth of IMC in Solid-State Reactions

In solid-state reaction of solder joints, the scallop-type $Cu_6Sn_5$ is converted into layer-type. Both $Cu_6Sn_5$ and $Cu_3Sn$ have the layer-type morphology, and their growth becomes diffusion-controlled. For simplicity, we consider the growth of a single layer of IMC, for example the $Ni_2Si$ between Ni and Si, as shown in Figure 5.1.

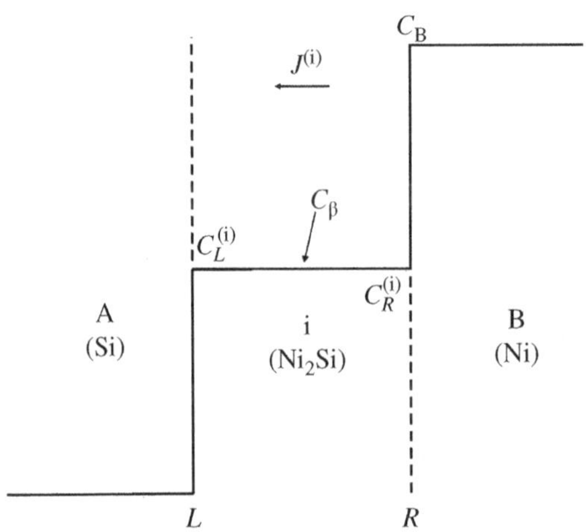


**Figure 5.1** The growth of a single IMC phase, the "i" phase, between two terminal phases of A and B with negligible solubility.

If we use Fick's first law to describe atomic flux, we encounter a serious difficulty because the concentration gradient across a layer of stoichiometric IMC is negligibly small. The rate of concentration change within the layer is close to zero. Then, in Fick's second law of $\partial C/\partial t = -\partial J_x/\partial x$, when $\partial C/\partial t = 0$, we have $J_x = const$. It means the interdiffusion flux under a constant concentration is almost the same in each point of the layer. Therefore, concentration gradient does not exist and cannot be the driving force. Explicitly, in Figure 5.1, we consider the growth of a single IMC phase, the "i" phase, between two terminal phases of A and B with negligible solubility. We have

$$J^{(i)} = -\tilde{D}^{(i)} \frac{C_R^{(i)} - C_L^{(i)}}{x_R^{(i)} - x_L^{(i)}} \cong -\tilde{D}^{(i)} \frac{\Delta C^{(i)}}{\Delta x^{(i)}}$$

If growth is diffusion-controlled, the concentrations at the two interfaces, $C_R^{(i)}, C_L^{(i)}$ are constants, corresponding to the concentrations in equilibrium phase diagram.

We have growth respectively at the left and the right boundaries of the "i" phase as,

$$\left(C_L^{(i)} - 0\right) \frac{dx_{Ai}}{dt} = -\tilde{D}^{(i)} \frac{\Delta C^{(i)}}{\Delta x^{(i)}} - 0$$

$$\left(C^B - C_R^{(i)}\right) \frac{dx_{iB}}{dt} = 0 - \left(-\tilde{D}^{(i)} \frac{\Delta C^{(i)}}{\Delta x^{(i)}}\right) \tag{5.1}$$

Rearranging them, we have

$$\frac{d\left(x_{iB} - x_{Ai}\right)}{dt} = \left(\frac{1}{C^B - C_R^{(i)}} + \frac{1}{C_L^{(i)}}\right) \tilde{D}^{(i)} \frac{\Delta C^{(i)}}{\Delta x^{(i)}}$$

Thus,

$$\left(\Delta x_i\right)^2 = 2 \frac{\left(C_L^{(i)} + C^B - C_R^{(i)}\right) \Delta C^{(i)}}{\left(C^B - C_R^{(i)}\right) C_L^{(i)}} \tilde{D}^{(i)} t = 2 \frac{C^B \Delta C^{(i)}}{C^{(i)}\left(C^B - C^{(i)}\right)} \tilde{D}^{i} t \tag{5.2}$$

In the last term of the above equation, we drop $(C_L^{(i)} - C_R^{(i)})$ because it is small as compared to $C^B$. Also, we take $C_L^{(i)} \approx C_R^{(i)} \approx C^{(i)}$.

However, we cannot determine $\Delta C^{(i)} \cong C_R^{(i)} - C_L^{(i)}$ in Eq. (5.2) because it is close to zero. In essence, this is the fundamental difficulty when the IMC phase has a stoichiometric composition. It has been a long-standing kinetic problem in analyzing a layered growth of IMC. To overcome it, we introduce Wagner diffusivity and follow Wagner's approach by using chemical potential gradient, rather than composition gradient. [4]

In Figure 5.2, we depict the free energy diagram of A (α-phase) and B (β-phase) and an IMC (i-phase). We begin by considering the mechanical mixing of A and B atoms, without chemical interaction. In mechanical mixing, the Gibbs free energy is

$$G = \mu_A N_A + \mu_B N_B$$

where $\mu_A$ and $\mu_B$ are chemical potential of A and B atoms in pure metals, respectively, so they are constants. However, in the analysis below for an IMC, they ($\mu_A$ and $\mu_B$) vary with the fraction of concentration of A and B in the mixing.

We assume $N_A$ and $N_B$ are number of A and B atoms, respectively, then $N_A + N_B = N$, which is the total number of atoms in the mixing.

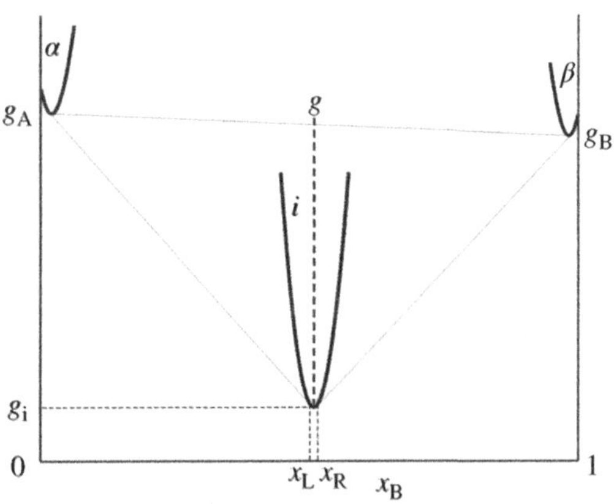


**Figure 5.2** Free energy diagram of A (α-phase) and B (β-phase), and an IMC (i-phase).

Thus, $X_A = N_A/N$ and $X_B = N_B/N$, which are the fractions of A and B in the mixing, respectively. We take

$$G = gN \text{ or } g = G/N$$

then

$$g = \mu_A X_A + \mu_B X_B \tag{5.3}$$

where "g" is the Gibbs free energy per atom. In Eq. (5.3), $\mu_A$ is a function of $X_A$, and $\mu_B$is a function of $X_B$. In Figure 5.2, when we draw a common tangent line to the α-phase and β-phase, the tangent meets the concentration line of $X_i$ at the point g. Because

$$dG = -SdT + Vdp + \sum_{i=1}^{2} \mu_i dN_i$$

we divide the above equation by $N$, and we have

$$dg = -sdT + \Omega dp + \mu_A dX_A + \mu_B dX_B$$

Since $X_B = 1 - X_A$, we have $dX_A = - dX_B$. At constant temperature and constant pressure, we have

$$dg|_{T,p} = (\mu_A - \mu_B) dX_A$$

or

$$\frac{\partial g}{\partial X_A} = \mu_A - \mu_B \tag{5.4}$$

Similarly, we have

$$\frac{\partial g}{\partial X_B} = \mu_B - \mu_A$$

Since $X_A + X_B = 1$, we obtain

$$\mu_A = g + X_B \frac{\partial g}{\partial X_A} \tag{5.5}$$

and

$$\mu_B = g - X_A \frac{\partial g}{\partial X_A} \tag{5.6}$$

Equations (5.5) and (5.6) indicate that if we know the function of g, we can calculate the chemical potential of A and B. Furthermore, we take differential of $\mu_A$ with respect to $X_A$,

$$\frac{\partial \mu_A}{\partial X_A} = \frac{\partial g}{\partial X_A} + \frac{\partial X_B}{\partial X_A}\frac{\partial g}{\partial X_A} + X_B \frac{\partial^2 g}{\partial X_A^2} = X_B \frac{\partial^2 g}{\partial X_A^2} = X_B g''$$

Now, we define a thermodynamic factor "φ" as below,

$$\phi = \frac{X_A}{kT}\frac{\partial \mu_A}{\partial X_A} = \frac{X_A X_B}{kT} g'' \tag{5.7}$$

In Figure 5.2, we draw the common tangent of the α-phase and the i-phase and the tangential point on the latter is $X_L$. Similarly, the common tangent of the β-phase and the i-phase has the tangential point on the latter at $X_R$. We define $\Delta X = X_R - X_L$, then we express

$$\frac{\left.\frac{\partial g}{\partial X}\right|_{X_R} - \left.\frac{\partial g}{\partial X}\right|_{X_L}}{X_R - X_L} \approx \frac{\partial^2 g}{\partial x^2} = g''$$

we have

$$g''\Delta X = \left.\frac{\partial g}{\partial X}\right|_{X_R} - \left.\frac{\partial g}{\partial X}\right|_{X_L} = \frac{g_B - g_i}{1 - X_i} - \frac{g_i - g_A}{X_i - 0} = \frac{\left[g_B X_i + g_A(1 - X_i)\right] - g_i}{(1 - X_i)X_i}$$
$$= \frac{g - g_i}{(1 - X_i)X_i} = \frac{\Delta g_i}{X_A X_B}$$

$$\Delta g_i = X_A X_B g'' \Delta X \tag{5.8}$$

In the above equation, $X_i$ is located roughly in between $X_R$ and $X_L$. The square bracket term is equal to "g" by the lever rule, and we can take $g - g_i = \Delta g_i$ to be the IMC formation energy per atom, which is known.

The purpose of the above derivation is to show that when the interdiffusion coefficient can be given as,

$$\tilde{D} = \left(X_A D_B^* + X_B D_A^*\right)\phi = \left(X_A D_B^* + X_B D_A^*\right)\frac{X_A X_B}{kT} g'' = \left(X_A D_B^* + X_B D_A^*\right)\frac{\Delta g_i}{kT\Delta X}$$

$$\tilde{D}\Delta X = \left(X_A D_B^* + X_B D_A^*\right)\frac{\Delta g_i}{kT} \tag{5.9}$$

where significantly, we note that while $\Delta X$ is unmeasurable, the product of $\tilde{D}\,\Delta X$, in terms of $\Delta g_i$, becomes measurable. Now, the interdiffusion coefficient is derived below.

We have in the moving coordination or moving frame,

$$j_B = C_B \frac{D_B^*}{kT}\left(-\frac{\partial \mu_B}{\partial x}\right) = -C_B \frac{D_B^*}{kT}\frac{\partial \mu_B}{\partial C_B}\frac{\partial C_B}{\partial x} = -D_B^*\left(\frac{X_B}{kT}\frac{\partial \mu_B}{\partial X_B}\right)\frac{\partial C_B}{\partial x} = -D_B^*\phi\frac{\partial C_B}{\partial x}$$

Similarly, we have

$$j_A = -D_A^*\phi\frac{\partial C_A}{\partial x}$$

Then, in laboratory frame or fixed frame, we have

$$\begin{aligned} J_B &= j_B + C_B v = -D_B^*\phi\frac{\partial C_B}{\partial x} - C_B\frac{1}{C}\left(j_A + j_B\right) \\ &= -\frac{C_A + C_B}{C} D_B^*\phi\frac{\partial C_B}{\partial x} - X_B\left(-D_A^*\phi\frac{\partial C_A}{\partial x} - D_B^*\phi\frac{\partial C_B}{\partial x}\right) \\ &= -\left(X_A D_B^* + X_B D_A^*\right)\phi\frac{\partial C_B}{\partial x} = -\tilde{D}\frac{\partial C_B}{\partial x} \end{aligned}$$

Thus, we have

$$\tilde{D} = \left(X_A D_B^* + X_B D_A^*\right)\phi. \tag{5.10}$$

About the thermodynamic factor, $\varphi$, we recall that Darken's interdiffusion coefficient [5] has been modified by the thermodynamic factor $\varphi$ when the ideal solution is changed to a regular solution because of chemical effect. Here, we have introduced $\varphi$ for IMC.

## 5.3 Wagner Diffusivity

In the above analysis, $\tilde{D}$ and $\Delta X\,(=\Omega\Delta C)$ appear together as a product, and the product is called "Wagner diffusivity" in the literature. By substitution, we have

$$\left(x_2 - x_1\right)^2 = 2\frac{C^B\Omega}{C^{(i)}\left(C^B - C^{(i)}\right)}\left(C^A D_B^* + C^B D_A^*\right)\left(\frac{\Delta g_i}{kT}\right) \tag{5.11}$$

We recall that the marker motion equation in a layered IMC growth of $A_iB$ is,

$$\frac{D_B^*}{D_A^*}=\frac{i\left(x_m-x_1\right)}{x_2-x_m} \tag{5.12}$$

By combining Eqs. (5.11) and (5.12), we can determine the intrinsic diffusivities in the "i" compound.

As an example of application of the above analysis, we consider the growth of $Ni_2Si$ between a Ni thin film and a Si wafer, because marker motion in $Ni_2Si$ was measured by using implanted Xe. [6] We take Si, $Ni_2Si$, and Ni to be A, i, and B phases, respectively, as depicted in Figure 5.1. Thus, $X_A = 1/3$, and $X_B = 2/3$. If we assume the atomic volume of Ni is Ω, we take $C^B = 1/\Omega$, and $C^{(i)} = (2/3)(1/\Omega)$, and $i = 1/2$. Experimentally, $\Delta x_i = x_2 - x_1 =$ 200 nm after $t$ = 40 min at $T$ = 300 °C. Substituting these values into Eqs. (5.11) and (5.12), we have

$$\left(\Delta x_i\right)^2=2\frac{1/\Omega}{(2/3)(1/\Omega)\left[(1/\Omega)-(2/3)(1/\Omega)\right]\Omega}\left[\frac{1}{3}D_{Ni}^*+\frac{2}{3}D_{Si}^*\right]\frac{\Delta g_i}{kT}t$$

$$\left(\Delta x_i\right)^2=\left[3D_{Ni}^*+6D_{Si}^*\right]\frac{\Delta g_i}{kT}t$$

The formation enthalpy of $Ni_2Si$ has been measured to be −10.5 kcal/gram-atom, which is the heat of formation of one mole of the compound divided by the number of atoms in the chemical formula of the compound. Thus, we have $\Delta g_i = (10.5 \times 3)/(23 \times 3) = 0.46$ eV/atom, and $kT = 0.05$ eV/atom at 300 °C. Marker analysis showed that Ni is the dominant diffusing species, and roughly the ratio in Eq. (5.12) is about 20/1. We have $D^*_{Si} = 0.1\, D^*_{Ni}$.

$$D_{Ni}^*=\frac{\left(2x10^{-5}\right)^2 cm^2}{2400x3.6x9.2\sec}=0.5x10^{-14}cm^2/\sec.$$

$$D_{Si}^*=0.5x10^{-15}cm^2/\sec.$$

The chemical effect is to slow down the tracer diffusivities in the silicide by one order of magnitude.

## 5.4 Kirkendall Void Formation in $Cu_3Sn$

To observe Kirkendall voids, a proper polishing of the product interface is crucial. The classical way of mechanical polishing tends to smear the interface and to make the voids invisible. Figure 5.3 shows very clearly the distribution of many voids in the layer of $Cu_3Sn$ by using focus ion beam polishing. [7] The formation of voids occurs at the interface between $Cu_3Sn$ and Cu. When the $Cu_3Sn$ layer grows, the moving interface does not drag the voids along, so they are buried in the $Cu_3Sn$.

The reaction to form $Cu_3Sn$ between $Cu_6Sn_5$ and Cu can be represented by the pair of chemical equations below:

$$Cu_6Sn_5 \rightarrow 2Cu_3Sn + 3Sn$$

$$3Sn + 9Cu \rightarrow 3Cu_3Sn$$

The last equation indicates that 9 Cu atoms will be needed to form $Cu_3Sn$. The diffusional flux of Cu will introduce a flux of vacancies moving in the opposite direction. Provided that these vacancies can be absorbed by the sinks in Cu microstructure or by the interface between Cu and $Cu_3Sn$, Kirkendall void will not form. Furthermore, the nucleation of the voids tends to be heterogeneous at the interface between $Cu_3Sn$ and Cu, so very low super-saturation of vacancy is needed, and void nucleation is easy. Thus, we must consider the sink of vacancies in order to avoid the growth

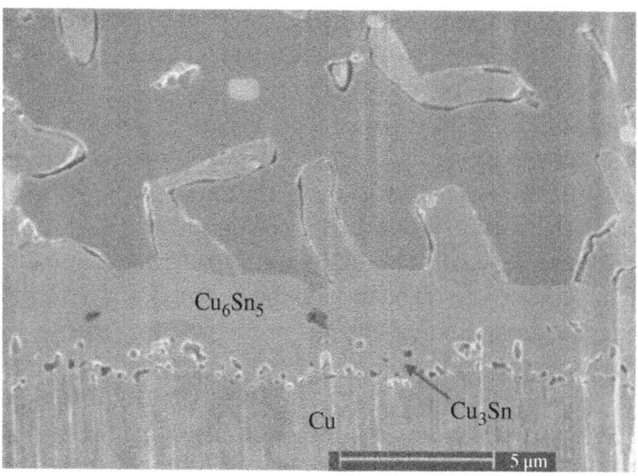


**Figure 5.3** SEM image of the distribution of Kirkendall voids in the layer of $Cu_3Sn$ by using focus ion beam polishing.

of Kirkendall voids. It is a reliability issue. We shall consider vacancy sink from the viewpoint of impurities and defects in the microstructure.

The question is whether the vacant lattice sites, left behind by the out-diffusion of Cu, can be absorbed by the Cu (or the IMC) lattice or not. We recall that in Darken's analysis of Kirkendall effect in interdiffusion, an important assumption is that vacancy is at equilibrium everywhere in the diffusion couple, so nonequilibrium vacancies are all absorbed and "lattice shift" is complete as revealed by the marker motion. The subject of lattice shift will be present in Section 13.2 of Chapter 13. Consequently, no void formation occurs in Darken's analysis because nucleation of a void needs super-saturation of vacancies. However, experimentally, Kirkendall voids (or Frenkel voids) can form because of a small amount of incomplete lattice shift.

In electroplating of the Cu film, if it has a high density of impurities, it can enhance heterogeneous void nucleation. If the grain size of $Cu_3Sn$ is very small, these grain boundaries can be good sinks to absorb vacancies to have a complete or near-complete lattice shift. However, the triple points in the grain boundaries could be excellent sites for void nucleation too. It has been found that when the Cu contains a high density of nanotwins, no Kirkendall void forms. It indicates that most likely the kinks in the incoherent twin boundaries can serve as good sinks for vacancies.

## 5.5 Sidewall Reaction to Form Porous $Cu_3Sn$ in μ-Bumps

In 3D IC technology, the size of μ-bump is about 20–10 μm. All the solder in the joint can be converted into $Cu_6Sn_5$ and $Cu_3Sn$ by reflow. Surprisingly, upon a longtime annealing above 170 °C, a porous-type of $Cu_3Sn$ is formed due to the depletion of Sn from $Cu_6Sn_5$. Porous $Cu_3Sn$ is mechanically weak, which is of reliability concern. [8]

Figure 5.4 shows a set of SEM cross-sectional images of μ-bumps after annealing at 220 °C for 20, 50, 100, and 300 hours, respectively. In Figure 5.4a, a symmetrical layered structure of $Cu/Cu_3Sn/Cu_6Sn_5/Cu_3Sn/Cu$ is seen, but no formation of porous $Cu_3Sn$. The thickness of $Cu_3Sn$ and $Cu_6Sn_5$ layers are about 3.5 and 14.6 μm, respectively. Figure 5.4b is the image of a μ-bump annealed for 50 hours, and the thickness of $Cu_3Sn$ and $Cu_6Sn_5$ layers has changed to 6.5 and 11.5 μm, respectively. In Figure 5.4c, the porous $Cu_3Sn$ began to form at the circumference of the middle $Cu_6Sn_5$ layer. The lateral growth of the porous $Cu_3Sn$ is in the radial direction, going from the edge toward the center. Figure 5.4d shows that the middle layer of $Cu_6Sn_5$ has been transformed completely into porous $Cu_3Sn$, and we have a layer of porous $Cu_3Sn$ sandwiched between two layer-type of $Cu_3Sn$. Accompanying the porous-$Cu_3Sn$ formation, a layer of $Cu_3Sn$ is formed on the sidewall of Cu UBM.

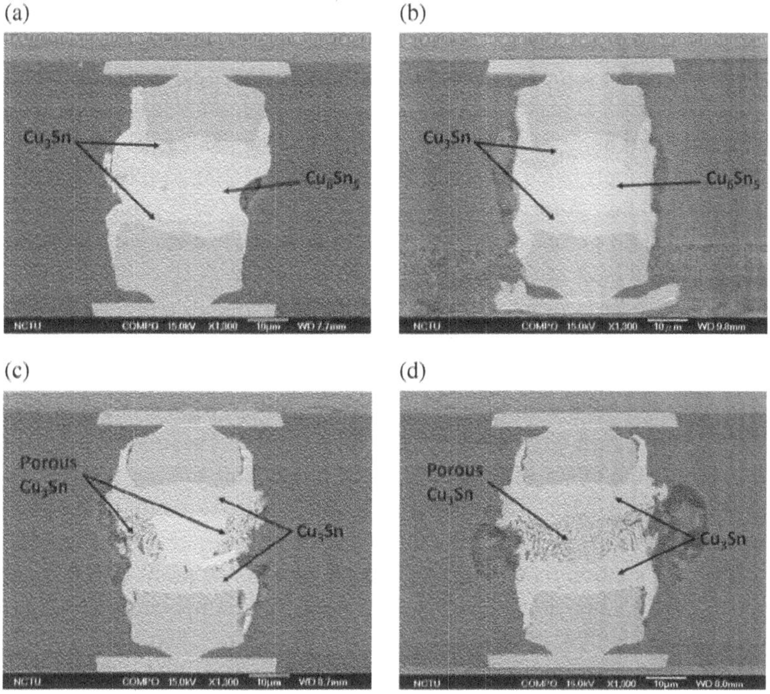


**Figure 5.4** A set of SEM cross-sectional images of micro-bumps after annealing at 220 °C for (a) 20, (b) 50, (c) 100, and (d) 300 hours, respectively.

We consider, respectively, the following two chemical formulae of the additive reaction and subtractive reaction in converting $Cu_6Sn_5$ into $Cu_3Sn$,

$$Cu_6Sn_5 + 9Cu \rightarrow 5Cu_3Sn(layer-type) \tag{5.13}$$

$$Cu_6Sn_5 - 3Sn \rightarrow 2Cu_3Sn(porous-type) \tag{5.14}$$

The first equation forms the layer-type $Cu_3Sn$ by adding Cu. The second equation forms the porous-type $Cu_3Sn$ by subtracting Sn. In principle, they are independent. Experimentally, they are coupled and in competition.

To add Cu to the reaction in Eq. (5.13), Cu is available from the thick Cu UBM in μ-bumps. To take Sn away from $Cu_6Sn_5$ in Eq. (5.14), there must be a sink for Sn. In a μ-bump, the nearby sidewall surface of the thick Cu UBM can do so. Thus, we need a third equation,

$$3Sn + 9Cu \rightarrow 3Cu_3Sn(sidewall) \tag{5.15}$$

Equation (5.15) means that the 3 Sn atoms go to the sidewall to form $Cu_3Sn$. If we add Eqs. (5.14) and (5.15), we obtain Eq. (5.13). Thus, thermodynamically we have the same transformation of $Cu_6Sn_5$ to $Cu_3Sn$ in Eq. (5.13), as well as in Eq. (5.14) plus Eq. (5.15), except that the morphology and kinetic paths are different.

When the supply of Cu across $Cu_3Sn$ from the Cu UBM is rapid and abundant, the reaction to form a layer-type $Cu_3Sn$ in Eq. (5.13) is dominant. If not, the reaction in Eqs. (5.14) and (5.15) may occur to form porous $Cu_3Sn$ plus a sidewall layer of $Cu_3Sn$.

However, there is an interesting question on the porous structure formation. How come the diffusion of 9 Cu atoms as shown in Eq. (5.13) only leads to some Kirkendall voids formation but no porous structure formation? Yet why does the diffusion of 3 Sn atoms in Eq. (5.14) leads to porous structure formation with about 41% of void volume? The answer is lattice shift. This is the major difference between Kirkendall void formation (lattice shift occurs) in the layer-type growth and the large amount of pore formation in porous-type growth of $Cu_3Sn$ (no or very little lattice shift occurs).

Before we analyze the kinetics of porous growth, we recall the classical cellular precipitation. [9–11] An example is the room temperature precipitation of Sn in an alloy of Pb – 7 at. % Sn. It takes place when the alloy was homogenized at 150 °C and quenched and kept at room temperature. The precipitation of lamellar Sn occurs via grain boundary diffusion because at room temperature the lattice diffusion of Sn in the alloy can be ignored. Morphology of the precipitation is depicted in Figure 5.5, where a moving cell boundary leads behind it a lamellar structure, which consists of the precipitated Sn and the remaining Pb(Sn) alloy. It is an incomplete and nonequilibrium precipitation because not all the super-saturated Sn can precipitate out. For comparison, the porous $Cu_3Sn$ formation is complete because the empty channels are open-ended and all the excess Sn can go to the sidewall.

In classical cellular precipitations, the diffusion of solute atoms occurs along a moving cell (or grain) boundary, so the time of diffusion is too short to allow all the super-saturated Sn to come out. [10] Historically, no complete cellular precipitation has been reported until the porous $Cu_3Sn$ formation in μ-bumps was observed. This is a unique and rare situation in phase transformations. Below a comparison between them is given. [12, 13]

To consider diffusion along a moving boundary, the divergence in a small volume of dxdydz, as shown by the small shaded area within the boundary in Figure 5.5, is given below. By taking $\Delta J_z = 0$, and $\Delta y = \delta$, we have

$$\frac{\partial C}{\partial t} = -(\nabla \cdot J) = -\left[\frac{\partial}{\partial x}\left(-D_b \frac{\partial C_b}{\partial x}\right) + \frac{\Delta J_y}{\Delta y}\right] = -D_b \frac{\partial^2 C_b}{\partial x^2} - \frac{V(C_0 - C)}{\delta}$$

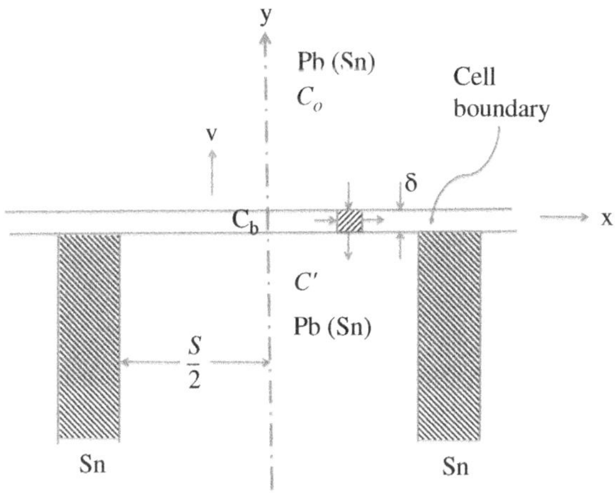


**Figure 5.5** A schematic diagram of cellular precipitation is shown. The cell boundary moves upward with a velocity of V along the vertical y-axis, and a lamellar structure of periodic Sn precipitates, and the remaining Pb(Sn) alloy is formed behind the cell boundary.

where $C_b$, $C'$, and $C_0$ are the concentrations in the boundary, in the transformed matrix, and in the untransformed matrix, respectively. Although $C_b$ is the concentration in the cell boundary, its unit is the same as that of $C'$ and $C_0$, which are concentrations in the bulk of the alloy. $V$ is the velocity of the moving cell boundary. Assuming steady state, Cahn derived the following equation [10],

$$D_b \frac{\partial^2 C_b}{\partial x^2} + \frac{V}{\delta}\left(C_0 - C\right) = 0 \tag{5.16}$$

The equation was solved by taking $C'/C_b = k$, which is the partition coefficient. The two boundary conditions are $dC_b/dx = 0$ at $x = 0$ and $C' = C_e$ at $x = S/2$, where $S$ is the lamellar spacing as shown in Figure 5.5. The solution is

$$\frac{C - C_0}{C_e - C_0} = \frac{\cosh\left(\sqrt{\alpha}\, x / S\right)}{\cosh\left(\sqrt{\alpha} / 2\right)} \tag{5.17}$$

where $\alpha = kVS^2/D_b\delta$, or $V = \alpha D_b \delta / kS^2$.

Cahn stated that if the diffusion is limited to the advancing cell boundary, the lamellae of cell cannot reach equilibrium composition at any nonzero growth rate. [10] In addition, the growth rate cannot be determined by the diffusion rate alone, but two controlling kinetic processes are required to describe the reaction.

Define Q to be the fraction of super-saturation that has been precipitated out, we have

$$Q = \frac{C_0 - C}{C_0 - C_e} = \frac{2}{\sqrt{\alpha}} \tanh \frac{\sqrt{\alpha}}{2} < 1 \tag{5.18}$$

Even we can measure $Q$ experimentally, we can only determine the product of $VS^2$ in $\alpha$, not $V$ and $S$ independently, meaning we have to measure one of them too. If we decrease the lamellar spacing $S$ to reduce the distance or the time of diffusion, the growth rate $V$ will increase, which will decrease the time of $\delta/V$. For a complete precipitation, we need $Q = 1$ and it requires $C' = C_e$, which is impossible because of the diffusion along a moving cell boundary. The above simple derivation is recalled here for the purpose of a direct comparison to the complete cellular precipitation of porous $Cu_3Sn$, to be given below.

Figure 5.6 depicts porous $Cu_3Sn$ formation. The cell boundary or the interface between $Cu_6Sn_5$ and porous $Cu_3Sn$ is moving into $Cu_6Sn_5$. If we project the three-dimensional figure in Figure 5.6 onto the x-y plane, the projection is similar to Figure 5.5. The morphological similarity between them is because we can regard the Sn lamellae in Figure 5.5 to be empty pores. Significantly, the empty pores are channels of open-end, so the time of out-diffusion of Sn is infinitive.

A detailed kinetic analysis of porous $Cu_3Sn$ formation has been published. [12] Here, a simple analysis of the growth of porous $Cu_3Sn$ on the basis of Figure 5.6 is given. The porous structure has a periodic $(a + b)$ layers, where "$a$" is the width of the lamellar $Cu_3Sn$ and "$b$" is the width of the lamellar pore. In a period of $\Delta t$, the interface of an arbitrary area of $W{*}W$ moves a distance of $\Delta y$ from the porous structure into $Cu_6Sn_5$, so the growth velocity is $V = \Delta y/\Delta t$.

The motion has transformed a volume of $W{*}W{*}\Delta y$ of $Cu_6Sn_5$ into the porous $Cu_3Sn$. The theoretical volume of voids in the porous structure is 41%. Therefore, the total number of Sn atoms depleted from $Cu_6Sn_5$ is

$$\# Sn = \frac{0.41 WW \Delta y}{\Omega} \tag{5.19}$$

where $\Omega$ is the atomic volume of Sn. By the law of conservation of mass, this number of Sn atoms must equal to the total flux of Sn atoms diffused out from the depletion of $Cu_6Sn_5$, which is

$$\# Sn = J_b A \Delta t = C_b \frac{D_b}{kT} \left( -\frac{\Delta \mu}{\Delta x} \right) (2nWd) \Delta t \tag{5.20}$$

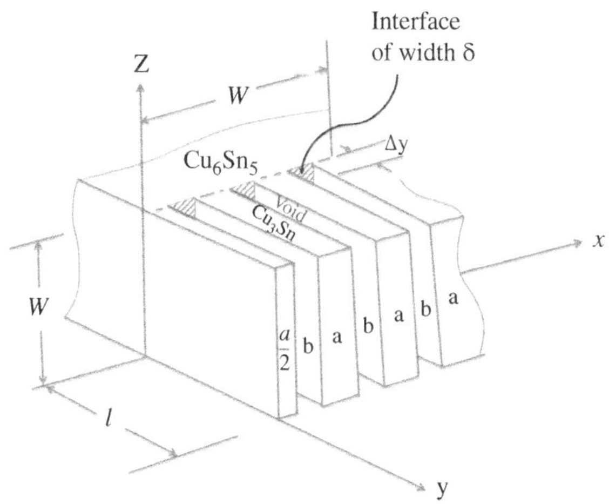


**Figure 5.6** A schematic diagram of the growth of porous $Cu_3Sn$ is shown, where the cell boundary or the interface between $Cu_6Sn_5$ and porous $Cu_3Sn$ is moving into $Cu_6Sn_5$. The porous structure has a periodic ($a + b$) layers, where "$a$" is the width of the lamellar $Cu_3Sn$ and "$b$" is the width of pores.

where $J_b$ is the interfacial flux of Sn diffusing in the x-direction on the interface. We have $J_b = C_b\langle v\rangle = C_bMF = C_b(D_b/kT)(-\Delta\mu/\Delta x)$, where $M = D_b/kT$ is mobility, and $F = (-\Delta\mu/\Delta x)$ is driving force. The area "$A$" is the cross-sectional area of the diffusing flux, and $A = 2nWd$, where $n$ is number of layer of $(a + b)$ in the width of $W$, so $n = W/(a + b)$, and d is the diameter of Sn atom. The factor of 2 is because each lamella of $Cu_3Sn$ has two surfaces. By equaling the above two equations, we have

$$V = \frac{\Delta y}{\Delta t} = \frac{\Omega}{0.41} C_b \frac{D_b}{kT}\left(-\frac{\Delta\mu}{\Delta x}\right)\left(2\frac{d}{a+b}\right) \tag{5.21}$$

To calculate $V$, we recall that in the derivation of Eq. 5.16, $C_b$ is the fraction of Sn atoms per unit volume, and $C'/C_b = k$, where $k$ is the partition coefficient. However, in decomposition of intermetallic compound (IMC), we take $C_b = X_{Sn}(1/\Omega_{\text{IMC}})$, where $X_{Sn}$ is the atomic fraction of Sn and $\Omega_{IMC}$ is volume per one atom in the IMC, respectively.

Comparing Figures 5.5 and 5.6, the basic similarity is the moving interface, which is also the reaction front where phase transformation occurs. In the Pb(Sn) case, only the diffusion of Sn along the interface was considered, but no lateral diffusion of Pb. In the porous $Cu_3Sn$ case, interdiffusion of Sn and Cu along the interface has been considered because of the phase change from $Cu_6Sn_5$ to $Cu_3Sn$. We recall the classic Darken's analysis of interdiffusion, where vacancy is assumed to be equilibrium everywhere in the sample, so lattice shift occurs and it leads to marker motion but no void formation. In the porous case, we have neither vacancy equilibrium nor lattice shift. The empty lattice sites left behind by the out-diffusion of Sn keep the void growing.

In the above, a comparison between the complete and incomplete cellular precipitation is given. [13] The similarity is that both have a moving interface, which is the reaction front where phase decomposition occurs. The difference is that the incomplete precipitation occurs in a closed system, where the diffusion is limited to the advancing cell boundary, so the lamellae of cell cannot reach equilibrium composition at any nonzero growth rate. On the other hand, the complete precipitation occurs in an open system, where the lamellar pores are empty channels of open-end, so the time of Sn diffusion is unlimited. Consequently, all the kinetic parameters can be defined in the complete case, but not in the incomplete case.

## 5.6 Effect of Surface Diffusion on IMC Formation in Pillar-Type μ-Bumps

The effect of surface diffusion on IMC formation in μ-bumps was studied by using the pillar-type samples as shown in Figure 5.7. One side polished polycrystalline Cu plates of 1 × 1 cm were etched and washed clean before they were electroplated with a layer of Sn about 5–10 μm. Focus

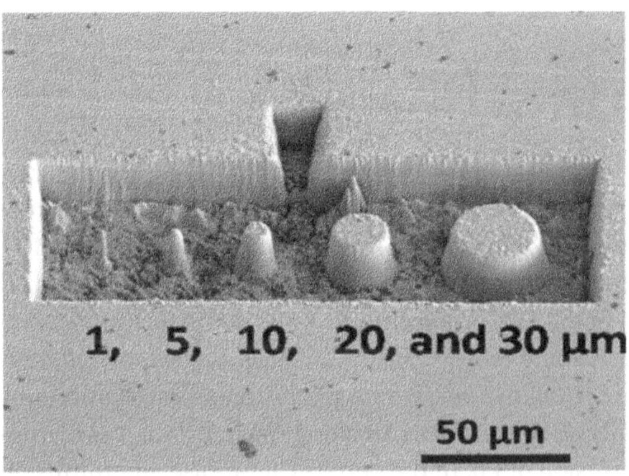


**Figure 5.7** The pillar-type samples are shown having various diameters, from 1, 5, 10, 20, to 30 μm.

ion beam (FIB) was used to pattern the Sn layer into pillars of various diameters, from 1, 5, 10, 20, to 30 μm. [14] The pillar structures were immediately placed into a vacuum oven and annealed at 185 and 195 °C for 30 minutes, and their SEM images were taken. Then, the samples were put back into the oven at the original temperature for another 30 minutes and repeated this process twice. In this way, the data of IMC thickness and its growth rate after annealing for 30, 60, and 90 minutes were obtained. Energy dispersive X-ray (EDX) analysis was used to confirm the IMC composition.

Figure 5.8 shows the SEM images of 1 μm pillars after annealing at 195 °C for 30, 60, and 90 minutes. Figure 5.9a and b show, respectively, the change of IMC thickness with different diameters when annealed at 185 and 195 °C. The trend of decreasing IMC thickness with increasing pillar diameter becomes obvious after 60 and 90 minutes annealing; actually, there seems to be a faster increase of IMC thickness for the pillars smaller than 30 μm after 90 minutes. The longer the annealing time, the more obvious of the IMC thickness increase in the smaller pillar.

Generally speaking, the diameter of $Cu_6Sn_5$ grains in C-4 solder joints is about 5 μm. Consequently, in 1 μm pillar joints, there should be only one single grain of $Cu_6Sn_5$. On the basis of this assumption, we built a simple model for IMC growth, ignoring lattice diffusion and grain boundary diffusion in small pillar joints. We consider surface diffusion alone. This is because we assume lattice diffusion is too small, and there is no grain boundary inside 1 μm pillar. However, when Cu atoms diffuse on the surface of the IMC to the Sn side, we must assume Cu atoms diffuse interstitially into Sn to grow the internal $Cu_6Sn_5$ layer. It is known that noble and near-noble metals diffuse interstitially in group IV elements.

Figure 5.10 illustrates the mechanism of Cu diffusion in 1 μm pillar, where we assume that the atomic diameter is d, and the circumference of

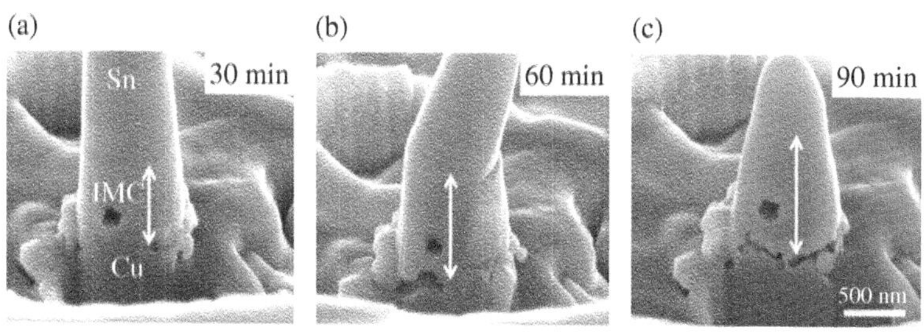


**Figure 5.8** SEM images of 1 1 μm pillars after annealing at 195 °C for (a) 30, (b) 60, and (c) 90 minutes.

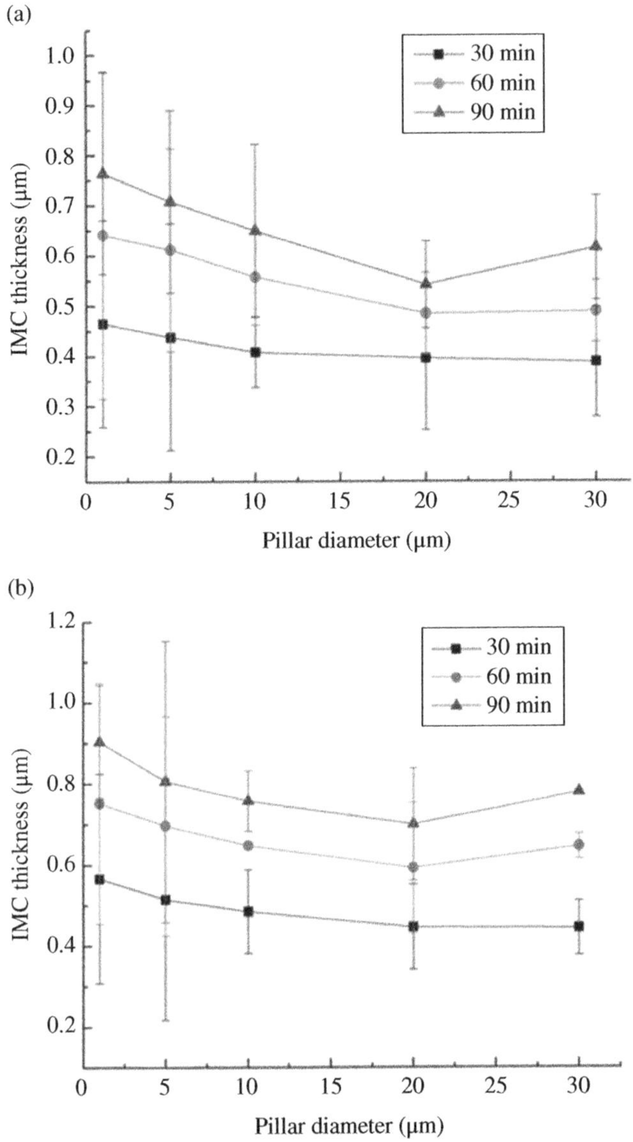


**Figure 5.9** (a and b) The change of IMC thickness with different diameters when annealed at 185 and 195 °C, respectively.

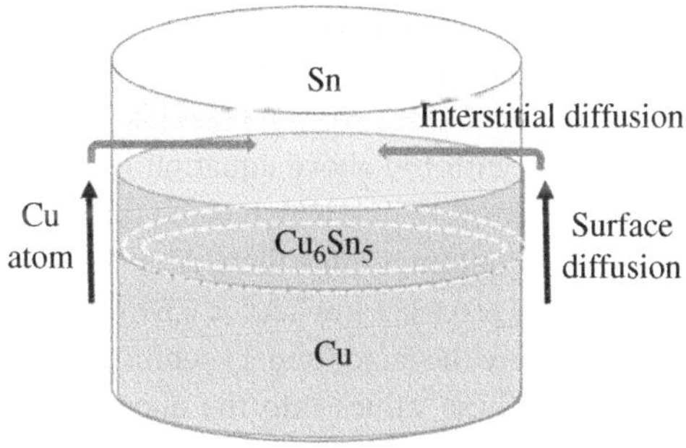


**Figure 5.10** The mechanism of Cu interstitial diffusion into 1 μm pillar is depicted.

pillar is $2\pi r$. In a period of "$\Delta t$," the amount of Cu atoms in the flux of surface diffusion is given as

$$\#\text{Cu}_{\text{atom}} = J_s \times 2\pi r \times d \times \Delta t$$

The $Cu_6Sn_5$ growth rate is $\Delta h/\Delta t$, as shown in Figure 5.10; therefore, in the period of "$\Delta t$," the $Cu_6Sn_5$ grows with a thickness of $\Delta h$. Take the unit cell volume of one $Cu_6Sn_5$ molecule to be $\Omega$. Then, the number of Cu atoms in a layer of $Cu_6Sn_5$ of thickness of $\Delta h$ is,

$$\#\text{Cu}_{\text{atom}} = \pi r^2 \Delta h \frac{6}{\Omega}$$

By combining these two equations in the above, we have

$$\frac{\Delta h}{\Delta t} = \frac{d\Omega J_s}{3r} \tag{5.22}$$

It means that in the period of $\Delta t$, the thickness of $\Delta h$ is inversely proportional to $r$, the radius of pillar. It agrees with the finding that the smaller the radius of pillar, the faster the growth of the IMC. By rearranging, we have an equation of the flux contributed by surface diffusion according to Fick's first law,

$$J_s = 3\frac{r}{d}\frac{\Delta h}{\Delta t}\frac{1}{\Omega} = -D_s\frac{\Delta C}{\Delta x}$$

Or

$$D_s = \frac{3r}{d}\frac{\Delta h}{\Delta t}\frac{1}{\Omega}\frac{\Delta x}{\Delta C} \tag{5.23}$$

For simplicity, we can roughly assume $\Delta C$ to be about 1% in the growth of $Cu_6Sn_5$, owing to the $Cu_6Sn_5$ composition range in the Sn-Cu equilibrium phase diagram. Since $\Delta x$ is the thickness of IMC, we can calculate the Cu surface diffusivity at 185 °C on $Cu_6Sn_5$ with the above equation by using values of $r = 0.5$ μm, $d = 0.256$ nm, $\Delta h = -0.12$ μm, which is the thickness growth from 60 minutes annealing to 90 minutes annealing, $\Delta t = 30$ minutes, and the unit cell volume of $Cu_6Sn_5$ is 0.0383 nm$^3$, $\Delta x = 0.64$ μm, and $\Delta C = (6/11) \times 0.1 \times 10^{22}$ atoms/cm$^3$, where the factor of 6/11 represents the fraction of Cu in $Cu_6Sn_5$. Substituting these values into the above equation, we have

$$D_s = \frac{3x0.5x10^{-4}cm}{2.56x10^{-8}cm}\frac{0.12x10^{-4}cm}{30x60\sec}\frac{1}{0.0383x10^{-21}cm^{-3}}\frac{0.64x10^{-4}cm}{(6/11)x0.1x10^{22}cm^{3}}$$

$$= \frac{3x0.5x0.12x0.64}{2.56x1800x0.0383x(6/11)x0.1}10^{-4-4-4+8+21-22}\frac{cm^2}{\sec} \approx 2x10^{-7}cm^2/\sec$$

We note that the theoretical Cu diffusivity on (111) plane is $9.42 \times 10^{-6}$ cm$^2$/sec, on (100) plane is $1.19 \times 10^{-9}$ cm$^2$/sec, and on (110) plane is $5.98 \times 10^{-11}$ cm$^2$/sec. By comparing to these data, it seems the Cu surface diffusivity on $Cu_6Sn_5$ is slightly slower than Cu diffusivity on (111) plane but much faster than on (100) plane.

## Problems

**5.1** In layer-type growth, what is the difference between diffusion-controlled growth and interfacial-reaction controlled growth?

**5.2** What is the definition of lattice shift? Please draw a schematic diagram to explain it.

**5.3** What is the difference between Kirkendall void in $Cu_3Sn$ and porous void in $Cu_3Sn$?

**5.4** Make a comparison between complete and incomplete cellular precipitation. Why certain cellular precipitation can be incomplete?

**5.5** What is Wagner's diffusivity? Why we need it to solve the layer-type IMC growth?

## References


1 D. A. Porter and K. E. Easterling, "*Phase Transformations in Metals and Alloys*," 2, Chapman & Hall, London, 1992.

2 P. G. Shewmon, "*Diffusion in Solids*," 2, The minerals, Metals and Materials Society, Warrendale, PA, (1989).

3 J. W. Christian, "*The Theory of Transformations in Metals and Alloys: Part I. Equilibrium and General Kinetic Theory*," 2, Pergamon Press, Oxford (1975).

4 C. Wagner, Theorie der Alterung von Niederschlagen durch Umlosen (Ostwald-Reifung), *Zeitschrift fur Elektrochemie*, 65, 581–591 (1961).

5 L. S. Darken, "Diffusion, mobility and their interrelation through free energy in binary metallic systems", *Transactions of AIME* 175, 184 (1948).

6 W. K. Chu, H. Krautle, J. W. Mayer, H. Muller, M. A. Nicolet, and K. N. Tu, *Applied Physics Letters*, 25, 454 (1974).

7 Kejun Zeng, Roger Stierman, Tz-Cheng Chiu, and Darvin Edwards, "Kirkendall void formation in eutectic SnPb solder joints on bare Cu and its effect on joint reliability", *Journal of Applied Physiology*, 97, 024508 (2005).

8 I. Panchenko, K. Croes, I. De Wolf, J. De Messemaeker, E. Beyne, and K. J. Wolter, "Degradation of $Cu_6Sn_5$ intermetallic compound by pore formation in solid–liquid interdiffusion Cu/Sn microbump interconnects", *Microelectronic Engineering* 117, 26 (2014).

9 D. Turnbull, "Theory of cellular precipitation", *Acta Metallurgica*, 3, 55 (1955).

10 J. W. Cahn, "The kinetics of cellular segregation reactions", *Acta Metallurgica* 7, 18 (1959).

11 K. N. Tu and D. Turnbull, "Analysis of kinetics of boundary diffusion limited cellular precipitation", *Scripta Metallurgica*, 1, 173 (1967).

12 A. M. Gusak, Chih Chen, and K. N. Tu, "Flux-driven cellular precipitation in open system to form porous $Cu_3Sn$", *Philosophical Magazine*, 96: 13, 1318–1331 (2016).

13 K. N. Tu and A. M. Gusak, "A comparison between complete and incomplete cellular precipitation," *Scripta Materialia*, 146, 133–135 (2018).

14 Yingxia Liu, Ying-Ching Chu, and K. N. Tu, "Scaling effect of interfacial reaction on intermetallic compound formation in Sn/Cu pillar down to 1 um diameter*Acta Materialia*, 117, 146 (2016).

# Part II

# 6

# Essence of Integrated Circuits and Packaging Design

## 6.1 Introduction

In this chapter, we introduce first the essentials for integrated circuits (ICs) and package designs for modern electronics products. Next, we describe the basic elements in IC and package scaling during the past development, and how they integrate. Then, many integrations and design styles will be reviewed, including SoC and multicore trends in IC designs; SiP and CoWoS, InFO in packaging. At the end, we summarize the current status of "More than Moore" 3D IC and the near-future electronics design integration.

ICs did not have a long history. Before ICs appeared, people have used vacuum tubes for electronic devices such as radios and radars. When transistors appeared, dramatic changes occurred. In 1946, Shockley at Bell Labs has proposed to develop semiconductors, and next year (1947), the **transistor** was created by him, Bardeen, and Brattain. They received the Nobel Prize in 1956. Then, the next key invention came in 1959 from Hoerni at Fairchild when the "planar" manufacturing process was realized. At the same time, Kilby at Texas Instruments and Noyce at Fairchild invented the IC based on the planar finding [1]. IC was recognized as a breakthrough for fabricating many transistors together on one plane. Furthermore, very thin metal lines can be deposited on the oxidized devices in order to connect all transistors, making all components to work in one production. Due to the arguments in the planar finding, at the end, Kilby got the patent for the invention of IC, but Noyce obtained the key patent for interconnection inside IC. Kilby was awarded the Nobel Prize in 2000 for this contribution (Noyce passed away in 1990).

Since then, we have started the roller-coaster trip in the development of ICs. In 1965, Moore at Fairchild had made the prediction for the future of

*Electronic Packaging Science and Technology*, First Edition. King-Ning Tu, Chih Chen, and Hung-Ming Chen.

ICs; it then became the "first rule" in semiconductor IC industry, and we usually call it "Moore's Law." The first version of the law is that the number of transistors on a silicon chip will increase exponentially, at least doubled every year in 10 years. Then, it became that the number will be doubled every two years and then corrected to be doubled every 18 months. However, what is usually ignored is that the increase can take place without the increase of production cost. Consequently, today the transistor is that cheap, so we can use as many as we want. Moore's law is not from natural evolution, it is economical and a prediction based on new technology, social science, and human habits in life. This law also has become a self-fulfilling prophecy because each company must keep up with its competitors.

In the path of past development, small-scale integration (SSI) has been named in the early 1960s; it has around tens of gates. Improvements in technique led to devices with hundreds of logic gates, known as medium-scale integration (MSI) in the late 1960s. Further improvements led to large-scale integration (LSI), i.e. systems with thousands of logic gates. With gates more than this level, we usually called very-large-scale integration (VLSI), which is widely used until now. Current technology has moved far past this mark and today's microprocessors have many millions of gates and billions of individual transistors. Around the year of 2008, billion-transistor processors became commercially available in the nanometer (sub-100 nm) era.

On the other hand, in order to house a semiconductor chip, we need to provide a "shell." Then, packaging comes into picture. In general, packaging supplies the chips with signals and power, at the same time can help to remove heat from the circuitry and provides physical support and protection [2]. To be more specific, it may be possible to directly use bare dies/chips (sliced from wafers) for targeted purposes; however, the chips may suffer from easy damage and contamination during transportation and heavy duty usage. As a result, we need a "shell" to protect the chips for longer usage. Structurally, a package is a plastic, ceramic, laminate, or metal seal that accommodates the die/chip inside [3]. In other words, it is a medium to host the die/chip on the printed circuit board (PCB) and provides better and safer connections between them. In an early definition [2] for the "perfect" electronics package, they still hold. It should be compact, and should supply the chips with a large number of signal and power connections, which have minute capacitances, inductances, and resistances; the wiring on the package should be very dense, and the interconnections are good transmission lines with low capacitance and high characteristic impedance. It should also provide a stable power-supply level even when the circuits are operating at high speeds and many output drivers are switching simultaneously. The thermal properties should match with chips well and avoid stress-induced cracks and failures. Last but not the least, it should cost much less than the chips it hosts.

During the IC packaging process, we perform the following operations: die attaching, bonding, and chip encapsulation. Die attachment is also called die mounting, which fixes the die to the package or the support structure. As aforementioned, the package has functions including delivering power from outer sources to chips, at the same time providing the signal communications in and out of the chips. They can be achieved by bonding: wire bonding, tape-automated-bonding (TAB), and controlled collapse bonding (C4 flip-chip) [2]. Wire bonding is to use thin wire (gold, aluminum, or copper) for the interconnection between the die and package, with a combination of heat, pressure, and/or ultrasonic energy. It is generally considered the most cost-effective technology. In TAB technology [2], a chip with its attached metal film leads is placed on a multilayer polymer tape. This film then can be flip-chip bonding, also called C4 (Controlled Collapse Chip Connection) bonding, that is the direct electrical connection between the face-down die and the substrate or PCB, through the conductive bumps on the chip's bond pads. The advantages include small size, high performance, great flexibility, and high reliability.

Encapsulation is sometimes called synonymously with the word "packaging." It means the die is encapsulated/accommodated with ceramic, plastic, metal, or epoxy to prevent the corrosion or damage from the outer world. At the end, there are two types of packages, depending on the ways to attach to PCB: pin-through-hole (PTH) (also called dual in-line package [DIP]) and surface-mount-technology (SMT). SMT is usually preferred for high-performance chips. Due to the rapid pace of increase of transistor density/capacity and the evolution of electronics systems, many novel packaging architectures have emerged, which will be discussed in Sections 6.5 and 6.6.

## 6.2 Transistor and Interconnect Scaling

As we already knew, the advance of electronics products come from the scaling of the transistor, or front end of the line (FEOL), and interconnect, or back end of the line (BEOL). In the miniaturization of FEOL, we can use the ideal scaling theory to describe the shrinkage of metal-oxide-semiconductor (MOS) transistors. In Table 6.1 [2], we can observe the effects when the scaling factor is applied. Assuming ideal scaling, all horizontal and vertical dimensions of transistors are scaled down by $1/S$ ($S > 1$); internal electric field strengths remain approximately unchanged. In conclusion, devices can get faster, thus both the speed and functional capability

**Table 6.1** Ideal scaling of MOS transistors.

| Parameter | Scaling factor |
|---|---|
| Dimension (W, L, $t_{gox}$, $X_j$) | 1/S |
| Substrate doping ($N_{SUB}$) | S |
| Voltages (VDD, $V_{TN}$, $V_{TP}$) | 1/S |
| Current per device ($I_{DS}$) | 1/S |
| Gate capacitance ($C_g$) | 1/S |
| Transistor on-resistance ($R_{Tr}$) | 1 |
| Intrinsic gate delay (T) | 1/S |
| Power-dissipation per gate (P) | $1/S^2$ |
| Power-delay product per gate ($P^*_T$) | $1/S^3$ |
| Area per device (A = WL) | $1/S^2$ |
| Power-dissipation density (P/A) | 1 |

of ICs get better without increasing power-dissipation density in such ideal world. However, a number of second-order effects undo the improvements brought by scaling, which we will discuss closely later.

On the other hand, interconnects play an important role in determining the speed, area, yield, and reliability of VLSI circuits, especially after we entered the deep-submicron (DSM) regime since 1990s. There are two facts we need to face, and they shape the current situation in physical design complexity. Just like aforementioned transistor scaling, Table 6.2 summarizes the scaling

**Table 6.2** Scaling of local and global interconnections.

| Parameter | Scaling factor |
|---|---|
| Cross sectional dimensions ($W_{int}$, $H_{int}$, $W_{sp}$, $t_{ox}$) | 1/S |
| Resistance per unit length ($R_{int}$) | $S^2$ |
| Capacitance per unit length ($C_{int}$) | 1 |
| RC constant per unit length ($R_{int}C_{int}$) | $S^2$ |
| Local interconnection length ($I_{loc}$) | 1/S |
| Local interconnection RC delay ($R_{int}C_{int}I_{loc}{}^2$) | 1 |
| Die size ($D_c$) | $S_C$ |
| Global interconnection length ($I_{Int}$) | $S_C$ |
| Global interconnection RC delay ($R_{int}C_{int}I_{loc}{}^2$) | $S^2S_C{}^2$ |
| Transmission line time of flight ($I_{int}/V_c$) | $S_C$ |

**Table 6.3** Intel's P856 Interconnect dimensions (0.25 um process).

| Layer | Pitch (in um) | Thickness (in um) | Final aspect ratio |
|---|---|---|---|
| M1 | 0.64 | 0.48 | 1.5 |
| M2 | 0.93 | 0.89 | 1.9 |
| M3 | 0.93 | 0.89 | 1.9 |
| M4 | 1.6 | 1.33 | 1.7 |
| M5 | 2.56 | 1.9 | 1.5 |

of local and global interconnection. The first fact we have observed is that the simple scaling does not apply to the global interconnects. There existed complications to the continuation of Moore's Law. We know that simple scaling expects to shorten the local interconnect length with scaling factor; however the RC constant (delay) per unit length is scaled inversely. At the end, we did not get benefits from the scaling, and this got worse when we encountered the massively larger chips. It is obvious that if we add a factor for die size (Sc), the global interconnect length will scale up, and finally impacts the RC delay of global interconnects to the maximum scale. The good old days when only dependence of gate count in circuit performance are gone. The second fact is about wire resistance evolution. Resistance, generally speaking, is inversely proportional to the cross-sectional area of the conductor. In Table 6.3, which was a rather old trend, we can still observe that the aspect ratio of the wires is above 1.5. The reason is that the scaling of the feature and wire sizes push this higher aspect ratio, in order to keep larger cross-sectional area. As a result, higher aspect ratio provides better interconnect performance on small widths, but introduces wall-to-wall capacitance, further worsening RC delay. We can somehow remedy this hard problem with the solution in modern packaging, to be illustrated in Sections 6.5 and 6.6.

## 6.3 Circuit Design and LSI

As mentioned in the introduction, Kilby and Noyce received the Nobel Prize in Physics in 2000 for their invention of ICs. Without ICs, we would not be able to have our modern lives today: starting from communicating with friends and colleagues, we need mobile phones with 5G technology, needing high-end communication chips and IoT environment; while working, we might need high-performance computing (HPC) servers for massive batch works or professional training purposes; during leisure time, we usually drive around and require autonomous driving or driving assistance, and high-end automotive in electric vehicles (EVs); at hospitals, we need biomedical and wearables electronics to help doctors make diagnosis and patients recover from their symptoms. In other words, the growing complication of applications has pushed very hard the design and manufacturing of ICs and electronics systems to newer levels we would never imagine before. [4]

From the nature of consumer electronics in different perspectives, we have different treatments in signals and data we receive, transmit, and process. In this end, the circuits are classified into digital and analog types. Since the natural world, most of it, is continuous and infinite, most of the signals are analogous. On the other hand, in order to store and process infinite data, computers are applied; therefore, transforming analog data to digital data is a must. Nowadays, due to the complexity of data processing and requirement of mobile devices, more and more analog circuits can be made by digital fashion; still the analog processing systems cannot be deserted.

From a common perspective in layout design style, we have the following types. The first type is full-custom style. It is an approach where layout elements are handcrafted and can be placed anywhere on the layout surface [5]. It is also the approach when IC came out, illustrated in Figure 6.1.

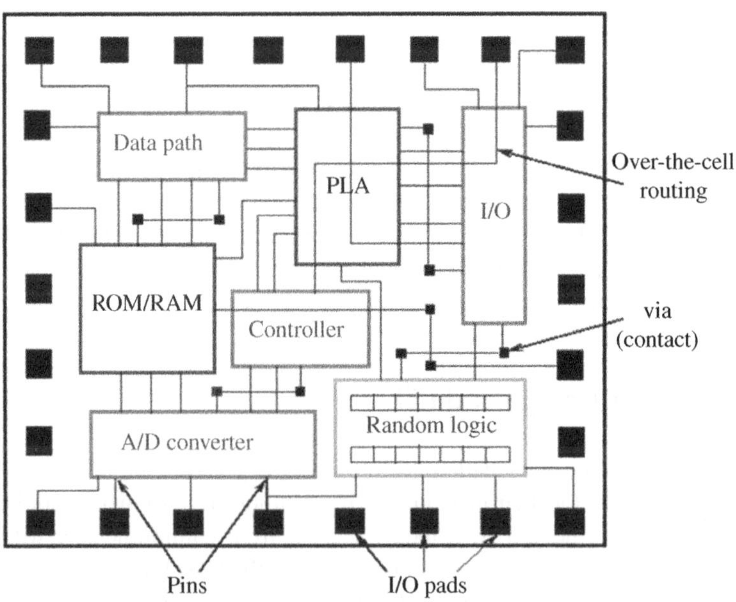


**Figure 6.1** IC design style: full-custom.

It is a free-style design, usually with no imposed constraints, except design rules from the manufacturing. However, the hard part is that the design takes longer time than other regular styles; usually it is used most in analog IC design. Another concern is that it cannot be applied when the number of transistors is large. The second type is standard-cell style, illustrated in Figure 6.2. In this design style, it is accomplished by using standard-cells. A

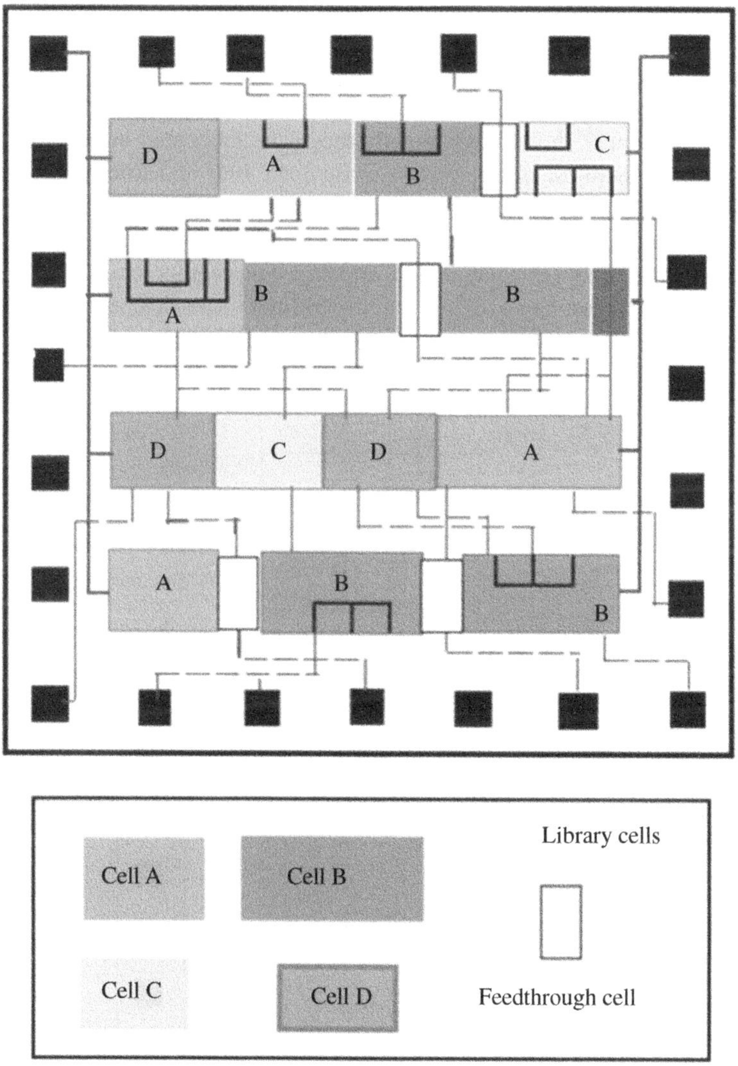


**Figure 6.2** IC design style: standard cell (row-based).

standard-cell is a logic block that performs a standard function. Usually, this style is for digital IC design, and it is supported by design tools, which will be described later. Because it is usually arranged in rows during layout, it is also called row-based design style. The third type is one of the common forms in programmable logic array, called field-programmable gate array (FPGA). In this style, two-dimensional array of logic blocks is the usual form, and each logic block can be programmed to implement any logic function of its inputs. It is flexible in comprising any design, and usually used as a prototyping of the designs-to-be in the market. Recently, it is also popular to be used for AI applications (Figure 6.3).

What is the current situation of circuit manufacturing? Who will manufacture the chips? In old days, usually they were fabricated by integrated device manufactures, so-called IDM companies. IBM and Intel were two major players, who fabricated CPUs as their main products. Through the years of development in consumer electronics, many more companies wanted to implement their creativities and innovations other than just CPUs, then a new business came up: the foundry business. The idea behind this is that not only IDMs can manufacture the chips, other small companies can also have their chips fabricated without owning fabs, this trend results in fabless companies. Then, the fabs which manufacture fabless companies'

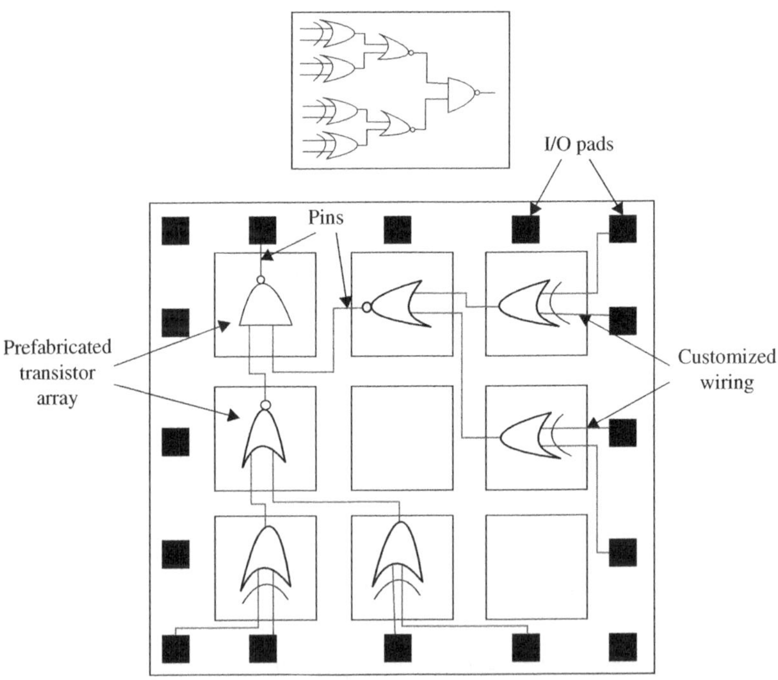


**Figure 6.3** IC design style: field programmable gate array (FPGA).

chips become foundries. The latter will only manufacture chips without IC design capabilities. This boosts up the IC designs big time. In Taiwan, Mediatech is a big IC design company, but Taiwan Semiconductor Manufacturing Company (TSMC), located in Taiwan as well as other places, is the leading foundry around the world. It is not only leading in IC manufacturing, but also impacting the packaging solutions, which will be discussed very soon.

Even with the foundries business taking care of IC manufacturing, the next question is with VLSI in circuit design, can we design ICs manually? Here are some numbers for consideration: now there are billions of transistors in one single chip (SoC chip, which will be elaborated in the next section), and the time-to-market (TTM) for such chips has been reduced from years to months or weeks. How can we do it? Through the advances in design methodology and computing facilities, it can be facilitated with design automation tools. The old name for this is computer-aided design (CAD); then it gets a new name: electronic design automation (EDA). Figure 6.4 presents the usual design methodology for IC. Just like Moore's Law, the self-fulfilling story happens here too: with better EDA tools, we can design better performance chips, which has led to better efficiency in designing ICs; in turn, it makes EDA tools to perform even faster. In the design methodology, each step has corresponding tools to form the supply chain. First, we are given the

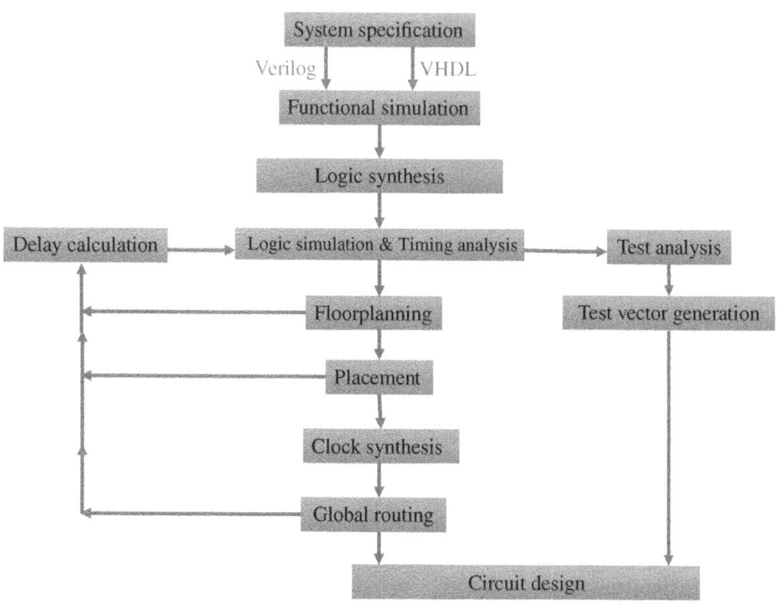


**Figure 6.4** VLSI design process and flow.

design specification of the chips, followed by the architectural and logic design, then we synthesize the circuits and perform the logic simulation. This is called the front end of the IC design. The output of this stage is called "netlist," the list of all the nets. Second, we enter the back end of the IC design: physical design, transforming the netlist to physical transistor placement and interconnect routing. Currently, it is impossible not to use EDA tools to design ICs. The major players are from the United States: Synopsys, Cadence, and Mentor Graphics (acquired by Siemens). It is interesting to mention that in 2020, the revenue of EDA tool vendors is around US$2.5 billion, whereas the revenue of the semiconductor is more than US$410 billion.

So far, it looks good that we can manufacture many chips with billions of transistors, and we have EDA tools to help. Is it the end of the story? The answer is far from it. Following the Moore's Law, we have scaled downward further and further but encountered many problems, mostly related to physics limit. When we approach the physics limits, we start to have process variation and it affects the yield. The manufacturing and yield problems that are used to be solved in the foundries now cannot be ignored during the stage of IC design. They are called the problems in design for manufacturability (DFM). As the process node shrinks below 130 nm, the issue of DFM has emerged as a serious challenge in obtaining the acceptable yield. One of the main issues is how to overcome the problem due to the fact that IC features are smaller than the wavelength of the light that creates them in lithography [3]. Figure 6.5 depicts this evolution of manufacturing

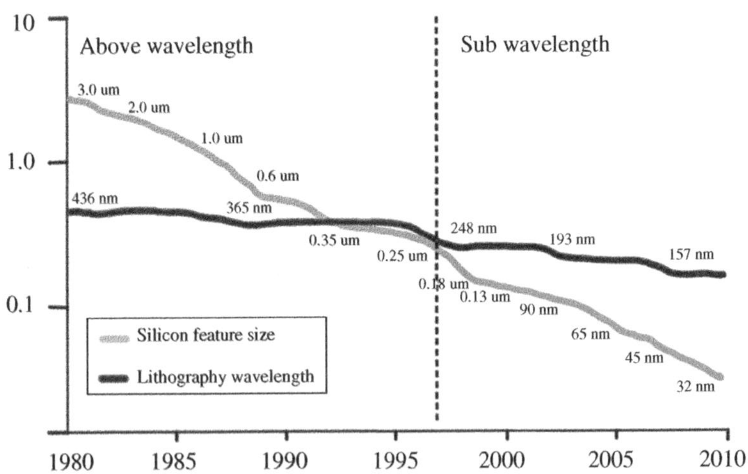


**Figure 6.5** Evolution of minimum feature sizes in semiconductor manufacturing relative to the exposure wavelength.

wavelength and transistor feature sizes. We can use the following analogy to describe that: it is impossible to use a wide brush to print a thin human hair. Although DFM problems are mostly related to manufacturing process, we need to mitigate the issues starting at IC design stage. Because they are related to the masks to fabricate the chips, the closest step is physical design (back end). Many researchers and semiconductor companies have worked on this topic since 2000, which keeps Moore's Law alive till now.

## 6.4 System-on-Chip (SoC) and Multicore Architectures

With the continuous advances of VLSI technology, and the main purpose is to put more and more transistors in a chip, finally it comes to how to place all transistors for an electronics system. They are usually called "System-on-Chip" (SoC). For a particular application (ASIC), a single chip/computer can include a CPU, bus, I/O devices, and of course memory. The basic economical reason for making SoCs is that they allow systems to be made at much lower cost than the equivalent board-level or package-level systems. Another reason, on top of the aforementioned, is that the performance can be much better and can achieve lower power than board or package-level equivalents. This is because on-chip connections are much more efficient than other levels of interconnections. The most common SoCs manufactured are communication SoCs since they are usually equipped with mobile devices. They have evolved from 3G to 4G LTE, to up-to-date 5G. Another well-known SoC is multiprocessor SoC (MPSoC) [6]. MPSoC is a system-on-chip with multiple processing elements. These processing elements may be general-purpose CPUs or digital signal processing units (DSPs) that are referred to as embedded CPUs or embedded DSPs that may also be specialized units. The reason for such SoC is that the irregular architecture can provide substantial benefits in real-time responsiveness and power consumption. Nonuniform memory systems can guard against irrelevant memory accesses from one processor delaying the time-critical accesses of another.

Another trend we cannot miss, and still ongoing, is multicore chips. The evolution to multicore is logical: with more and more transistors for computing, it is getting harder and harder to put them all together on one processor; in alternative approach, we put more processors/cores on one die. It is also called CMP (chip multiprocessor architecture). Multicore processors are widely used across many application domains, including general-purpose, embedded, network, DSP, and graphic processing unit (GPU). The most up-to-date trend is wafer-scale chip from Cerebras [7]. The fabless design house, Cerebras, has developed a novel computer system which is powered by a 21.5 cm by 21.5 cm wafer-scale processor with 400,000 programmable compute cores. It is structured as a regular array of 633 by 633 processing elements, each with its own local high bandwidth SRAM memory and direct high bandwidth connection to its neighboring cores. Furthermore, each tile has a programmable execution core, memory, and a router for interconnection with other tiles.

## 6.5 System-in-Package (SiP) and Package Technology Evolution

Here, we shall switch gears to another way of generating circuit systems. Because SoCs suffer from higher cost in manufacturing (the yield of such chips is usually not high enough), one not-so-advanced-way is to put all components/chips/devices in one package, also called "System-in-Package" (SiP). It may be easy to say to put all things in a package, but it is not easy to conclude that SiP or SoC which will be a better solution. It has been proposed that, from some prior discussions, the dependence of these two options involve four factors: time to market, cost, size, and performance. [8]

**Time to market** should be an advantage for SiP over SoC due to obvious reasons. With the SiP approach, the design complexity is contributed from specialized dies and the substrate. Because those specialized dies have shorter product design cycle time, we can expect to design them in parallel, and at the same time design the substrate vehicle with less complexity compared with an integrated SoC solution. At the end, a quicker cycle time can be reached, and it has much tunability in the first prototype pass with changeable discrete components and possible layout change of these components. On the other end, **lower-cost** solutions are believed to be out when taking the SoC approach. SoCs put all the circuitry in one die, while SiP can accommodate two or more dies and multiple passives. As mentioned in the beginning of this section, in order to make SoCs more competitive in price, the yield should be higher and enough to compensate for the loss of multiple higher-cost dies in SiP. In addition, the cost structure of SoC is a lot different from SiP. In testing, SoC only needs one test for the single die solution, whereas SiP must test each die before assembly and then test the completed module after assembly. SoCs also suffer from the higher non-recurring engineering (NRE) cost. Therefore, it is still arguable which will be at lower cost.

About **size**, SoCs apparently enjoy the advantage as well, but SiP is reducing the size and starting to come back with advanced packaging schemes such as die stacking. Although the advanced packaging schemes increase the assembly cost in SiP solution, the trend chooses the one due to form factor of the packages.

The last factor is **performance**. Indeed, shorter interconnects can bring better performance, which only happens in very specific purpose (or regular) chips such as CPU or computation-centric ASICs. It has been pointed out that a good SiP design would use the best die technology for cost and performance in each functional block, while SoC could require the entire die to use the highest-performing functional block [8]. We need to pay attention to the fact that we also need to use the best technology for passive components.

Before we see more SiPs in modern packaging, we should ask in general how packaging evolves. It is also the history of integration; in the following, we use the numbers (transistor density) in [9] to reflect the advances. The conventional packaging usually happens after getting the bare dies from the wafer sawing. This covers the periods from PTH/DIP, then Quad Flat Pack (QFP) due to higher I/O count of the die, Ball Grid Array (BGA) wire-bonding (Figure 6.6) and flip chip packages (Figure 6.7), and multichip module (MCM). The first single chip microprocessor in 1971 had the density of around 19000 transistors/cm$^2$. The routing density of the PCB basically determined the size of the system, the ratio of IC to package level integration was roughly around 2000:1. In early 1990s, the transistor density is around 265000 transistors/cm$^2$, this translates to a ratio of around 5000:1 between transistor and system component density. In 1998, Intel's Pentium III introduced the ratio of IC to package level integration that grew to more than 0.5M:1; at that time, BGA packages helped reduce the package size. It is worth noting that in the 1970s and 1980s, computing was the primary driver, while in the mid-1990s, the driver has shifted to

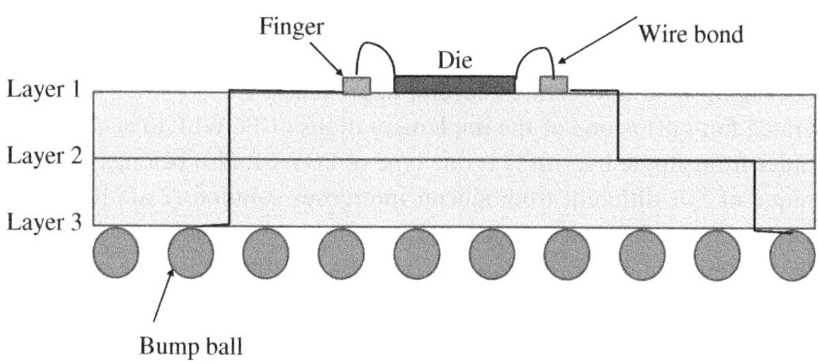


**Figure 6.6** Wire bonding in a BGA.

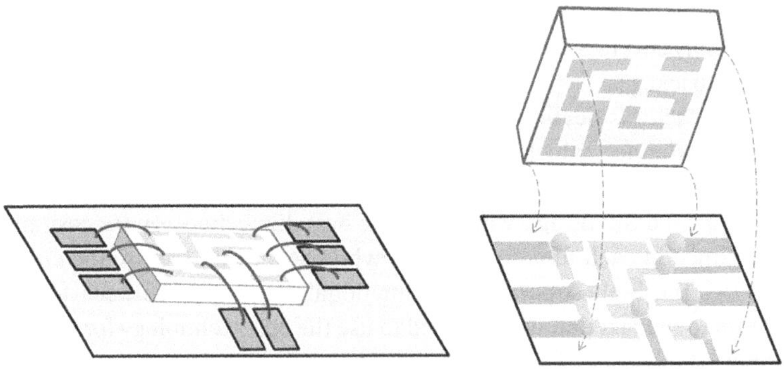

**Figure 6.7** Flip chip bonding packages.

consumer electronics. Mobility became an essential requirement as well as the driver.

When we enter wafer level chip scale packaging (WLCSP), the story becomes different. The reason for using WLCSP/WLP is that it can contain more input/output (I/O) pins due to the exploding need of I/O because of finer and finer digitization. It also runs the opposite against the conventional packaging: we package first, then saw the packaged wafer. In WLCSP, the size of package is almost the same as the bare die, single package size can be effectively controlled. At first, fan-in style WLP is applied. However, due to the decrease of the chip area, the number of pins is also decreased; therefore, not enough for big I/O number design. In order to solve this, fan-out WLP or FOWLP has come to the rescue, providing the area surrounding the chip for connecting more I/Os with redistribution layer (RDL). This has become the mainstream of the modern packaging.

The story gets more interesting: with more efficient packaging to help boost the chip performance. Foundries such as TSMC have entered the packaging field and created several approaches, InFO (wafer level integrated fan-out) is one of the implementations of FOWLP to acquire large order from Apple Inc. InFO is one type of FOWLP, also belongs to a technique of SiP, different from silicon interposer solution (2.5D IC, will be introduced in the next section). InFO technology not only provides a system scaling but also complements the chip scaling, and helps sustain the Moore's Law for the smart mobile as well as Internet of Things (IoT) applications [10], will be detailed in Section 8.8. Figure 6.8 shows the evolution of packaging from MCM to current heterogeneous integration (HI).

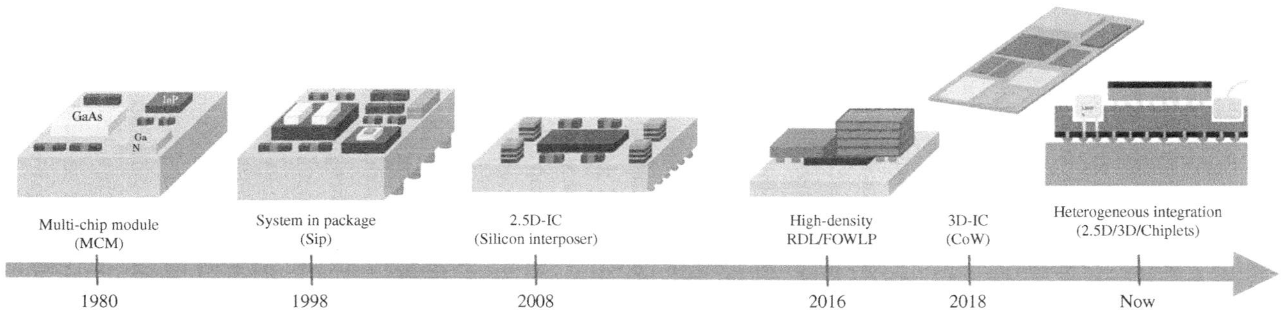


**Figure 6.8** The package evolution: from MCM to HI.

## 6.6 3D IC Integration and 3D Silicon Integration

3D integration solutions have been called for in the semiconductor market for a long time. In a broader sense, 3D integration usually means to place the components or devices in a vertical direction instead of placing them side by side. The reason of going vertical is very simple: we try to shrink the form factor of the chips, since we still need many transistors with our forthcoming computing tasks, horizontal communication will limit the area being used. It consists of 3D IC packaging, 3D IC integration, and 3D silicon integration [11]. 3D IC packaging has been in the market, such as die stacking with bonding wires and package on package stacking (PoP). As for 3D IC and silicon integrations, through silicon via (TSV) plays a very essential role because it is needed in vertical communication.

In 3D IC packaging, comparing stacking ICs with 2D ICs, the planning of I/Os and modules are complex because I/Os and modules should be planned into several tiers. Due to the planning of I/Os and modules in stacking IC, all modules and I/Os in the system partitioning step should be placed in a suitable tier to minimize the nets cut, and then the 2-D floor planning tool is used for each tier. If also considering PoP, the form factor will be smaller, but the performance of this setup is not boosted greatly. Other 3D integration styles should get better performance due to the effective usage of TSVs.

3D silicon integration has several definitions, in [12], they showed the thinned silicon layers with TSV and silicon interconnections, and silicon chips mounted on the thinned silicon interposer. In order not to confuse with 3D and 2.5D IC integration, we adopt the classification in [11]: 3D silicon integration is to stack up wafers with TSV for electrical feedthrough. When people were starting to go vertical in the 1980s, it was favored back then. Through the years of development and research, it is still not easy and to be commercially in mass production. The major reasons include the material and technology difficulty, design issues such as alignment and not-fixed models, and bumpless characteristics. As for now, wafer-to-wafer (W2W) bonding is still difficult; chip-to-wafer (C2W) is the current research focus and hope to achieve this integration.

3D IC integration, on the other hand, is considered the savior in modern 3D integration. It contains many types: C2W, W2W, and chip-to-chip (C2C) with microbumps. Among them, we can place the chips either stacked up *or* side by side, or other substrates with much smaller versions of bumps called microbumps. If those substrates (or so-called interposers) carry the active components or devices (transistors), we can still treat them as 3D IC integration. However, if the interposers only contain passive devices such as decoupling capacitors, system houses called them 2.5D IC integration, it is considered to be the most effective 3D IC integrator. Another effort made by TSMC for another big customer Xilinx Inc. is presented/called chip-on-wafer-on-substrate (CoWoS). It is the representative implementation for 2.5D IC integration. Xilinx Inc. is an FPGA vendor which generates high-performance programmable devices, where the development of TSV interposer technology for a high-performance 28 nm logic die is mounted on a large silicon interposer with Cu TSV. [13] Figure 6.8 also presents some of 2.5D package design. More design and package details will be presented in later chapters.

## 6.7 Heterogeneous Integration: An Introduction

It has been years of effort to achieve many aspects of the production of consumer electronics, communication devices, and the integration of them. For the semiconductor industry, the paradigm shifted from a continuous focus on extending Moore's Law to a new horizon where productivity advances can be reached through HI is no small endeavor. Following semiconductor industry association (SIA)'s closure of ITRS activities in 2015, the Heterogeneous Integration Roadmap (HIR) has taken over to strive with the new plan, mainly sponsored by the IEEE Electronics Packaging Society (EPS) and SEMI. In general, HI refers to the integration of individually manufactured components into a higher level assembly (such as SiP) that provides enhanced functionality and improved operating characteristics [14]. The technical working groups (TWGs) in this roadmap workforce include High Performance Computing and Data Center, IoT, Medical, Health and Wearables, Automotive, Aerospace & Defense, Mobile, Integrated Photonics, MEMS, and Sensor Integration.

About the central idea of HI, we ask why we need it? From the current definition, HI means to combine the components or devices manufactured by different technologies, or by different companies. As an example of homogeneous integration, SoC and multicore chips are presenting the computing power for modern applications. HI is to have higher level of assembly to achieve more powerful functionalities, and of course to improve the quality of operation. Now, we are very close to the limit of Moore's Law; SoC and multicore solutions show almost the end of homogeneous integration. We can watch closely the memory system evolution and then we see the value creation from HI [15]. DRAM (dynamic random access memory) is an important component in computer system, and it has three major standards: DDR (double data rate), LPDDR (low power DDR), and GDDR (graphic DDR). Starting from 2010, people start to use packaging schemes to improve the performance of DRAM, and it officially enters the HI regime. HMC (hybrid memory cube), wide I/O, and HBM (high bandwidth memory) are proposed, and they correspond to DDR, LPDDR, and GDDR, respectively. HMC is proposed from Micron and Intel, but it is discontinued. Wide I/O is proposed by Samsung; however, the mass commercial application is not seen yet (compared with HBM). HBM is the one which becomes modern HI memory and is widely used. Now, we see that HI is being used in memory manufacturing, and will be in other places soon.

In short, the revolution about semiconductor technology starts with Moore's Law, yet this revolution will continue with the advances in packaging technologies. As aforementioned, SiP has opened the door for achieving HI. In Chapter 8, we will discuss more about them in detail.

## Problems

**6.1** What is Moore's Law? Is it a real physics law in science field? What has been keeping this law going and what will be the staller to keep it going?

**6.2** As technology advances, interconnects dominate the circuit performance. What are the interconnect issues handled during the traditional VLSI design flow? How do we use package technology to resolve interconnect issues in electronic systems?

**6.3** As 3D integration is becoming a modern trend in building electronics systems, what are the usual categories based on the technology and materials?

**6.4** What is the major difference between IDM companies and fabless companies? Can foundries sell chips?

## References

**1** Nenni, D. and McLellan, P. (2013). *Fabless: the transformation of the semiconductor industry*. Semiwiki.com.

**2** H.B. Bakoglu, *Circuits, Interconnections, and Packaging for VLSI*, Addison-Wesley, 1990

3 L. Xiu, *VLSI Circuit Design Methodology Demystified: A Conceptual Taxonomy*, IEEE Press, 2008
4 J. Lienig and H. Bruemmer, *Fundamentals of Electronic Systems Design*, Springer, 2017
5 S.M. Sait and H. Youssef, *VLSI Physical Design Automation: Theory and Practice*, World Scientific, 1999
6 W. Wolf, *Modern VLSI Design*, 4, Prentice Hall, 2009
7 M. James, M. Tom, P. Groeneveld, and V. Kibardin, (2020). *ISPD 2020 physical mapping of neural networks on a wafer-scale deep learning accelerator*. ACM International Symposium on Physical Design.
8 M. P. Gaynor, *System-in-Package: RF Design and Applications*, Artech House, 2007
9 M. Swaminathan and K. J. Han, *Design and Modeling for 3d ICs and Interposers*, World Scientific, 2014
10 C.-F. Tseng, C.-S. Liu, C.-H. Wu, and D. Yu, (2016). *InFO (wafer level integrated fan-out) technology*. IEEE Electronic Components and Technology Conference.
11 J. Lau, (2011). *Evolution, challenge, and outlook of TSV, 3D IC integration and 3D silicon Integration*. IEEE International Symposium on Advanced Packaging Materials (APM).
12 J.U. Knickerbocker, P.S. Andry, B. Dang, et al. (2008). *3D silicon integration*. IEEE Electronic Components and Technology Conference.
13 B. Banijamali, S. Ramalingam, K. Nagarajan, and R. Chaware, (2011). *Advanced reliability study of TSV interposers and interconnects for the 28 nm technology FPGA*. IEEE Electronic Components and Technology Conference.
14 *Heterogeneous Integration Roadmap*, SEMI, 2019 Ed. https://eps.ieee.org/technology/heterogeneous-integration-roadmap/2019-edition.html.
15 Y.-C. Lin, "*The Value Creation of Heterogeneous Integration: DRAM as an Example (Chinese)*," DIGITIMES Taiwan, 2020

# 7

# Performance, Power, Thermal, and Reliability

## 7.1 Introduction

In this chapter, we introduce the metrics we usually apply when designing ICs, and they are conformed from generations to generations of designs. When we design the corresponding packages, they will be the first to be referenced. After the introduction of general guidelines in those metrics, we will describe them in detail. We first discuss the issues in performance and power dissipation, two of the most important metrics. We introduce some popular and effective low-power architectures such as DVFS and voltage islands. Then followed by thermal management and reliability concerns, they are closely related to the prior two major metrics, thus will be connected to the package designs in modern electronics.

Before we detail the metrics, we need to introduce the basics for transistors, especially the field-effect transistor, so that we know how we make the transistors work. In order to make all computations happen, except for the compute circuits (CPU), we need a larger component to store the data ready for computation: the memories. Afterward, we can discuss the challenges we face. Among all the challenges in IC design, performance is always the first-order effect to consider. Since the invention of IC and the planar manufacturing, we have been in a roller-coaster trip in fabricating chips for decades. In order to conform the protocols of evolving communication systems or to fulfill the functions in various electronics systems, the speed of the chips is usually the target to meet, we also call it "performance." When we see in the market the speed of the CPU, it is usually marked by the unit of GHz. It is the running frequency of the CPU chip, also stands for the reciprocal of the longest propagation time in sequential circuits (from one register to another). We have more detailed discussion in Section 7.3.

*Electronic Packaging Science and Technology*, First Edition. King-Ning Tu, Chih Chen, and Hung-Ming Chen.

With the advances of technology, the transistors have become smaller and smaller, and the speed/delay became faster/smaller, we are able to put more and more transistors in a system, as "predicted" by Moore's Law. Eventually, we have more powerful system chips, and we can obtain the expected performance; however, we have encountered the biggest problem ever: power consumption. As we knew, if we put so many transistors (billions of transistors in modern SoC or CPU), the power dissipation has accumulated to the level that we cannot hold a handheld device. Figure 7.1 shows another famous (or notorious) power trend when IC evolves. If we do not think of thermal management solutions, it will be like holding rocket nozzles when launching the rockets. Sections 7.4 and 7.7 have more discussion on how to resolve this issue. Section 7.5 presents the relationship/trade-off between performance and power, and Section 7.6 depicts the delivery of supply voltage and clock signal.

Power dissipation and power density have become the first-order effect in IC design, and the second-order effect comes from thermal management. Fast and highly integrated devices/circuits usually generate a large amount of heat. The heat must be removed efficiently [1]. The reason is that virtually all failure mechanisms are enhanced by increased temperatures. The thermal problem should be solved in both chip placement and packaging. And based on different package cost, the thermal solutions are also differentiated. We will look into it in Section 7.8.

Robustness of the VLSI is composed of two major factors: reliability and variability. They are usually categorized as design for reliability (DFR), design for variability (DFV), and overall design for manufacturability (DFM). Reliability is a fundamental requirement in chips to guarantee correct operation over their lifetime. But the continuous scaling downward of

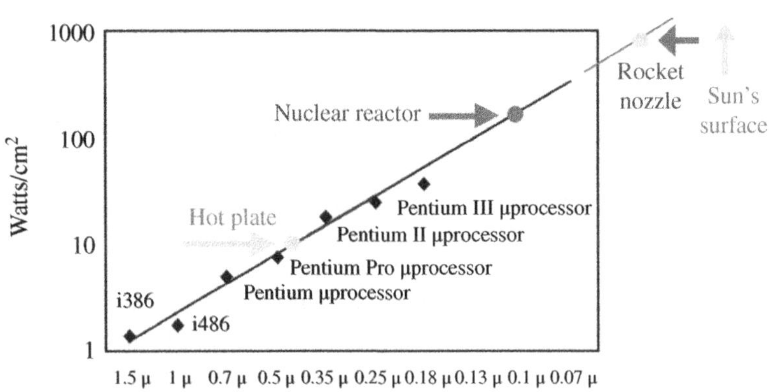


**Figure 7.1** The trend of power density along the process advances.

both the device and the interconnect has severe consequences for a reliable operation of an IC [2]. There are various issues in DFR, such as hot-carrier effect and bias temperature instability (BTI) in devices, and electromigration (EM) in interconnects. Most of them are related to materials, doping of transistor, and temperature. On the other hand, variability mostly comes from the manufacturing process, electrical and environmental variations. As we already knew, the extreme scaling worsens the variability, causing the design flow changes and fundamental mindset evolution. The details will be presented in the last part of this chapter.

## 7.2 Field-Effect Transistor and Memory Basics

We have introduced the invention of transistors and scaling trends along the years in Chapter 6, Sections 1 and 2. In this section, we discuss the basics for transistors and a larger component consisted of transistors, the memory cell. To be more exact, in modern semiconductor-based electronics, we usually use MOSFETs as our transistors. In the following, we explain how field-effect transistor works. Figure 7.2a depicts the cross section of an n-type field-effect transistor, it uses the electric field applied on the gate electrode to control the switch property. The gate, source, and drain contacts of $TiSi_2$ silicide are shown in dark color. This switch property is used to represent the data we transmit or store: 0 or 1. If we apply the positive voltage on the gate, the inversion layer will be formed on the surface, we start to have current. When we have much larger voltage, this intersection becomes a channel, it connects to the two n-type wires which form the source and drain. In Figure 7.3 (n-type MOS transistor), it shows more complete transistor figure, the size of channel region is labeled relative to the direction of current flow [3]: the channel length (L) is along the direction of current flow between source and drain, while the width (W) is perpendicular to current flow. The amount of the current flow is a function of the W/L ratio; therefore, we usually choose W and L when we draw the layout to reflect the strength of the transistor. In the later years, the transistors suffered from "short channel effect" and many other problems/issues occurred.

The working principle of the DRAM (dynamic random access memory) cell, consisting of one field-effect transistor and one capacitor, is as follows. Under the gate oxide below the poly-Si, we can form an n-channel between the n+ source and n+ drain by applying a positive charge to the gate (along the word line [WL] in the memory cell, shown in Figure 7.2b), which attracts electrons to the inversion layer to form the n-channel below the gate oxide, allowing electric current to flow across the gap from the source to the

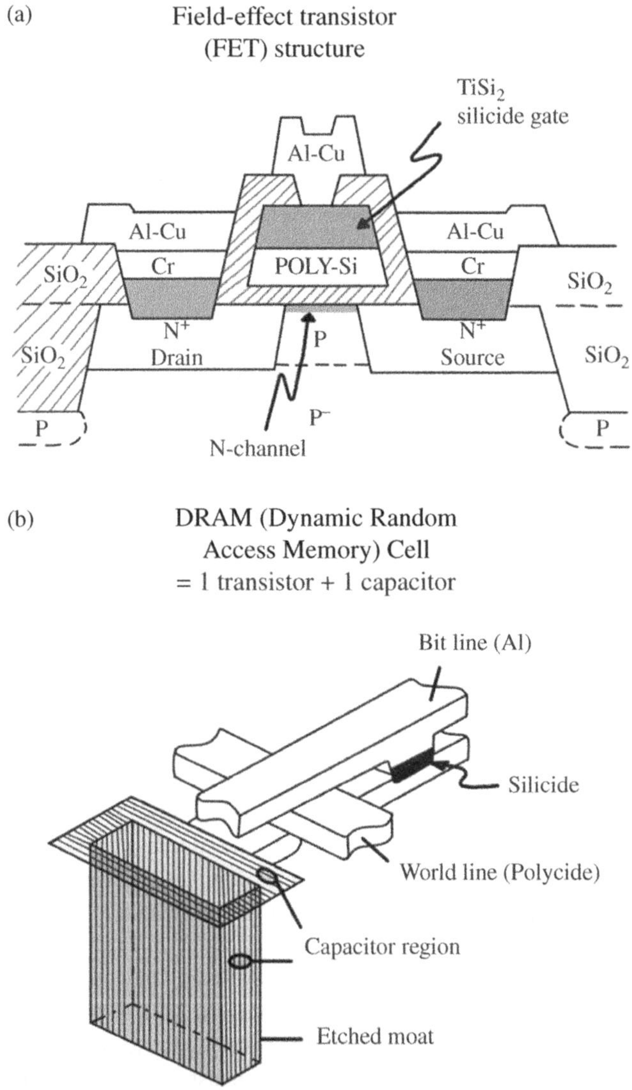


**Figure 7.2** (a) Cross section of an n-type transistor. (b) Basic DRAM cell.

drain. The stored charges in the capacitor, which is connected to the source, will flow out along the bit line (used in the memory cell) when the gate is open. When it is detected that a pulse of charge has flowed out from the capacitor, it is "1." On the other hand, when the gate is open and no charge flows out from the capacitor, it is "0." Furthermore, a negative charge stops

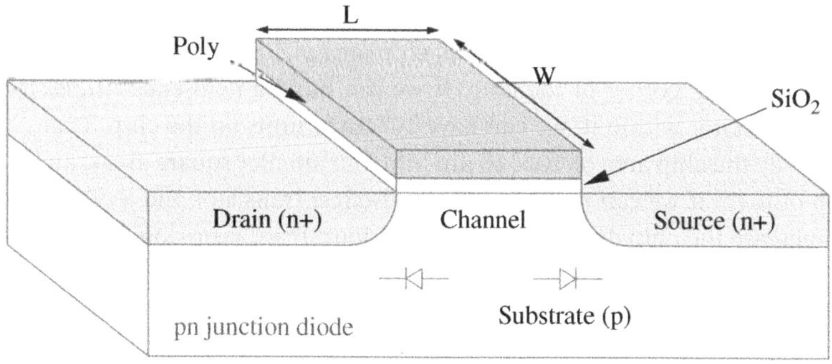


**Figure 7.3** An n-type transistor with channel length (L) and width (W) [3].

current and closes the gate because of the p-type region between the n+ source and n+ drain. Figure 7.4a show a schematic sample/core circuit for an aforementioned memory cell/unit at the point of ($X_1$, $Y_2$), also shown in Figure 7.2b. The WL along $X_1$ is used to turn-on or turn-off the transistor, and the bit line along $Y_2$ is used to transmit charges coming out from the capacitor. Figure 7.4b depicts a square network of X and Y lines. At each point of (Xi,Yj), we build a field-effect transistor and a capacitor, then we have the random access memory (RAM).

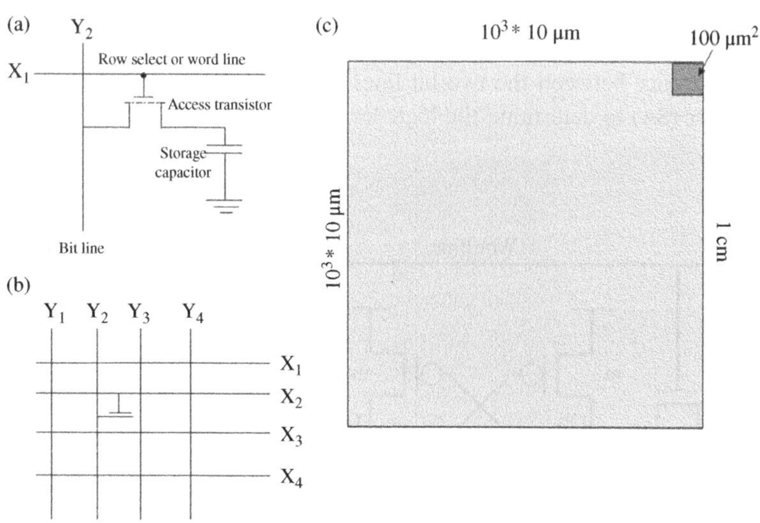


**Figure 7.4** DRAM cell and block configurations [4]. (a) Schematic circuit for a memory cell. (b) Square network of X and Y lines. (c) Translating the circuits to a Si chip.

Now, we translate the circuits to a Si chip as shown in Figure 7.4c. On a Si chip of $1 \times 1$ cm, we consider a small chip area of $10 \times 10$ μm, as shown in the upper right corner of the chip. If we can build a field-effect transistor and a capacitor within it, we can have $10^6$ RAM units on the chip. Then, if we divide the chip area of $10 \times 10$ μm into four smaller square areas, and in each of them if we can again build a field-effect transistor and a capacitor, we increase the chip density by a factor of four. If we keep doing so, this is the physical picture of Moore's Law of miniaturization.

What about another kind of RAM? Figure 7.5 shows a sample/core circuit for a static memory cell/unit (6T Static RAM). With this, we can generate a larger memory sub-circuit for bigger storage for on-chip computation. The value is stored in the middle four transistors (M1–M4), which form a pair of inverters connected in a stable loop. The other two transistors (M5 and M6, acting as switches) control the access to the cell by the bit lines (BL and BLb, the opposite state of BL). Bit lines are used to read and write the data value. There is another control: WL, if it is 0, the inverters reinforce each other to store the value. The Read/Write can be performed when the WL is selected (=1).

When we perform Write, we will set the BL (the write data) and BLb to the desired values to force Q and Qb (the prior stored value in bit-cell) to change states, then set the WL to 1 (Vdd). Then, the write operation is complete through the transistors. When performing Read [2], first both bit lines are pre-charged to Vdd, then set the WL to Vdd and contents of the cell cause a slight discharge on one of the pre-charged bit lines. This small voltage difference between the two bit lines is sufficient for an SRAM sense amplifier (SA) to determine the logic level in the memory cell. There are

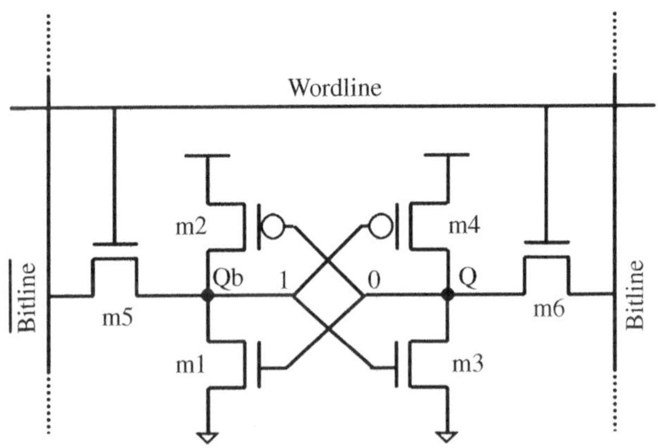


**Figure 7.5** A typical 6T Static RAM cell.

many other peripheral circuits like SA in an SRAM subsystem for column read/write, row circuitry, and so forth. In summary, with the recent trend of AI applications and in-memory computing (IMC), the architectures of future mobile and high-performance computing (HPC) devices will have much diversified appearances.

Here, we supplement on how we make the transistors (then ICs) and memories work as expected. We need to deliver the clock signals and supply voltage on a chip, which will be described in Section 7.6. Then, the rest is how we deliver those from outside package to the chip. We usually use voltage regulator module (VRM) to supply power through Vdd and Gnd planes in the PCB and packages/interposers. The planes are connected through bonding (described below) or vias (described in Chapter 8) to the on-chip power network. In Section 1 of Chapter 6, we have mentioned that the package functions include delivering power from outer sources to chips, and providing signal (including clock) communications in and out of the chips, these are provided by the bonding between chip and package frame. Early bonding techniques include wire bonding (still popular though in low-cost solutions), through thin gold wires bonded to metal pads. The problem for this bonding is that it cannot accommodate large number of I/Os, we called it pad-limited. The chip size is then determined by the pad frame rather than by the logic within the chips. Of course, the power of delivering many signals and supply is limited since they only come from the periphery of the chip. Since late 1990s, flip-chip connections were adopted, also called controlled collapse chip connection (C4 flip-chip). In this way, the chip is in nearly direct contact with the package [5], eliminating the inductance associated with the bond wires. It provides better power delivery and much more I/O distribution, so it has been used for mainframes and HPC machines. In Chapter 8, we will see more modern chip and package designs.

## 7.3 Performance: A Race in Early IC Design

Chip performance is considered as a golden metric among all design considerations. If the chips manufactured are below the targeted performance, they will be tagged lower prices in the market. Because of this, performance has been becoming the race in IC development history, and we can also consider this as another version of Moore's Law. Since performance is one of the most important issues, we hope that the performance/timing can be consistent throughout the design phase. However, it is usually not the case, we call this problem as "timing closure," part of the design closure problem sets. Due to this reason, the products cannot be in the market as scheduled. Timing closure is often the most difficult task in designing a chip due to the fact that the timing behavior of logic gates will vary at different PVT (process, voltage, and temperature) conditions [6]. In addition, the closure issue also comes from the delay variation in the speed of transistors and wires. This is because logic gate speed is affected by the drive and load environment surrounding the gates. Before 1990, the chips were so small that the logic gate's delay dominated the overall performance, and the performance is easy to evaluate. When the chips are getting larger, the loading of the devices/gates become extremely larger, we can no longer ignore the already existed interconnect delay. We use a commonly followed delay model to describe such a situation.

In general, most circuits can be represented as an RC tree. The root of the tree is the voltage source and the leaves are the capacitors/devices/transistors at the end of the branches. The Elmore delay model estimates the delays from the source to the leaves/sinks by taking the sum of RC products, whereas each resistance R is multiplied by the sum of all the downstream capacitors, and the formula $t = \sum RC$ simply shows the relationship. Here, we can see that if downstream capacitors include very long interconnects (typical phenomena in DSM and nanometer designs), the loading becomes very obvious and non-negligible, and causing large RC delay. In conclusion, performance becomes nontrivial to evaluate and achieve.

In order to understand how we evaluate the performance, we summarize the primary performance/timing analysis [3, 6] as follows. Clocking disciplines help us construct synchronous electronic systems that operate at some speed. A clock is an electric signal that oscillates between a high state and a low state, usually a square wave with a predetermined period (frequency). We use the clock signals to coordinate the actions of all the circuit components on a chip, either rising edge or falling edge of the waves will trigger the components for synchronization. Without the clocking principles, it is impossible to construct the complicated and extremely fast SoC chips nowadays.

How do we use clock signals to coordinate the actions of the components? We know that there are two kinds of machines in a system: combinational logic network and sequential machine. The combinational logic includes simple data path calculation such as adders, comparators, and counters. They do not need clock signals to perform their functions since it is not necessary to memorize the previous states. However, if the output values depend not only on the present input values but also on the prior states of inputs, we need to use sequential models for greater functionalities. The structure of a generic sequential system is just like a machine with many

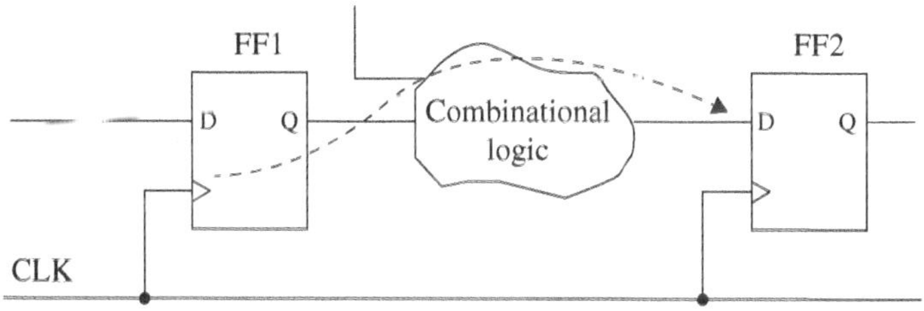


**Figure 7.6** Illustration of the performance/timing analysis in digital designs.

states, when the system starts at some initial state, it will move on to another state depending on the input or prior states, then it will stop at some terminating state(s). Since there are only a finite number of states in such system, we also called it finite state machine (FSM). In such systems, memory elements (flip-flops, registers, and latches) hold the machine's state; the machine's inputs and outputs are called primary inputs (PIs) and primary outputs (POs). Between PIs and POs, there are usually many signal paths, one path from PI to PO may contain several memory elements. In between memory elements, there are usually one combinational logic network. Figure 7.6 shows a simple illustration in such a setup. EDA tools usually apply static timing analysis (STA) algorithms to determine the propagation time in the combinational logic between clocking elements. Among all the propagation times in all combinational "segments," the maximum propagation time will determine the clock period. It is because if we use smaller propagation time as clock period, other paths containing the maximum propagation time segment will fail the synchronization. We use this as the speed of the chip: $f = \frac{1}{p}$, where f is the frequency of the chip and p is the maximum propagation time. Clocking is complicated when there are many clock domains in an SoC, we need different clocks (clock networks) in order to make the functions work correctly. In summary, we control the sequential machines by delivering clock signals to all sequential elements, those signals will arrive at the same time in order to implement the "synchronous" systems for massive computation nowadays.

## 7.4 Trend in Low Power

Ideally, when we come to design the chips, we hope that the chips can operate by using the least power possible without affecting or sacrificing chip performance. As aforementioned, it is not acceptable to hold the power density in rocket nozzle level. To be more detailed, modern applications (such as AI) need tremendous data movement between processors and memory in one computing device, and also the compute density is too high to handle. As a result, lower power solutions are absolutely necessary.

In order to learn low-power techniques, we need to know where the power comes from. There are two principal sources of power dissipation in modern CMOS-based VLSI: dynamic and static power. If we check the general power formula: $P = IV = CV^2$, we can see that the first half is the power dissipation with flowing current, while the second half shows the power with capacitive loads. We skip the switching activities' rate and frequency for simplicity, but actually they are just the coefficients. It is worth mentioning that low power actually means low entropy production or low waste heat, which will be discussed in Chapter 10.

First, we look into dynamic power. It consists mostly of the switching power. To estimate this power [5], we can consider each node of the circuit: the capacitance of the node is the sum of the gate, diffusion, and wire capacitances on the node. Thus, the switching power depends on the sum of the effective capacitances (the capacitance multiplied by the activity factor, which can be measured by logic simulations) of all nodes. As we see in the formula, Vdd is a quadratic term, it is good to select the minimum Vdd possible that can support the required frequency of operation. It is the reason why we see such low supply voltage in modern IoT and mobile devices. Another source comes from short-circuit current. Short-circuit power dissipation occurs when both pull-up and pull-down networks are partially ON while the input switches in CMOS technology. Unless the input edge rate is much slower than the output edge rate, short-circuit current is a small fraction of current to the load and can be ignored in calculations.

Next, let us get closer to the sources of static power, which gives us more headache. Static power exists even when a chip is not switching [5]. In nanometer processes, the threshold voltage (to control power and performance, simply say) is lower and the gate oxides are thinner, leakage can account for as much as a third of total active/switching power. Prior to the 90 nm node (before nanometer process), leakage was only of concern when the circuit is in sleep mode since it was negligible compared with dynamic power. The sources of leakage include subthreshold leakage, gate leakage, and junction leakage in transistors (or to be more exact, field-effect transistors, FETs). Subthreshold leakage current flows when a transistor should be in the OFF state. Raising the source voltage or applying a negative body voltage (to change the threshold voltage) can decrease leakage. Gate leakage occurs when carriers tunnel through a thin gate dielectric while the gate is ON, also called tunneling leakage. It also depends on the voltage across the gate. Junction leakage happens when a source or drain diffusion region is at a different potential from the substrate. It is often minor compared with the other leakages. Due to the serious leakage issues in the nanometer process, novel structures of FETs were invented; among all the structures, the industry has chosen the fin-like structure FETs to suppress the leakage, which is called FinFET, invented by Prof. Chenming Hu from UC Berkeley. The idea is to use more coverage of gates to "turn-off" the leakage current. It has been applied until sub-10 nm process in major foundries such as Intel, Samsung, and TSMC.

## 7.5 Trade-off between Performance and Power

In the previous section, we have discussed the importance in low-power solutions; however, better performance and higher speed require greater power consumption [6]. Thus, we come to a common sense/law in IC design: we always trade-off between chip performance and power dissipation. First, we adopt the perspective in [5] to consider three scenarios in trade-off and optimization: minimizing the energy, minimizing the energy-delay product (EDP), and minimizing energy under a delay constraint. Second, we discuss a common methodology in realizing performance and power trade-off: cell library synthesis.

The power-delay product (PDP) is simply the energy [5]: the product of the power of an operation and the time for the operation to complete is the energy consumed. PDP can also be recognized as energy consumed by a gate per switching event. The minimum energy point is the least energy that an operation could consume if delay were not important, usually happens in subthreshold operation where Vdd < Vt, where Vt is the threshold voltage which decides the mode of transistor operation. We will discuss how to control Vt in order to trade-off between leakage power consumption and high performance in the next paragraph. As for EDP, similar to PDP, it is a popular metric that balances the importance of energy and delay. Figure 7.7 shows a sample illustration of EDP. If we neglect leakage, we can solve for the supply voltage that minimizes EDP. However, when considering leakage, the best supply voltage is slightly higher. On the other hand, the designers also generally face the problem of having minimum energy under a delay constraint. For a given supply voltage and threshold voltage, the designers can make logic and sizing decisions that affect delay and energy. As shown in Figure 7.7, such curves can be determined using a logic synthesizer or transistor sizing tool constrained to various delays. This can be realized through the implementation in the standard cell library, and utilized in the digital design flow.

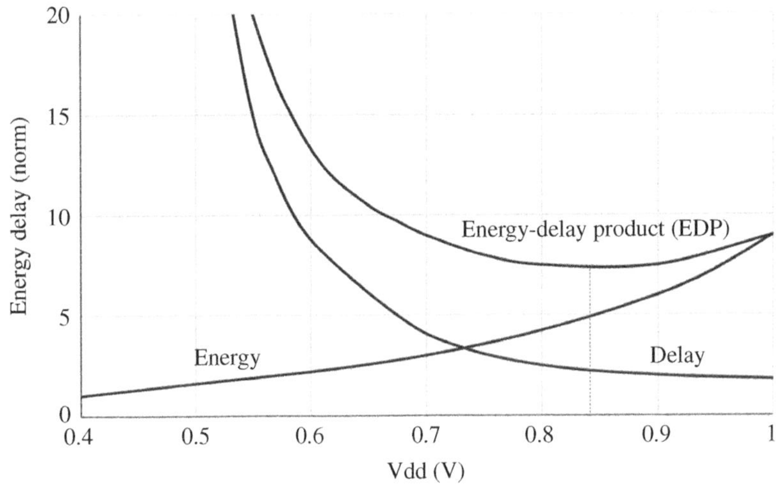


**Figure 7.7** Illustration of energy-delay product (EDP) for the trade-off.

In practice, a scheme we usually apply in this trade-off is the consideration of multi-objective standard cell libraries. When we realize a digital IC design, in addition to employing EDA tools for synthesizing circuits and layouts, we need to prepare "technology files" and cell libraries for the final chip synthesis. Technology files mean the design and process parameters in targeted technology node. As for cell libraries, for a single technology node, there are several libraries for implementation. Take TSMC 28 nm node as an example, there are six logic libraries: 28HPC/HPC+/LP/HP/HPL/HPM. They have different objectives: 28HPC/HPC+ focus on HPC, which means the supply voltage is higher; 28LP is for low-power design and 28HPM process has multiple gate lengths for a greater range. In other words, with different targets for the products (high performance or low power), we can use this approach to dealing with the trade-off.

## 7.6 Power Delivery and Clock Distribution Networks

We switch gears to implementation detail in realizing power and clock: how do we deliver the power and how do we achieve the target speed of the chips? They are critical in determining the power dissipation of chips and the grade of package we should apply for our chips. In order for transistors to perform their assigned tasks, a certain level of voltage must be delivered at the transistors' Vdd terminals [6]. Thus, a power network is a physical metal structure whose function is to supply the needed power (or current) to all of the transistors/components on the chip. Figure 7.8 shows a sample illustration of power network on a chip, usually it is spread at higher metal routing layer since the wire size is wider than normal signal nets.

Power network is complicated. Every macro block and memory cell has its own power network, which must be connected to the main chip power network. Power network synthesis today faces several challenges due to technology advancement [7]. Generally, conventional power network includes one set of P/G network and power ring surrounding the chip. However, to achieve more cost-effective fabrication and better performance, multiple macro blocks are generally integrated in one die to reduce the fabrication cost and to increase the performance due to reduction in interconnect delay. More macro blocks in one die imply more power consumption per unit area. To optimize power consumption for power demanding design, the concept of multiple supply voltage (MSV) is proposed (described in the next section). Integrating multiple macro blocks in one die also introduces the concept of multiple power domain (MPD). Since not every macro block operates at same driving voltage, each macro block has its own unique power network and driving voltage to efficiently control power consumption. Thus, constructing a reliable power network is the most tedious and labor-intensive work in physical design phase [6]. The main concerns for a power network are IR drop and EM problems, which will be detailed in Sections 7.9 and 7.10.

Just like we describe the power network, a clock network is a physical structure whose function is to deliver the clock signals to all sequential

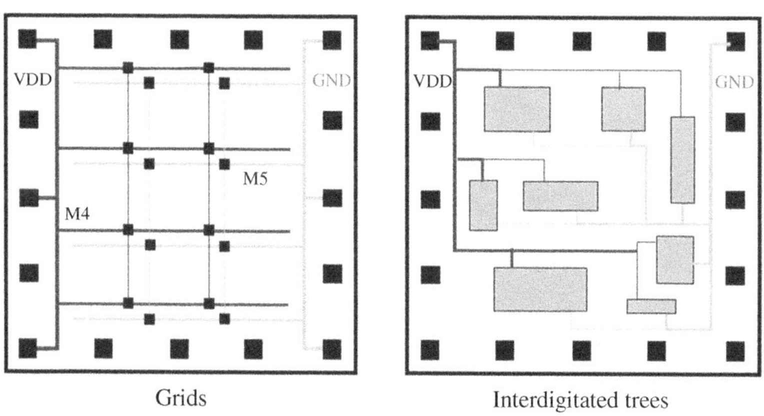


**Figure 7.8** A power network illustration.

elements on the chip. During the old days of chip planning, we usually used "clock tree synthesis" (CTS) to name this stage. There are two stages in clock generation: first one is clock generation from the pads and second one is the on-chip clock network generation. Many chips require clock signals of frequencies much too high to be driven onto the chip from the pads; therefore, the high-frequency clock must be generated on-chip from a lower-frequency input. The phase-locked loop (PLL) is then used to generate the on-chip clock signal [3]. It compares the input clock with the internal clock to keep the internal clock in the proper phase relationship. After setting up PLL, we should synthesize the clock tree/network. The mission of the clock network is to deliver the signals so that they should arrive at all sequential elements simultaneously in order to "synchronize" the system. If the signals cannot arrive simultaneously, the clock skew will occur and will cause the performance or function degradation. Figure 7.9 shows the usual architecture for clock network.

It is worth mentioning that clock network costs the most power for a synchronous design and directly affects circuit speed. With technology downward scaling, variations result from manufacturing and the operating environment. These variations require more guard band in the design process. To achieve lower power dissipation, a voltage scaling technique is broadly adopted in modern design, which introduces severe voltage variation. One way to reduce clock skew uncertainty is to improve the delay correlation between clock paths by using a multiple-driving-paths clock network such as mesh, cross link, and multilevel tree. However, [8] adopts the other way, that is, to reduce skew uncertainty by reducing path delay variability.

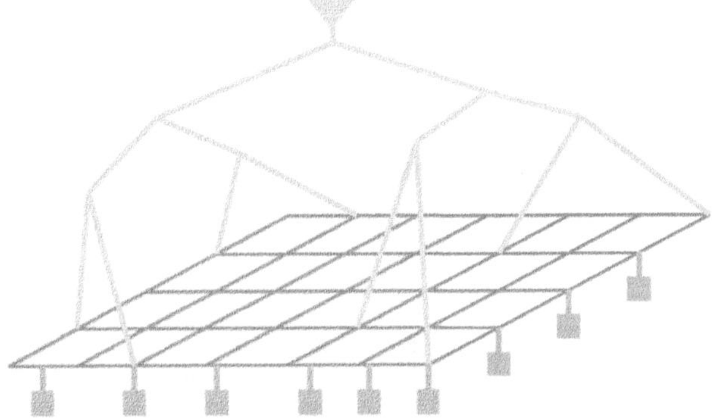

**Figure 7.9** A clock mesh/tree hybrid architecture for clock network.

## 7.7 Low-Power Design Architectures

After understanding the basics in performance and power in IC designs, we need to know more about how we design lower power chips nowadays. In this section, we introduce three important techniques and architectures, more details can be found in [9]. The first one is dynamic voltage frequency scaling (DVFS), and it is based on a simple logic: If something does not need to operate at a high voltage, then lower it; if something needs to be alive in order to respond to data when it arrives but is not processing anything, lower the frequency as much as possible [10].

DVFS was first developed to control the power consumption of microprocessors [3]. It is a dynamic technique: both the clock frequency and power dissipation vary during operation/runtime. DVFS relies on the fact that processors do not always need to run at full speed in order to finish all the tasks. If the processor's workload does not require all available performance, we can slow down the processor to the lowest available performance level that meets the demand. In other words, DVFS schemes are particularly good for variable workloads [10]. In the architecture of DVFS, there is a power controller to control the processor power consumption through a combination of clock frequency and power supply changes. It needs an adequate algorithm to determine the proper frequency and power supply settings from a load estimate. An interface is required to tell the controller the needed performance and to communicate between the system-level load estimate and the commands to the clock generator and power supply. Another component that affects the efficiency of DVFS power management is voltage regulator [11]. A voltage regulator converts the input voltage from the noisy power supply into one or more desired voltage levels to be used by the processor. Recent industry trends are toward integrating voltage regulators on chip, such that the granularity of DVFS in space (more voltage domains, will be discussed in the next paragraph), time (faster response time), and voltage levels (smaller voltage steps) can all be improved, thereby enabling greater opportunities for more efficient power management. Figure 7.10 shows the system relationship in DVFS architecture.

The second architecture for low power is MSV, also called voltage island (VI). As aforementioned, dynamic power is proportional to $Vdd^2$, lowering Vdd on selected blocks helps reduce power significantly. Consider an example in Figure 7.11 [9], there are three different voltages in a simplified design. The cache RAMS (random access memory) are operating at the highest voltage since they are on the critical timing path. The CPU runs at a rather high voltage since it stands for system performance. The rest of the

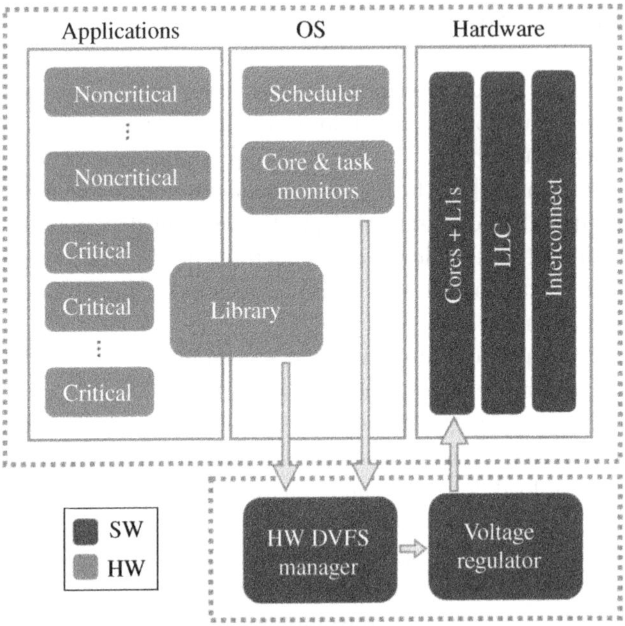


**Figure 7.10** An example of DVFS architecture chip.

chip can operate at a lower voltage, without jeopardizing the overall system performance. In Figure 7.12, we describe a little bit more details in implementation [12]: the needed level shifter for the signal traveling between different voltages. We must use some level shifters to maintain signal integrity while a low Vdd device drives a high Vdd device. If we use a tree-based

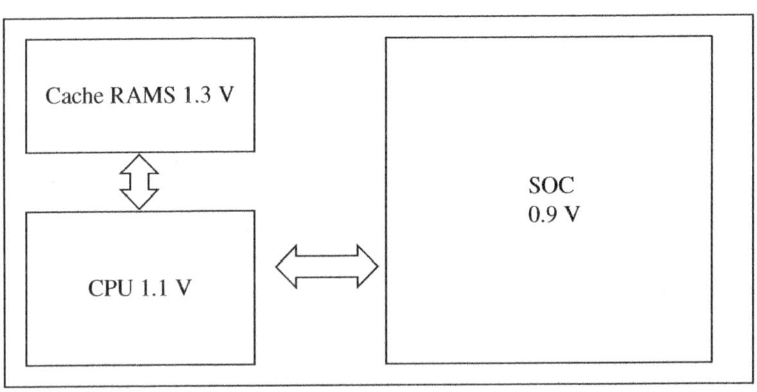


**Figure 7.11** Illustration of MSV designs.

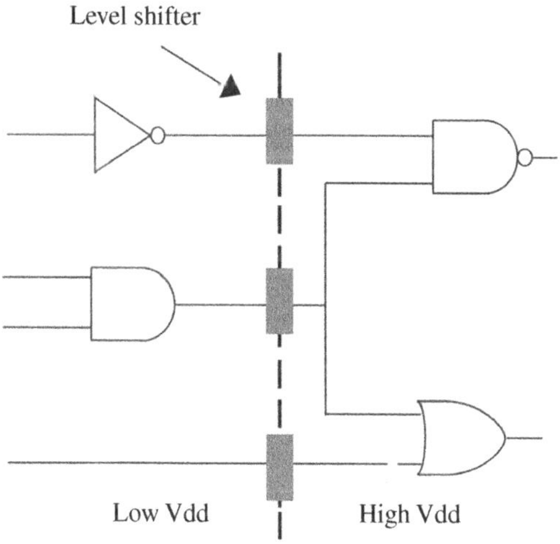


**Figure 7.12** In MSV designs, level shifters are needed for the signal traveling between different voltages.

algorithm to perform the buffer insertion and level shifter assignment, there will be two issues. First, the generated routing tree may not have feasible level shifter location. Therefore, signal integrity will not able to be maintained. Second, if we regard level shifter as a standard cell and place it during placement stage, signal integrity can be maintained though, the location of level shifter will affect total wire length, delay, and power. Those detailed issues should be accounted for in the implementation.

The third architecture is based on the concept of Gating: to "gate" the power dissipation for the logic elements and blocks. It is also based on another simple logic: If something is not needed, shut it down [10]. It has two branches in application: power gating and clock gating. The purpose of power gating is to solve the problem of **leakage** power dissipation, which belongs to static power. To reduce the overall leakage power of the chip, it is highly desirable to add mechanisms in turning off blocks that are not being used. It is done by inserting sleep transistors to shutting off the idle blocks [13]. Since sleep transistors are rather large, due to early planning in power delivery for area-array design style, it is essential to consider the power gating techniques in early SoC physical design stage. Moreover, a low-power floor planning framework is developed in [13] by considering sleep transistor insertion in the presence of power supply noise. Furthermore, every interface of a power-gated region needs to be managed [9]. The outputs of the power-gated block are the primary concern because they will cause electrical or functional problems in other blocks. The usual approach to controlling the outputs is to use isolation cells to clamp the output to a specific and legal value.

On the other hand, clock gating technique has been applied for a long while. Because a significant portion of dynamic power in a chip is in the distribution network of the clock, turning the clocks off when they are not required becomes an effective method. Clock gating ANDs is a clock signal which is able to turn off the clock in the idle blocks [5]. It has been highly effective because the clock has such a high activity factor (high switching), thus gating the clock to the input registers of a block prevents the registers from switching. In this way, it stops all the activity in the downstream combinational logic. Now, modern design tools support automatic clock gating: they can identify circuits where clock gating can be inserted without changing the function of the logic.

## 7.8 Thermal Problems in IC and Package

A considerable amount of heat energy is typically generated in electronic systems due to Joule heating and power dissipation [14]. When current flows through a transistor or resistor, Joule heating is generated by the friction between electrons. Usually, if we need HPC capability in our applications/solutions, we definitely need to consider the thermal problem directly from high power. We can use abovementioned techniques for low power, thus lowering the temperature concern. However, in many HPC requirements such as data center and AI-driven applications, we cannot avoid the thermal issue, which mainly comes from the high supply voltage. When supply voltage is higher, the performance is boosted up, yet the high leakage causes large power dissipation and temperature rise. As a result, approved ambient temperature should be defined and designed in order to assure component and system functions and reliability.

The problem of degradation on chip performance due to temperature issue becomes more critical in advanced node technology [15]. The benefit of smaller transistor comes with the cost of higher power density per unit area. Higher power density translates to more dissipation of heat and this exorbitant amount of heat rapidly increases on-chip temperature. Studies have shown that chip performance and reliability can deteriorate significantly due to temperature variation. Evidence has shown that a 10° variation in temperature can induce 5% increase in interconnect delay, 25% increase in crosstalk-induced noise, and 50% reduction in component life time. Thus, reducing the maximum on-chip temperature and eliminating hot spots are critical to chip's reliability and performance.

Shrinking in interconnect dimension also has an adverse effect of increase in higher power and heat dissipation due to increase in interconnect resistance. Viewed in this light, the effect of chip temperature and variation must be addressed when considering chip performance and reliability. Figure 7.13 shows the thermal aware placement to reduce on-chip temperature [15].

On the other hand, in package level, we need to consider how to remove the heat from IC very rapidly in order not to damage the system. Heat transfer [1, 14] is the transportation of thermal energy due to a temperature difference. There are three heat transfer modes: conduction heat transfer (thermal conduction), convection heat transfer (thermal convection), and radiation heat transfer (thermal radiation). Only thermal conduction occurs in solids, whereas it is found that thermal conduction coupled with convection in liquids and gases. In low-cost applications, such as PCs or workstations, the chip power dissipation is relatively low, the chip packaging is simple. In high-end systems, chips dissipate more heat, therefore the packaging is more complicated. As another example from [1], dice are flip-chip mounted on the substrate via control-collapsible solder balls or pressure contact bumps. Then the major thermal path from the die to the substrate is through the solder balls and contact bumps. If a large amount of power dissipation is required, a thermal path from the backside of the die has to be established, the corresponding packaging solution should consider these guidelines. In addition, there are methods to increase heat transfer [14]: heat sinks, thermal interface materials, fans, and heat pipes. The packaging should be designed accordingly.

(a) (b) (c)

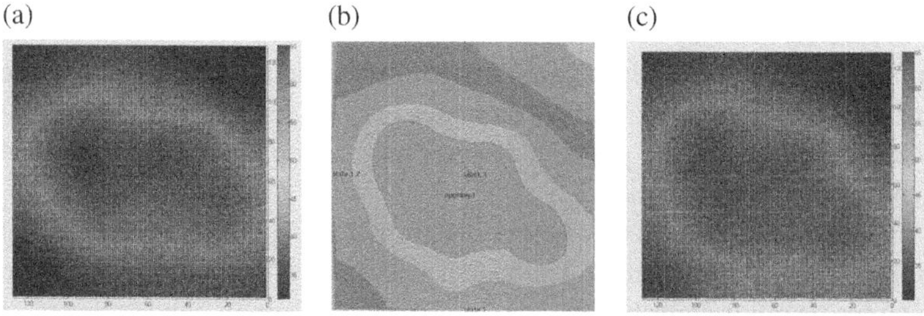

**Figure 7.13** (a) Temperature distribution of a testcase using the thermal model in [15] before thermal aware placement. (b) Temperature distribution of a testcase using Icepak (a commercial thermal simulation tool) before thermal aware placement. (c) Temperature distribution of a testcase after thermal aware placement. Peak temperature in (c) after thermal aware placement is reduced and spread out more evenly compared with (a).

## 7.9 Signal Integrity and Power Integrity (SI/PI)

In this section, we will discuss two issues when designing chips and packages, even boards. They are part of design integrity problems: signal integrity (SI) and power integrity (PI). They will affect the quality of the designs, and are related to the supply voltage decision and delivery. As aforementioned, in order to achieve lower power (dynamic and static), supply voltage must be downgraded. The side effect is twofold. One is causing more leakage, then thermal problem emerges. Another issue is the shrinkage of supply noise margin, causing signal integrity problem. SI is a set of measures of the quality of an electrical signal. If SI is jeopardized, the function of the chip will be affected or become faulty. The usual problem in SI is crosstalk: it is a phenomenon in which a signal transmitted on one circuit, or one channel of a transmission system, creates an undesired effect in another circuit or channel [6]. The crosstalk occurs when two metal wires are close to each other, the signals traveling through each of the wires potentially interfere with each other through the coupling capacitor between them (also mentioned in Chapter 6, Section 2). As a result, the impact of the aggressor net (the net that activated the interference) on the victim net (the net that receives the interference) becomes the delay addition (or subtraction) in the victim net's signal propagation time, also called crosstalk delay.

Another effect caused by crosstalk is noise. Continuing on the scenario of aggressor and victim nets, the switching of the aggressor signal could cause a glitch on the victim. If this glitch exceeds the noise margin, it could result in a logic error and thus functional failure. There are two ways to remedy crosstalk delay and noise. First, we can increase the space between two neighboring wires to reduce the coupling and interference. However, this will possibly increase the layout area. Second, we can size the buffers/drivers in the design to boost the signal strength in the victim net, or replace the stronger buffers/drivers with weaker ones in the aggressor net. These solutions are located in the physical design phase, a good and robust physical synthesizer should do the trick.

PI is an analysis to check whether the desired voltage and current are met from source to destination. The usual problem we encounter is IR-Drop (voltage drop): it describes the phenomenon of a drop in voltage potential when a current flows through a resistor. In a chip, all interconnections are metal segments of aluminum or copper, and they are resistors. Due to the law of $V = IR$, when the supply voltage is delivered through power network, it is possible that the supply voltage level is greatly reduced at the end of the segments. It is because of the loss of voltage delivery in the resistive network.

If, at some locations inside the chip (usually in the center), the amount of voltage degradation is over a certain limit, the cells at those locations could experience speed loss, or even worse, cannot operate properly [6]. IR-Drop becomes an important and serious issue in nanometer design with CMOS process scaling down to 20 nm. As the advance of technology, many challenges come up to satisfy the multiple constraints. For example, the leakage current would increase the power of IC and ruin the functionality within it. Furthermore, to increase the speed of signal transmitting in the circuit, we need to lower the driving voltage of each transistor as aforementioned. However, the driving voltage of each transistor may not be sufficient due to IR-Drop. Therefore, to optimize the power delivery network while considering IR-Drop is essential and unneglectable [16]. To construct a power delivery network, there are several concerns to be addressed. First, optimization of power delivery network has strong connection to the power distribution of standard cells, static random-access memory (SRAM), and other functional IPs on the chip. Therefore, the instance must have shorter distance (this implies net's resistance) from the nearest power stripe (backbone of power network) than the one with lower requiring power. Second, the position and the potential power loss of the power pads are critical in optimization of power delivery network so that the accuracy of IR-Drop's analysis will not be omitted.

## 7.10 Robustness: Reliability and Variability

When we enter into the deep end of Moore's Law, we find that many unexpected effects emerge, and they get worse. Usually, they should be taken care of in the manufacturing side, or foundry in the fabless regime, but it lost control after the node of 0.25 um. We call these unexpected or uncontrolled problems of "deep submicron" (DSM) effects. This name gets advanced to "nanometer" effects. Because they cause the degradation of VLSI/SOC performance and yield, we can no longer ignore these issues and should consider them into the design phase of IC planning. We classify these consideration and techniques to resolve the issues as the category of "design for manufacturability" (DFM). The word "Manufacturability" comes from "manufacture" and "ability," meaning the impact on the yield and performance on VLSI designs. As aforementioned in the introduction, DFM includes the robustness consideration in VLSI design: DFR and DFV.

We describe those DFR concerns in twofold. In devices, we have the following wearouts (just list the popular ones): oxide wearout, overvoltage failure, and latchup. They are usually called hard errors, in comparison with another kind of reliability problem of soft errors. As gate oxides are subjected to stress, they gradually wear out, causing the threshold voltage to shift and the gate leakage to increase [5]. Eventually, the circuit fails when transistors become too slow. Here, we list the primary mechanisms for oxide wearout: hot carriers, negative bias temperature instability (NBTI), and time-dependent dielectric breakdown (TDDB). Hot carriers are the carriers that gain enough energy to jump from the silicon substrate into the gate oxide. As these carriers accumulate, they create a space charge in the oxide that affects the transistor's threshold voltage and other parameters. NBTI is particular to p-MOS devices (while PBTI is with n-MOS). It is a result of a negative bias applied to the gate of a p-MOS with respect to the bulk, and refers to shifts in threshold voltage, drive current, and trans-conductance. TDDB occurs when the electric fields across gate oxides induce stresses that damage the oxide, the gate current gradually increases. As for overvoltage failure, it may be triggered by excessive power supply transients or by electrostatic discharge (ESD) from static charges entering the I/O pads. Overvoltage at the gate node accelerates the oxide wearout. The last phenomenon about device reliability we want to mention is latchup. In the CMOS process, if low-resistance paths between Vdd and Vss are developed, catastrophic meltdown will occur. It happens when parasitic bipolar transistors (not real bipolar) are formed by the substrate, and diffusion turns into ON. In summary, with advanced gate engineering and proper layout procedures, these problems can be avoided or alleviated.

In wire reliability, we have mainly the migration problems of EM and stress migration [3]. High current densities lead to the electron wind that causes metal atoms to migrate over time, we call it EM. It causes the wearout of the wires, forming two effects: the depletion of atoms (voids) in the cathode and deposition of atoms (hillocks/whisker) in the anode, they will eventually make the chip fail. EM majorly depends on the current density J, it is shown in Black's equation, which expresses the mean time to failure (MTTF): MTTF $\propto J^{-n} e^{\frac{Q}{kT}}$, where Q is activation energy, and n is constant of 2, see Chapter 10. From this relationship, we can see that EM is also related to temperature T. Stress migration, another wire reliability problem, is caused by mechanical stress and occurs even when no current flows through the wire. The stresses are caused by the different thermal expansion coefficients of the wires and the materials. Failures occur with a time-dependent creep in moderate temperatures (150–200 °C) or with short-time stress at very high temperatures. EM can be resolved by careful physical design since it mainly depends on current density, while stress migration needs proper interconnect engineering.

On the other hand, if we discuss about the problems in DFV, we might rather express them as a kind of reliability issues in nanometer technologies [3]. As we move into technologies beyond 90 nm, variations in many important parameters become huge. With this we can no longer treat reliability as a deterministic problem. Furthermore, design parameters such as power supply voltage and operating temperature introduce additional factors that can cause failures. Use timing analysis as an example [2]. In worst case of timing analysis, it is assumed that the worst-case path delay equals the sum of the worst-case delays of all individual logic gates from which it is constructed. This, however, produces pessimistic results, incorrect critical paths, and results in overdesign in modern framework. Overdesign usually means a conservative approach, but also the waste of design efforts. The reason is that the aforementioned deterministic values for gate and wire delays may be appropriate for inter-die parameter variations, but does not account for intra-die (inside the die) variations. With all inter-influential parameters and faults by noise sources, delay becomes unpredictable. Statistical static timing analysis (SSTA) is therefore needed in order to cope with the local variations. An objective of SSTA is to find the probability density function (PDF) of the signal arrival times at internal nodes and POs. With these efforts, researchers hope to minimize the overdesign problem and regain the design cost control in nanometer design era.

## Problems

**7.1** We have listed one solution for modern IC design trade-off between performance and power. Can you find other popular solutions?

**7.2** Find the examples from the current solutions of packaging on low-power chips and high-performance chips. What factors will affect the decisions?

**7.3** The designed ICs will affect the solutions of packages due to power and I/O demands. Analyze the relationship between them and list the guidelines.

**7.4** How is package decision affecting the power delivery of ICs? Show your research.

## References


1 H.B. Bakoglu, *Circuits, Interconnections, and Packaging for VLSI*, Addison-Wesley, 1990

2 H. Veendrick, "*Nanometer CMOS ICS*," Springer, 2008

3 W. Wolf, *Modern VLSI Design*, 4, Prentice Hall, 2009

4 S. M. Sze, "*VLSI Technology*," McGraw Hill, 1988

5 Weste and Harris, *Integrated Circuit Design*, 4, Pearson, 2011

6 L. Xiu, *VLSI Circuit Design Methodology Demystified: A Conceptual Taxonomy*, IEEE Press, 2008

7 C.-J. Lee, S.-Y. Liu, C.-C. Huang, et al. (2012). Hierarchical power network synthesis for multiple power domain designs. ISQED-12.

8 C.-K. Wang, Y.-C. Chang, H.-M. Chen and C.-Y. Chin, "Clock tree synthesis considering slew effect on supply voltage variation," *ACM Transactions on Design Automation of Electronic Systems*, 20(1) Article 3, 2014

9 M. Keating, D. Flynn, R. Aitken, A. Gibbons and K. Shi, *Low Power Methodology Manual*, Springer 2007

10 B. Bailey, (2020). Is DVFS worth the effort? Semiconductor Engineering. weblog, September 2020.

11 Y. Bai, V.W. Lee, and E. Ipek, (2017). Voltage regulator efficiency aware power management. ACM ASPLOS, 825–838.

12 B. Tseng, and H.-M. Chen, (2008). Blockage and voltage-island-aware dual-Vdd buffered tree construction. ACM ISPD, 23–30.

13 C.-Y. Yeh, H.-M. Chen, L.-D. Huang, et al. (2007). Using power gating techniques in area-array SoC floorplan design. IEEE SOCC, 233–236.

14 J. Lienig and H. Bruemmer, *Fundamentals of Electronic Systems Design*, Springer, 2017

15 S. Aroonsantidecha, S. Liu, C.-Y. Chin and H.-M. Chen, "*A Fast Thermal Aware Placement with Accurate Thermal Analysis Based on Green Function* IEEE ASP-DAC, pp. 425–430, 2012

16 C.-C. Huang, C.-T. Lin, W.-S. Liao, et al. (2014). Improving power delivery network design by practical methodologies. IEEE ICCD-14.

# 8

# 2.5D/3D System-in-Packaging Integration

## 8.1 Introduction

On the brief history of computing hardware, it has moved from vacuum tubes to discrete devices, then to integrated circuits (ICs). The chips are getting larger and larger, following the Moore's Law, so the chip capacity and performance increases; they are driven by the demands of applications. In the near future, these application demands are driving the advance to vertical stacking in chips and memories. We have mentioned the vertical trending in Section 6.6 for modern chips and packages: 3D IC packaging, 3D IC integration, and 3D silicon integration. With the practical 3D solutions consideration in modern technology, we shall focus on the category of 3D IC integration in this chapter.

Unlike 3D Si integration, 3D IC integration stacks up chips in the vertical direction with through silicon vias (TSVs), thin chips/interposers, and micro-bumps to achieve high performance, low power, wide bandwidth, small form factor, and hopefully low cost [1]. Among 3D IC integration solutions, the so-called 2.5D IC utilizes a piece of dummy silicon (or glass), people call it interposer, with TSVs/TGVs for the communication of signals and power delivery. Such designs have become the savior in high-performance computing devices due to its lower cost compared to genuine 3D manufacturing chips. We shall discuss the concept and format of 2.5D IC designs through Sections 8.2–8.4.

On the other hand, the genuine 3D IC integration lies in real 3D structure: die-to-die direct communication. In this way, we should be able to reach lower power with higher communication bandwidth ideally. Bandwidth is usually defined as the amount of data (bits) transferred per second. The reasons are as follows. The communication bandwidth between logic and

*Electronic Packaging Science and Technology*, First Edition. King-Ning Tu, Chih Chen, and Hung-Ming Chen.

memory is dictated by the number of I/O terminals used to transfer data [2]. As the number of I/O terminals increases in modern designs, the communication bandwidth increases as well. Sending data back and forth between two chips dissipates power, which is proportional to the parasitic of the interconnects used to communicate between dies. Because the long interconnect wires have larger capacitance and hence large amount of power is wasted, making wires shorter will reduce power consumed. Therefore, 3D IC technology is considered very useful in increasing the communication bandwidth between the processor and memory while reducing power, which will be detailed in Sections 8.5–8.7.

Besides 2.5D IC and 3D IC solutions, we shall discuss the modern solutions provided by foundries such as InFO and SoIC in Section 8.8. They are called for from ever-increasing market demands on higher computing efficiency, wider data bandwidth, higher functionality packaging density, lower communication latency, and lower energy consumption per bit data [3]. The new trending in chiplets designs and packaging will be presented as well.

## 8.2 2.5D IC: Redistribution Layer (RDL) and TSV-Interposer

Through this section to Section 8.4, we discuss in detail about interposer/2.5D IC designs. Figure 8.1 presents a typical packaging hierarchy, with an interposer as first-level package, which is inserted between the die stack and the

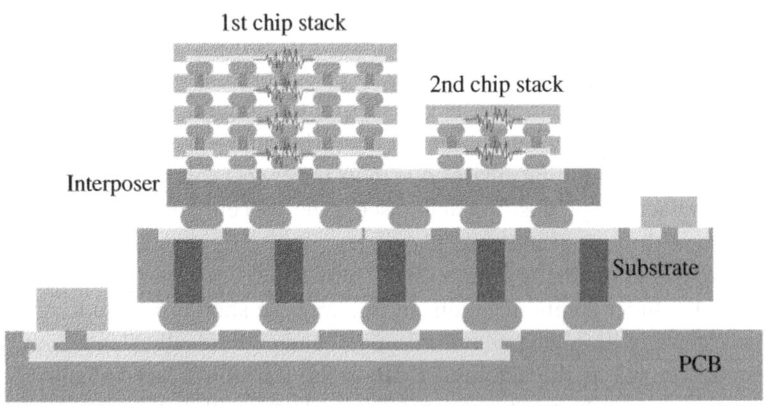


**Figure 8.1** Packaging hierarchy in interposer-based 3D IC. CoWoS is one of the representative implementation.

second-level package [2]. Why do we need interposers when we already have packages? We can understand it from the amount of different levels of C4 solder bumps. The number of micro-bumps between the die stacks and the interposer is around the range of 100K, while that of C4 solder bumps between the interposer and package (second-level) is around 10K. The number of BGA balls between package and PCB is around 1K. Of course, the pitches of these "interconnections" are in the similar scale. They provide the different order of magnitude improvements in integration.

The interposer owns the following electrical functions/advantages when applied to wiring to connect dies (stacking or side-by-side) to each other [2]. First, the short interconnections reduce parasitic and thus improve communication speed and bandwidth, compared to the presence of package only. Second, it provides better noise management and improves the ability to decouple the power supply by increasing the effectiveness of the decaps (decoupling capacitors) mounted on the interposer due to its thinness. Third, it improves the heat dissipation (depending on the interposers used), providing the effective thermal management solution. Fourth, it improves the mechanical reliability by minimizing the coefficient of thermal expansion (CTE) mismatch between die stacks and the interposer (when silicon is used as interposer material).

Coming back to more convincing reasons to apply interposers, the following embodiments of interposer can further elaborate, mentioned in [2]. The major reason of applying interposers is that it can serve as a **space transformer** through redistribution by connecting the micro-bumps to the C4 bumps. With an order of the magnitude increase in communication capacity, we can effectively deliver signals and power supply, much better than merely using second-level package. On the other hand, since the interposer is a passive structure (no transistors in them), we can use the vias in the interposer to connect dies to each other in the form of a stack to increase the transistor density. This can be done when the dies are mounted on either side of the interposer and connected to each other using the through vias in the interposer (TSV-interposer), seen as the second embodiment. This approach works for some low-power applications. The third embodiment is realized by Xilinx, Inc., and TSMC called CoWoS, where the interposer is used to support the wiring between the smaller ICs, will be depicted next.

The 2.5D IC integration used by Xilinx in [4] consists of a passive silicon interposer and three active dies. The silicon interposer consists of four high densities (~1 um pitch) wiring layers, TSVs, and micro-bumps. The four layers are as follows: one layer for redistribution, two for signal routing, and the last one for ground reference layer, which separates the two signal routing layers. Redistribution layer (RDL) is an integral part of 3D IC integration, especially for 2.5D IC designs with a passive interposer. The RDL allows for circuitry fan-outs of and allows for lateral communication between the chips attached to the interposer [1]. There are at least two ways to fabricate the RDL, namely, (i) polymers to make the passivation and Cu-plating for the metal layer, and (ii) semiconductor BEOL Cu damascene process. Two kinds of signals are routed in the interposer: signals that connect the micro-bumps to the C4 bumps through the TSVs, and signals that connect dies to each other in the interposer, which do not pass through the TSVs. This technology is called chip on wafer (actually interposer) on substrate (actually package), namely, CoWoS. It was debuted in 2011 [4] and updated in 2017 [5]. Now this technology is seen as a major driver in AI chip development.

There is another version of 2.5D IC integration from Intel. The Embedded Multi-die Interconnect Bridge (EMIB) is an elegant and cost-effective approach to package a high-density interconnect of heterogeneous chips [6]. The industry refers to this solution as 2.5D IC package integration. Instead of using a large silicon interposer typically found in other 2.5D IC approaches, EMIB uses a very small bridge die, with multiple routing layers. This bridge die is embedded as part of the substrate fabrication process. It is proposed in 2008 [7] and debut in 2014.

## 8.3 2.5D IC: Silicon, Glass, and Organic Substrates

We have seen the silicon interposer in prior section, yet we ask is silicon the only material for the interposer? With the ultimate advantage of CTE mismatch minimization, i.e. Si on Si, it is the usual choice to use silicon. The problem for CMOS grade silicon is that it has a resistivity of 10 Ω-cm and therefore lossy [2]. Although it can be improved through doping to 100 Ω-cm to lower coupling, the cost is also increased. Glass as an alternative material is a possible solution. It is an insulator as opposed to silicon, which is a semiconductor, the through glass vias (TGVs) can be filled with copper without the need for an insulator lining. Because the resistivity of glass is between $10^8$ and $10^{13}$ Ω-cm, it is much higher than silicon and therefore the coupling between vias is greatly reduced, as compared to that with silicon. The problem of glass interposer comes back to CTE mismatch: the mechanical reliability is a big issue. Furthermore, the thermal conductivity of glass is much lower than that of silicon; it has no heat spreading and creates more hotspots.

We can see other characteristics of adopting glass substrates. It can provide the lower cost due to cheaper material cost. Another good part is regarding the possible benefit in power integrity (PI). TSVs can be lossy and can produce excessive coupling in a low-resistivity (high conductivity) silicon substrate. Glass is a low-loss dielectric medium. TGVs can have superior electrical performance. As of now, the silicon interposer still has an edge due to the following reasons [2]. First, it can reduce the amplitude of the resonances between power and ground planes, and that between chip and interposer. Second, it has indirect impact on signal integrity (SI) by eliminating return path discontinuities, which affect the target impedance (will be introduced in Section 8.9). Third, since power is at the end converted to waste heat, silicon substrate can perform better in thermal solution. It reacts favorably to temperature increase at higher frequencies.

The organic interposer [8], like the glass interposer, is one of the alternate types of interposers being explored to realize the cost benefits. Organic interposers are certified to be cheaper due to a well-established supply chain and the ability to be manufactured using traditional processes such as wet etching. The challenge with organic interposers is in their mechanical properties due to their flexibility. They also have a lower finer-pitch I/O density than silicon and glass substrates. Organic interposers are suited for logic-memory integration, large CPUs/GPUs, and certain types of ASICs. Interestingly, certain high-performance RF applications have been demonstrated in using organic interposers. However, their mass applicability to next-generation high-performance application is still unclear.

## 8.4 2.5D IC: HBM on Silicon Interposer

We have mentioned that going vertical is driven by the density of transistors, and the dense level is driven by application demands. But the trend to 3D IC has to start with memory due to its repetitive devices characteristic [9]. With early adoption of vertical memory, AMD has initiated the stacking memory solution in 2008; in 2013 high bandwidth memory (HBM) is adopted as industry standard. HBM is a new type of memory chip with low power consumption and ultrawide communication lanes. It uses vertically stacked memory chips interconnected by microscopic wires (TSVs). Through years of development, [10, 11] show the efforts from memory makers SK Hynix and Samsung. In [10], they adopted a logic-interface chip between the interposer and stacked-DRAM with thousands of TSVs. The logic-interface chip in the base level of HBM decreases the capacitance in I/O, repairs the chip-to-chip connection failure, and supports better testability and improves reliability. Figure 8.2 illustrates the configuration of the stacked-DRAM (HBM) on the interposer.

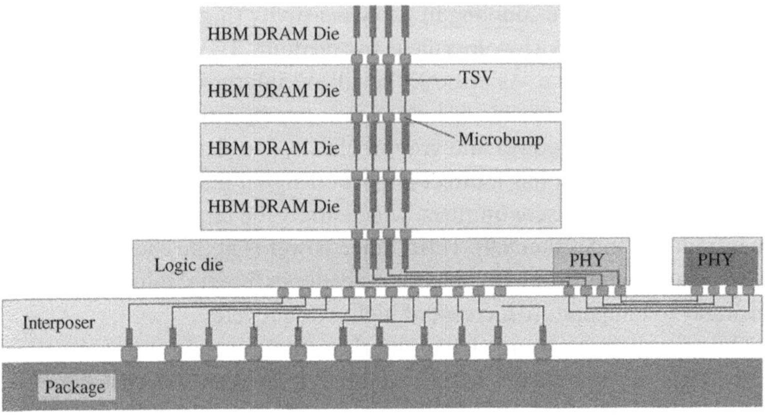


**Figure 8.2** An illustration for high bandwidth memory (HBM) with interposer.

This trend also brings the reform in CoWoS. As aforementioned in [5], for the first time CoWoS containing the HBM2 has been applied in fabricating high-performance wafer-level SiP. An ultra-large Si interposer, up to 1200 $mm^2$ made by a two-mask stitching process, is used to form the basis of the second-generation CoWoS (CoWoS-2) to accommodate chips of logic and memory and to achieve the highest possible performance. On the other hand, due to rapidly evolving artificial intelligence (AI) technology, many applications have been successfully deployed, such as image recognition, health care, and autonomous driving. Such rapid evolution has been possible, owing to the emergence of neural network (NN) accelerators, such as GPUs and TPUs, which have a higher data rate of communication throughput. All these (mentioned in [11]) require an enhanced memory system with large capacity and high bandwidth: HBM. Other details of HBM are presented in the next section.

## 8.5 3D IC: Memory Bandwidth Challenge for High-Performance Computing

Besides 2.5D IC designs, we mentioned in prior sections, the following three sections will discuss genuine 3D designs in 3D IC integration. What we mean by genuine 3D designs is that the chips are stacked up without passive interposers. As aforementioned, due to the features of mobile devices, they drive performance and density, and eventually we have "Mobile DRAM" in the mobile product market. Knowing the evolution of volatile mobile DRAM in mobile products will help understand the current utilization roadmap of 3D IC integration. Note that we will not discuss 3D memory devices such as 3D/vertical NAND and 3D ReRAM [9] here because they are stand-alone storage-class memory products with no 2.5D/3D integrated computing systems involved.

Mobile DRAM consumption in smart phones is growing with the growth of personal computer (PC) DRAM [1]. In 2011, the DRAM demand of smart phone has officially surpassed that of PCs. Therefore, the wider bandwidth at reduced power becomes the minimum (and continuous) requirement in modern and future mobile DRAM. Another minimum requirement is higher densities in thinner packages. In nonvolatile memories such as Flash, they also have similar requirements: faster and denser storage solutions with embedded Multi Media Card (eMMC) and small solid-state drive (SSD) form factors for thin and light-weight devices. They even combine mobile DRAM with NAND Flash into a multichip module and put them into multichip package (MCP).

Coming back to the wider bandwidth, bandwidth is usually defined as the amount of data (bits) transferred per second. Because each I/O or solder bump can transfer only a certain amount of data. Thus, for the transfer of a large set of data, a large number of I/O solder bumps is needed. DRAM has 4-, 8-, 16-, or 32-bit data width to communicate with CPU/logic/SoC. For example, DDR3 (double data rate, type 3)-1600 chip has a speed rating of 1600 Mb/s per I/O. If this chip has 32-bit I/O data width, it has total memory bandwidth of $32 \times 1600 = 51{,}200$ Mb/s = 51.2 Gb/s or 6.4 GB/s. Another example: DDR4-3200 chip's memory bandwidth (with 64-bit I/O data width) is up to $8 \times 3200 = 25600$ MB/s = 25.6 GB/s. On top of it, Wide I/O standard can provide 512-bit I/O data width, which we will explain next why it becomes the standard.

How to achieve such high I/O data width for high-performance computing (HPC)? Unfortunately, the data width is limited by IC packaging technology. If using wire-bonding technology with larger pad sizes and pitches, we need much larger chip size to accommodate. With TSV technology, which can provide very small via size and pitch (much smaller than wire-bonding), a much wider I/O data width such as 512-bit is possible. If we use the same DDR4-3200 chip for 4-DRAM stacking in Wide I/O, we can achieve the memory bandwidth of $512 \times 3200 = 1{,}638{,}400$ Mb/s = 204.8 GB/s, very attractive for HPC. Note that this DRAM stack needs to be interconnected to the logic die directly. Thus, we have recent implementations of standardized Wide I/O and related standards for real 3D IC integration.

There are currently three major implementations related to Wide I/O DRAM in the market [12]: Samsung's Wide I/O series, HMC from Micron and Intel, and HBM from SK Hynix. Wide I/O is a standard from Joint

Electron Device Engineering Council (JEDEC). For over 50 years, JEDEC has been the global leader in developing open standards and publications for the microelectronics industry. Wide I/O and Wide I/O 2 have been backed by companies like Samsung and are designed to provide mobile SoCs with a maximum amount of bandwidth at the lowest possible power consumption. This series is explicitly designed to be 3D interfaced. However, 2.5D configurations are also possible. The second effort is Hybrid Memory Cube (HMC) from the joint Intel-Micron standard. HMC is designed to emphasize massive amounts of bandwidth at higher power consumption and cost than Wide I/O 2. Intel and Micron have claimed that up to 400 GB/s of bandwidth may be possible via HMC; however, it is not a JEDEC standard yet. It is explicitly designed to respond to multicore scenarios, but at the same time it is more expensive than the counterpart. The last one we introduce is High Bandwidth Memory (HBM), as planned by SK Hynix, AMD, and Nvidia. HBM is explicitly designed for graphics, but it can be seen as a specialized application of Wide I/O 2. Both AMD and Nvidia are adopting it for next-generation GPUs. HBM is the compromised version in cost and bandwidth: it is not as cheap as Wide I/O, or as power efficient, but it was designed for high-performance GPU environments. It is still considered cheaper than HMC [12]. HBM stacks up DRAM dies, including an optional base die with a memory controller, which are interconnected by TSVs and micro-bumps (upper half of Figure 8.2).

## 8.6 3D IC: Electrical and Thermal TSVs

The TSVs consist of vertical holes that are drilled or etched in the silicon, oxidized, and then filled with a conducting metal, which is usually copper [9]. The obvious advantages are short interconnects and lower power. TSV technology is then the most important key enabling technology of 3D Si integration and 3D IC integration (Section 6.6) [1]. Although we know that we also have TSV in interposer/2.5D IC technology, they are slightly different: in 2.5D's TSV, the difficult level of manufacturing is less than that in 3D IC's because in the former, the interposer/substrate only contains passive components. Therefore, in this section we only focus on 3D IC part. In fact, TSV is not a new concept; it was invented more than 50 years ago by the 1956 Nobel Laureate in Physics: William Shockley.

There are six key steps in making a TSV [1]: (i) via formation by either deep reactive-ion etch (DRIE) or laser drilling; (ii) dielectric layer deposition by plasma-enhanced chemical vapor deposition (PECVD); (iii) barrier and seed layer deposition by physical vapor deposition (PVD); (iv) copper

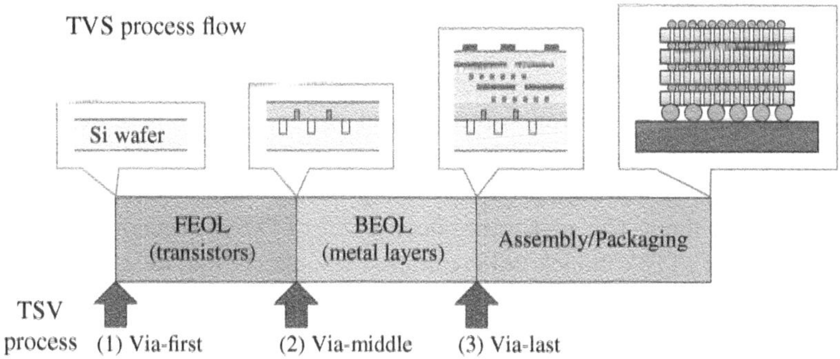


**Figure 8.3** TSV fabrication options in 3D IC integration: via-first, via-middle, and via-last.

plating to fill the vias; (v) chemical and mechanical polishing (CMP) of copper plating residues; and (vi) TSV revealing. From the fabrication order for TSV process in 3D IC integration, we have via-first, via-middle, and via-last processes. Figure 8.3 presents those different TSV fabrication options in the manufacturing process stages. Via-first process means that the vias are fabricated before the FEOL (transistors), which only can be done by the fabs. Via-middle process has face-to-face and face-to-back configurations. The vias are fabricated right after the FEOL and before the tiny vias (inside dies) and metal layers. Usually they are done by the fabs. Face-to-face configuration means that the die is "connected" using transistor face to another die's transistor face. TSVs have shorter distances compared to face-to-back configuration. Actually, these two configurations have the same FEOL, TSV, and BEOL process, just different in the process after BEOL. Via-middle process is usually done by the fabs. Lastly, the vias are fabricated after the BEOL in via-last process. Via-last process also has face-to-face and face-to-back configurations, and it is usually performed by the fabs with small vias and fine RDLs. For larger vias, the assembly house (or outsourced semiconductor assembly and test, OSAT) can do the job.

TSVs are widely used in 3D ICs and at the same time can provide passive cooling due to high-thermal-conductivity metal vias. While signal TSVs (STSVs) are TSVs that carry both electrical signal and heat between dies, thermal TSVs (TTSVs) are dummy TSVs that facilitate heat transfer across 3-D ICs [13]. Applying TTSVs uniformly across the die or specifically near hotspots can reduce the peak temperature. There were some prior works about the planning of TTSVs, but they underestimated the thermal resistance of micro-bump interconnections, which is measured to be 10x higher than that of metal vias. In [13], TTSV placement optimization is performed early in the design phase, an effective medium theory (EMT)-based model to capture the direction-dependent thermal conductivity of TSVs is developed. On the other hand, the metal inside TTSVs can conduct heat more effectively than the silicon substrate, and the metal bumps underneath TTSVs can help heat penetrate through the interlayer thermal interface material (TIM). However, the surrounding silicon dioxide blocks the heat flowing into them. This makes the effectiveness of TTSVs questionable [14]. Some existing TTSV models fail to capture these effects, and the experimental results in [14] based on finite element simulations confirmed that the temperature reduction was indeed brought by the metal bumps underneath the TTSVs rather than the TTSVs themselves.

The last thing but not the least we should be aware of is the reliability issue of TSVs. Due to CTE mismatch between TSV and substrate materials, large thermo-mechanical stress builds up during fabrication. Especially, highest stress occurs at TSV/dielectric liner interface, which might initiate mechanical instability at this interface. In [15], they focused on the interfacial crack at TSV/liner interface, which may induce both mechanical and electrical failure. Crack can initiate in any locations and propagate to any directions in TSV, if the stress is high enough, there is a good chance that the crack will initiate at this TSV/liner interface, and causing the failure of the vias.

## 8.7 3D IC: 3D-Stacked Memory and Integrated Memory Controller

Continuing the discussion in Section 8.5, here we describe the implementations in reflecting the efforts in 3D heterogeneous dice integration. Since the performance of the classical memory hierarchy can be improved using 3D processor cores on one chip stacked using TSV with cache and DRAM on another chip(s), we introduce two practices on this: memory on logic. The study in [16] demonstrated an implementation of a four-layer-stacked chip with 45-nm DRAM dice and a 65-nm logic controller die, which are interconnected by via-last TSV processes. Figure 8.4 shows the process flow of DRAM/logic die-stacking. A thicker top DRAM die was presented, then the middle DRAM die was bonded onto the top DRAM die with thermo-compression bonding using the flip-chip C-4 solder joint method. As three middle DRAM dice were stacked on the top DRAM die, the logic die was bonded onto the last middle DRAM die. In this work, the stacked DRAM/logic performance and system verifications have shown the manufacturing capability in high-performance 3D heterogeneous integration.

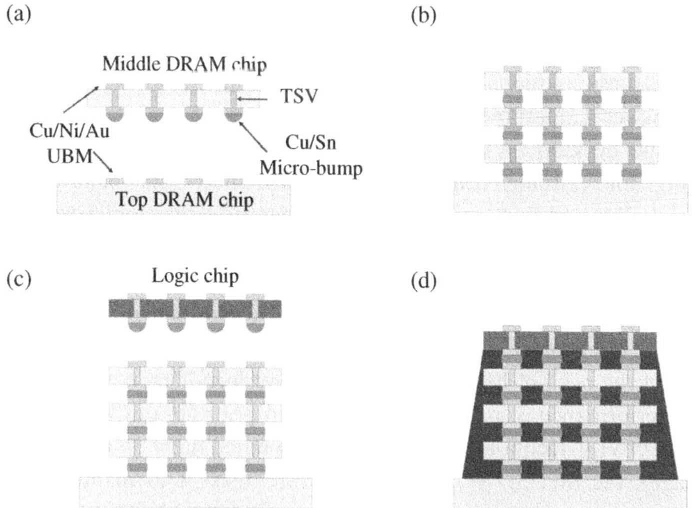


**Figure 8.4** Process flow of DRAM/logic die stacking module in [16]: (a) a thin middle DRAM die stacked on the thick top DRAM die, (b) three middle DRAM dice stacked on the top DRAM die, (c) a thin logic die stacked on the DRAM dice, and (d) underfill encapsulation for the die-stacking module.

On the other hand, Samsung's X-Cube (eXtended-Cube) has presented a new technology to allow chip-stacking of SRAM dice on top of a base logic die through TSVs, debuted in 2020 [17]. Quite different from existing technologies such as intermediate interposers and silicon bridges (EMIB), the logic die is designed with TSV pillars, which then connect to micro-bumps with only 30 μm pitch, allowing the SRAM-die to be directly connected to the main die without intermediate mediums. Stacking more valuable SRAM instead of DRAM on top of the logic chip would possibly represents a higher value and efficiency, as this would allow smaller die footprints for the base logic dies, with larger SRAM cache structures being able to "reside" on the stacked die.

## 8.8 Innovative Packaging for Modern Chips/Chiplets

3D IC chips stacked and connected with vertical TSV entered the market in about 2013 led by stacked chips of homogeneous memory and followed in 2014 by early stacked heterogeneous memory and logic chip systems [9]. Nevertheless, in this section we discuss modern packaging solutions, mostly created from the foundries due to customized needs from the leading customers regarding the concern of manufacturing cost.

Among those needs for modern electronics market, the most famous development is integrated fan-out wafer-level system integration (**InFO**). What drives these solutions emerging from various and diversified packaging? The answer is smart mobile devices. Mobile computing devices, different from the desktop or laptop computing, are power sensitive while demanding for extreme functionality, performance, and bandwidth. Transistor scaling and chip scaling alone no longer fulfill the mobile device's demanding requirements on multifunctional integration in a smaller form factor with good power efficiency, thermal dissipation, and electrical performance [18]. Due to the high pin count and high-density integration, InFO allows chip I/O signals to be fanned-out to a region larger than silicon die area to meet the need of high I/O counts, smaller form factor, and higher electrical performance. It is much different from prior fan-out wafer-level package (FO_WLP) technologies. In 2016, [18] debuted InFO_PoP (package on package), the first 3D FO_WLP, and TSMC used it to manufacture Apple's A10 processor. Figure 8.5 shows the cross-sectional view of InFO_PoP. Flip-chip PoP (FC_PoP), flip-chip PoP for high memory bandwidth (FC_HMB_PoP), and 3D IC with TSV are compared in [18], and the conclusion is that InFO_PoP can achieve satisfaction in terms of performance, power and form factor, and cost for targeted applications. After this, many variations/extensions have emerged for different targets: InFO-oS (on substrate), InFO-MS (memory substrate), InFO_AiP (antenna in package), and InFO_SoW (system on wafer).

Keeping the evolution of the packaging solutions for further improvement in high-performance computing (HPC), 5G/AI, and edge computing applications, leading foundry TSMC continues to provide another attempt to further overcome the challenges in thermal, power delivery, and yield: system on integrated chips (SoIC) [3]. It was proposed in 2018,

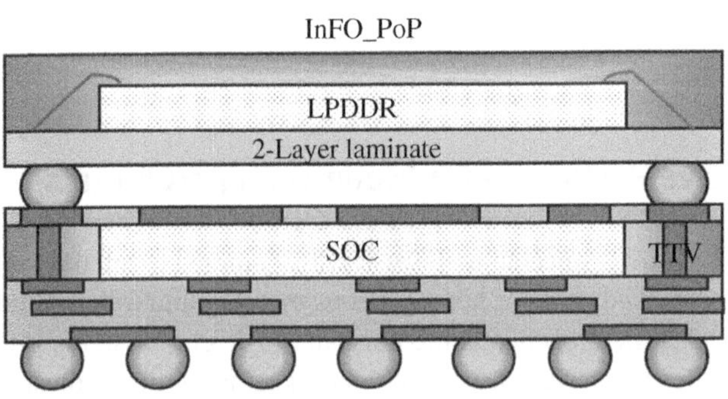


**Figure 8.5** Cross-sectional view of single chip InFO_PoP with TIV (Through-InFO-Via) [18].

and it can be holistically integrated into advanced packaging technology platforms such as flip-chip (FC), integrated fan-out (InFO), 3D IC, and 2.5D IC with Si interposer (CoWoS). According to [3], with the innovative bonding scheme (bumpless hybrid bonding), SoIC enables the strong bonding pitch scalability for chip I/O to realize a high-density die-to-die interconnects.

Why is SoIC necessary from TSMC's point of view? Intel actually has a similar solution named Foveros. There is a rich set of packaging technologies to make things even more interesting because they can be mixed-and-matched, Intel therefore introduces "co-EMIB," which is a combination of EMIB and Foveros. First introduced in 2019, Foveros plays a role in advanced 3D face-to-face (F2F) die-stacking packaging process technology. The key feature of Foveros is the F2F chip-on-chip bonding through extremely fine-pitched micro-bumps. Coming back to the necessity, the reason for SoIC and Foveros to emerge is chiplet design. Chiplets usually mean the independent components, which make up a large chip built out of multiple smaller dies, and the dice are not necessarily manufactured by the same nodes. It is expected that turning to chiplets can significantly reduce cost, compared to a more complex SoC [7].

Major foundries such as TSMC, UMC, Global Foundries, and Samsung are building advanced packaging capabilities into the backend of manufacturing [19]. As aforementioned, TSMC is planning to add chiplets into the frontend using SoIC. All of these will likely require significant changes across the industry, from EDA tools to test and post-silicon diagnosis. All of this is necessary as the cost of putting everything onto a single die continues to rise. Modularity allows chipmakers to customize chips relatively quickly based upon a platform type of approach. CPU, GPU, and FPGA chip designers figured this out years ago and have since started the march to disaggregated implementation by going to multi-die and letting the interposer/packaging take care of the integration. Take the hybrid CPU platform from Intel as another example. Lakefield (debut in 2019), which consists of a 10-nm processor core and four 22-nm processor cores inside a package, is a typical chiplet-style product.

In conclusion, we emphasize that we have come to the age that we need advanced packaging solution for boosting IC-level performance [7]. The inflection point was the rise of disruptive AI architectures based on neural networks (NNs). That was an important point; it showed the feasibility of NNs, and it gave rise to accelerators and high bandwidth memory inside the package. As a result, innovative packaging solutions from foundries and IDMs are indeed providing satisfactory performance for those applications and will continue to evolve.

## 8.9 Power Distribution for 3D IC Integration

With all possible scenarios in stacked ICs such as F2F or F2B connections, stacked dice side by side on a logic die or an interposer, one of the most important things we should face is how to have effective power distribution for various configurations. The reasons are two-fold. First, due to the stacking of ICs, the current density in the package will increase [2]. Second, with the huge communication bandwidth between the logic and memory IC (described in Section 8.5), it is expected that the power consumed by I/O will increase, thus increasing the importance of overall power distribution.

In Figure 8.6 [20], we can obtain all necessary components in a 3D IC model. First, consider a packaged chip mounted on a PCB, the chip is usually powered by a voltage regulator module (VRM) mounted on the PCB. The VRM supplies current through the P/G planes in the PCB and package to the P/G terminals of the chip. Since the functionality and reliability of transistors rely on the voltage fluctuation levels between P/G terminals, the fluctuations should not be allowed to exceed a maximum limit, which is determined by calculating the *target impedance* [2]. The target impedance as a design parameter has been adopted by both industry and academia, described as follows. From Ohm's law, the voltage fluctuation (ΔV) due to a current source (I) can be calculated as $\Delta V = ZI$, where $Z$ is the target

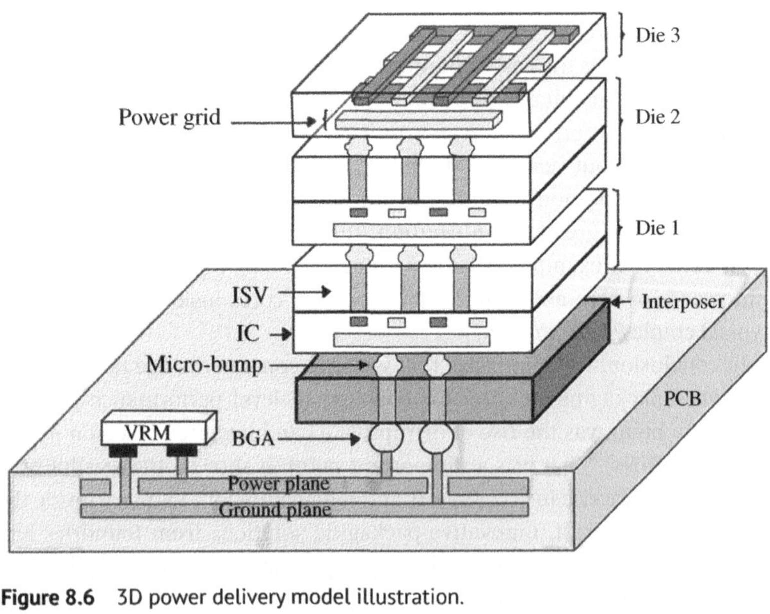


**Figure 8.6** 3D power delivery model illustration.

impedance. Here $Z$ is expressed in the frequency domain, as the current $I$ varies with frequency, the target impedance $Z$ also varies with frequency. At the end of the day, low target impedance is always required and preferred to minimize power supply fluctuations in any system. The design goal is to ensure that the power distribution impedance is always less than the target impedance.

Here, we show one example in integrated design power network co-simulation. One major problem remains in design synthesis and flow, i.e. how to model heterogeneous die(s) with major logic die for power synthesis and signoff. In [21], it provides a realistic model and principle for heterogeneous dies power network for 3D ICs. It is based on a given abstract or an early stage information like bump location and power consumption from the provider. It also uses this model to synthesize power network with bottom logic die in the design flow. The result is DRC clean power network without IR and EM violation for all power domains. First, it analyzes the location and power consumption of power bump for heterogeneous die(s). Second, according to this prior analysis, the stripe location and power sink location of heterogeneous die(s) model can be decided. After the initial model is synthesized, it is converted to a node graph with corresponding resistance of via and metal layer, also nodal voltages. Third, the model is optimized to adjust power stripe width. It will improve the model iteratively until the target IR Drop is met. A real case from design house containing a 3D DRAM stack is demonstrated for the effectiveness of this cross-layer realization.

## 8.10 Challenge and Trend

The obvious challenge is the problem of 2.5D/3D integration practice. Although we have seen many implementations under various architectures with the consideration in design complexity and cost, still it won't be too useful if there is no help from the tools. Design issues for TSV stacking include methods for repartitioning processor chips and their memory in 3D IC internally to improve performance and reduce power [9]. It is shown in [22] that one of the recent efforts in providing a unified perspective on the fundamental opportunities and challenges posed by 3D ICs, especially from the context of design tools and methods. Still, commercial 3D IC design tools are called for to effectively and efficiently generate 2.5D/3D designs.

It is worth mentioning that the modern computing devices evolve to a new direction [23]. Memory-centric designs deploy computation to storage and enable efficient in-memory computation (IMC) while avoiding massive amount of data movement. The IMC schemes have shown distinct advantages and concerns when applying to different types of memory technologies, from conventional SRAM, DRAM to emerging ReRAM. To attain an efficient design within short design cycle, it is imperative to have an integrated design framework with automated tools to support hybrid memory systems and perform effective optimization across design stages. We can say that there will be possibly new style of packaging solutions for such smart AI edge devices.

## Problems

**8.1** What is the difference between the TSV designs on interposer and on genuine 3D IC? Are the thermal issues the same between those two styles?

**8.2** Why do the foundries develop advanced packaging technologies? Try to list the advantages/profits by developing those solutions.

**8.3** Try to list the difficulties in developing the design tools for 2.5D and 3D ICs.

**8.4** What is the difference between 2D IC, 2.5D IC, and 3D IC power delivery networks?

## References


**1** Lau, J.H. (2013). *Through-Silicon Vias for 3D Integration*. McGraw Hill.

**2** Swaminathan, M. and Han, K.J. (2014). *Design and modeling for 3D ICs and interposers*. World Scientific.

**3** Chen, F.C., Chen, M.F., Chiou, W.C., and Yu, D. (2019). *System on Integrated Chips (SoIC) for 3D Heterogeneous Integration*. IEEE ECTC.

**4** Banijamali, B., Ramalingam, S., Nagarajan, K., and Chaware, R. (2011). *Advanced Reliability Study of TSV Interposers and Interconnects for the 28nm Technology FPGA*. IEEE ECTC.

**5** Hou, S.Y., Chen, W.C., Hu, C. et al. (2017). Wafer-level integration of an advanced logic-memory system through the second-generation CoWoS technology. *IEEE Transactions on Electron Devices* 64 (10): 4071–4077.

**6** Embedded multi-die interconnect bridge. https://www.intel.com.tw/content/www/tw/zh/silicon-innovations/6-pillars/emib.html.

7 Santo, B. (2020). *Explainer on Packaging: Interposers, Bridges and Chiplets.* SoC DesignLine https://www.eetimes.com/explainer-on-packaging-interposers-bridges-and-chiplets/?utm_source=newsletter&utm_campaign=link&utm_medium=EETimesDaily-20201111&oly_enc_id=6133B9781801C5P.

8 Overview of the interposer technology in packaging applications. https://ww2.frost.com/frost-perspectives/overview-interposer-technology-packaging-applications/ (2018).

9 Prince, B. (2014). *Vertical 3D Memory Technologies.* Wiley.

10 Lee, D., Kim, K.W., Kim, K.W. et al. (2014). *A 1.2V 8Gb 8-Channel 128GB/s High-Bandwidth Memory (HBM) Stacked DRAM with Effective Microbump I/O Test Methods Using 29nm Process and TSV.* IEEE ISSCC.

11 Oh, C., Chun, K.C., Byun, Y.-Y. et al. (2020). *A 1.1V 16GB 640GB/s HBM2E DRAM with a Data-Bus Window-Extension Technique and a Synergetic On-Die ECC Scheme.* IEEE ISSCC.

12 Hruska, J. (2015). *Beyond DDR4: The Differences Between Wide I/O, HBM, and Hybrid Memory Cube.* ExtremeTech https://www.extremetech.com/computing/197720-beyond-ddr4-understand-the-differences-between-wide-io-hbm-and-hybrid-memory-cube.

13 Ren, Z., Alqahtani, A., Bagherzadeh, N., and Lee, J. (2020). Thermal TSV optimization and hierarchical floorplanning for 3-D integrated circuits. *IEEE Transactions on Components, Packaging and Manufacturing Technology* 10 (4): 599–610.

14 Chou, C.-H., Tsai, N.-Y., Yu, H. et al. (2013). *On the Futility of Thermal Through-Silicon-Vias.* IEEE VLSI-DAT.

15 Jung, M., Liu, X., Sitaraman, S.K. et al. (2011). *Full-Chip Through-Silicon-Via Interfacial Crack Analysis and Optimization for 3D IC.* IEEE ICCAD.

16 Shen, W.-W., Lin, Y.-M., Chen, S.-C. et al. (2018). 3-D stacked technology of DRAM-logic controller using through-silicon via (TSV). *IEEE Journal of the Electron Devices Society* 6: 396–402.

17 Frumusanu, A. (2020). X-cube 3D TSV SRAM-logic die stacking technology. AnandTech. https://www.anandtech.com/show/15976/samsung-announces-xcube-3d-tsv-sramlogic-die-stacking-technology.

18 Tseng, C.-F., Liu, C.-S., Wu, C.-H., and Yu, D. (2016). *InFO (Wafer Level Integrated Fan-Out) Technology.* IEEE ECTC.

19 Sperling, E. and Mutschler, A. (2020). New architectures, much faster chips. Semiconductor Engineering. https://semiengineering.com/new-architectures-much-faster-chips/.

20 Zhang, D.C., Swaminathan, M., Keezer, D., and Telikepalli, S. (2014). *Characterization of Alternate Power Distribution Methods for 3D Integration.* IEEE ECTC.

**21** Liao, W.-H., Lin, C.-T., Fang, S.-H. et al. (2017). *Heterogeneous Chip Power Delivery Modeling and Co-Synthesis for Practical 3DIC Realization*. IEEE/ ACM ASP-DAC.

**22** Lu, T., Serafy, C., Yang, Z. et al. (2017). TSV-based 3-D IC: design methods and tools. *IEEE Transactions on Computer-Aided Design of Integrated Circuits and Systems* 36 (10): 1593–1619.

**23** Chen, H.-M. et al. (2020). *On EDA Solutions for Reconfigurable Memory-Centric AI Edge Applications*. IEEE ICCAD.

# Part III

# 9

# Irreversible Processes in Electronic Packaging Technology

## 9.1 Introduction

Electronic packaging technology is wafer-based and flux-driven. Its applications occur in the operation of microelectronic devices built on semiconductor chips or wafers. The basic steps in processing of these devices consist of adding or subtracting a monolayer of atoms on or off a wafer surface, respectively, or in joining two materials together as in solid–liquid interdiffusion (SLID) reaction to form a solder joint. In these processes, we are not dealing with equilibrium states, rather we deal with kinetic states in going from one state to another or we deal with the steady state of a flux of atoms, or electrons, or heat.

Furthermore, if we consider a p-n junction in a transistor, it is not at an equilibrium state. If we anneal the junction at a high temperature for a long time, it will disappear by interdiffusion of the p-type and n-type dopants. At device operation near 100 °C, the dopants are super-saturated but frozen in place in the semiconductor in order to produce the electrical potentials needed to guide the transport of charges. In doping a semiconductor, we need to diffuse or to implant a flux of atoms into the semiconductor to obtain the desired concentration profile of dopant.

During the operation of electronic devices, we can have a flow of matter, a flow of energy (heat), a flow of charge particles, or all of them in a device system, which is an open system. Most importantly, from the point of view of device reliability, it is the interaction among them or their cross-effects that can lead to device failure. In electromigration in Al and Cu interconnects, it may involve the interaction between atomic flow driven by a high current density, as well as by a high stress gradient. In thermomigration in eutectic flip-chip solder joints, a concentration gradient may lead to phase separation under a temperature gradient in the eutectic structure.

*Electronic Packaging Science and Technology*, First Edition. King-Ning Tu, Chih Chen, and Hung-Ming Chen.

In simple cases, the flows are steady. But they may not go from an equilibrium state to another equilibrium state. This is because the flow can keep on going steadily. Without equilibrium, we cannot use the condition of minimum Gibbs free energy to describe the end states. In a steady state of flow process, we are in the domain of nonequilibrium or irreversible thermodynamics. [1–4]

In thermodynamics, the kinetic process going from one equilibrium state to another equilibrium state can be reversible or irreversible. Theoretically, it can be reversible. Practically, all of them tend to be irreversible. If the actual system is kept in a closed system and in a homogeneous boundary conditions, i.e. at constant temperature and constant pressure, it goes irreversibly to the equilibrium end state, for example in a classical phase transformation. On the other hand, if the system is kept in inhomogeneous boundary conditions, e.g. there is a temperature gradient or a pressure gradient, it becomes an open system and tends to go irreversibly to a steady state instead of an equilibrium state. Usually, we call the steady-state processes to be "irreversible processes."

To illustrate the difference between reversible and irreversible processes, Figure 9.1a depicts a closed container that contains water and vapor at constant temperature and pressure. The water and the vapor are at equilibrium in a closed system, meaning that across the interface between them, there is an exchange of molecules of water and vapor, and the exchange is at micro-balance. Or we can say the exchange is reversible. Figure 9.1b depicts an open system when we open a hole in the upper piston of the container to allow vapor molecules to flow away, and also we applied a constant heat source at the bottom of the container to keep the outgoing vapor at a constant rate, it becomes a steady state process, and it is an irreversible process.

Furthermore, we consider, for example the bilayer thin film reactions between Rd and Si or between Ni and Zr, where amorphous alloy of Rd-Si or Ni-Zr was formed, respectively, upon slow heating rather than rapid quenching of the samples. Because crystalline intermetallic compound phases of Rd-Si or Ni-Zr must have a lower free energy than the amorphous alloy, it is clear that we cannot use maximum free energy change to describe the formation of these amorphous alloys. However, we may be able to use maximum rate of free energy change (or gain) in a short period of time to explain it. We have

$$\Delta G = \int_0^\tau \frac{\Delta G}{\Delta t} dt = \int_0^\tau \frac{\Delta G}{\Delta x}\frac{\Delta x}{\Delta t} dt = \int_0^\tau -Fv dt \tag{9.1}$$

where $\Delta G/\Delta t$ is the rate of Gibbs free energy change, $F$ is the driving force of the reaction or chemical potential gradient, and $v$ is the reaction velocity

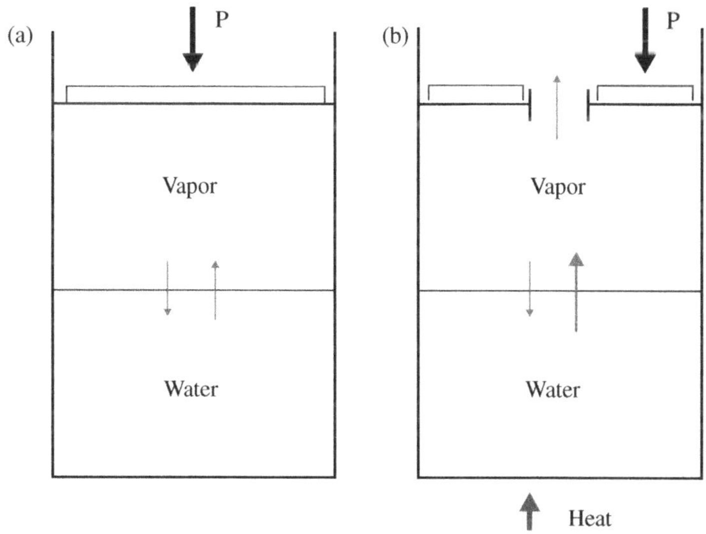


**Figure 9.1** (a) It depicts a closed container keeping a fixed amount of water and vapor at constant temperature and pressure. The water and the vapor are at equilibrium, meaning that on the interface between them, there is an exchange of water molecule and the exchange is at micro-balance. Or the exchange is reversible. (b) It depicts that when we open a hole in the upper part of the container to allow vapor molecules to flow away, and also we applied a constant heat source at the bottom of the container to keep the outgoing flux of vapor at a constant rate, it becomes a steady-state process, or an irreversible process.

or rate of formation of the phase under consideration. For the amorphous alloy, its driving force of formation will be lower than that of the competing crystalline phases. But if its rate of formation (in a short period of time, $\tau$) is greater than that of the crystalline phases, it is possible that the product of $Fv$ for the amorphous alloy is greater than that of the crystalline phase, so the amorphous phase forms.

We can express Gibbs free energy change in a rate process as,

$$\frac{dG}{dt} = -S\frac{dT}{dt} + V\frac{dp}{dt} + \sum_{i}^{j} \mu_i \frac{dn_i}{dt} \tag{9.2}$$

Indeed, we can use rapid quenching (large change in $dT/dt$), rapid mechanical milling (large change in $dp/dt$), and ion implantation (large change in $dn_i/dt$) to produce amorphous or meta-stable materials. On the other hand, when time becomes infinitive, the equilibrium crystalline phase will win.

## 9.2 Flow in Open Systems

In irreversible processes where the flow of matter, heat, or charge is under steady state in open systems, they have been characterized by the rate of entropy change rather than free energy change to be discussed below by an example of heat transfer. In a field-effect-transistor, we pass an electric current or a flow of charges through the device to turn on or turn off the transistor, so it is an open system. While the charges in and out of the device are conserved, the entropy production is not.

We shall use heat flow to illustrate entropy change in a flow process. In Figure 9.2, we consider two heat chambers at temperatures of $T_1$ and $T_2$, where the temperature of $T_1$ is higher than $T_2$. If we connect these two chambers, heat will flow from $T_1$ to $T_2$. We assume a quantity of heat of $\delta Q$ flows from $T_1$ to $T_2$, the change in entropy in the two chambers will be

$$dS_1 = -\frac{\delta Q}{T_1}$$

$$dS_2 = +\frac{\delta Q}{T_2} \tag{9.3}$$

The net change in entropy is equal to

$$dS_{net} = dS_1 + dS_2 = \delta Q\left(\frac{1}{T_2} - \frac{1}{T_1}\right) = \delta Q\left(\frac{T_1 - T_2}{TT_{21}}\right) \tag{9.4}$$

which is positive, so the heat flow has generated a certain amount of entropy. Because the heat flow takes time to go from one chamber to the other, it is a rate process of entropy generation. We shall consider entropy production in

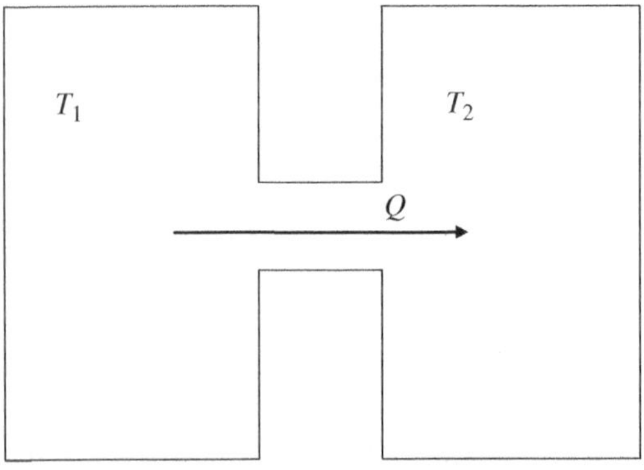


**Figure 9.2** Two heat chambers at temperature of $T_1$ and $T_2$, where the temperature of $T_1$ is higher than $T_2$.

the flow of matter, heat, and electrical charges below. Before we do so, we need to define the flux and its driving force, as well as the cross effects.

There are three kinds of flux or flow, and they are governed by the three well-known phenomenological laws: the flow of matter (Fick's law), the flow of heat (Fourier's law), and flow of electric charge (Ohm's law). In one-dimension, they are given below.

$$J = -D\frac{dC}{dx} = L^{diff}\left(-\frac{d\mu}{dx}\right)$$

$$J_Q = -\kappa\frac{dT}{dx} = \kappa T^2\frac{d(1/T)}{dx} = L^{heat}\frac{d(1/T)}{dx} \tag{9.5}$$

$$j = -\sigma\frac{d\phi}{dx} = \left(\frac{\sigma}{e}\right)\left(\frac{d(e\phi)}{dx}\right) = L^{ele}\left(-\frac{d(e\phi)}{dx}\right)$$

where the flow of matter, $J$, is equal to diffusivity times the gradient of concentration (proportional to the gradient of chemical potential, $\mu$). The flow of heat, $J_Q$, is equal to thermal conductivity times the gradient of temperature. And the flow of charge carrier, $j$, is equal to electrical conductivity times the gradient of electric potential (voltage) or the electric field. The last equation can be written as $E = j\rho$, where $E = -d\phi/dx$ is electric field, $\phi$ is voltage, and $\rho$ is resistivity.

We can consolidate the above three equations into a single equation of

$$J = LX \tag{9.6}$$

For each force $X$, there is a corresponding conjugate primary flow of $J$, and $L$ is the proportional constant or parameter. Typically, we take the negative gradient, $-d\mu/dx$, or $-d(1/T)/dx = -(1/T^2)(dT/dx)$, or $-d(e\phi)/dx$ as the driving force in these flow processes. Onsager called them conjugate forces, and the pairs of conjugate force and the corresponding conjugate flow (or flux) will be given below.

We can use matrix to represent Eq. (9.6),

$$J_i = \sum_j L_{ij} X_j \tag{9.7}$$

where $L_{ii}$ are the coefficients between force and its primary flow, and $L_{ij}$ are the coefficients of cross-effect when $i \neq j$.

In heat conduction, it has been found that temperature gradient can also drive a flow of electric charges, as well as a flow of atoms. These are defined as cross-effects. When a temperature gradient induces an electric charge flux, it is called the thermal-electrical effect or Seebeck effect and the application of this effect in thermal couple for temperature measurement is well known. Reversely, when an electric field induces a heat flow, it is called Peltier effect, which may affect interfacial reactions at the hot end of the device. When a temperature gradient induces an atomic flux, it is called thermomigration, or Soret effect. Indeed, when thermomigration occurs in a solder alloy, it is a concern of solder joint reliability. Another cross-effect which is of keen interest here is electromigration, which means atomic diffusion driven by an electric field under a high electric current density, and it can lead to void formation or open circuit at the cathode end.

When we consider heat flow, temperature is not a constant and is a variable. Thus, in considering thermal-electrical effects or thermomigration, the temperature should be treated as a variable. However, atomic flow and charge carrier flow can occur at a constant temperature, so when we consider the cross-effect between them as in electromigration, we can assume a constant temperature process.

## 9.3 Entropy Production

To consider entropy production in a flow process, we shall deal with inhomogeneous systems. The main thermodynamic supposition for such a system is a postulate of quasi-equilibrium for physically small volumes in the system. Each physically small volume or cell can be considered to be in quasi-equilibrium and its entropy can be determined from the relations of equilibrium thermodynamics by using thermodynamic variables and parameters. According to the first law of thermodynamics, we express a change of the internal energy of an ideal binary system, for example the mixture of an element and its isotope, as

$$dU = TdS - pdV + \sum_{i=1}^{2} \mu_i dn_1 \tag{9.8}$$

We note that when positive work is done by applying a pressure to the system, the volume decreases, so we have the –pdV term which has the negative sign. Then

$$TdS = dU + pdV - \sum_{i=1}^{2} \mu_i dn_i = dH - Vdp - \sum_{i=1}^{2} \mu_i dn_i \tag{9.9}$$

Here $dS$, $dU$, $dH$, $dV$, and $dn_i$ are, respectively, the changes of entropy, internal energy, enthalpy, volume, and number of particles of i-species in the fixed physically small cell. The whole system's entropy (and its change) is defined here as the sum of entropies (and their changes) of all cells.

The change of entropy in each cell can be represented as a sum of two terms: $dS = dS_e + dS_i$. The first term, $dS_e$, is due to divergence of entropy flux (describing the redistribution of already available entropy between cells); it corresponds to the difference between the entropy coming to a cell through its boundaries and the entropy going out from the cell through the boundaries. The second term, $dS_i$, means entropy production within the cell and it is always positive (or zero when the system is in equilibrium and at zero fluxes). In a steady-state process, the sum $dS$ is zero in every cell, but $dS_e$ or $dS_i$ may not be zero. This is because a steady state means that the parameters of each cell are constant in time. So entropy, which is a state function, will remain constant, and $dS$ will be zero. In a steady-state process, when the entropy production, $dS_i$, is not zero, it is compensated by the divergence of entropy flux, $dS_e$, in the cell. For example, in a steady-state process, when heat is generated in a cell, the heat must be transported away or dissipated, otherwise the temperature will increase and it cannot be kept in a steady state.

Onsager showed that in Eq. (9.10) below, the first term is the divergence term and the second term is the rate of entropy production term, which can always be represented as a product of flux $J_i$ and its conjugate force $X_i$.

$$\frac{\partial S}{\partial t} = -div\left(J_S\right) + \frac{1}{T}\Sigma J_i X_i \tag{9.10}$$

It is worthwhile making a comparison between Eq. (9.10) and the atomic diffusion equation or the Fick's second law, which is

$$\frac{\partial C}{\partial t} = -\nabla \cdot J$$

The second law is based on conservation ofmass. On the other hand, if there is a source or sink of atoms, we can add a source/sink term, which is optional and rare, but the last term in Eq. (9.10) on entropy production is build-in.

We consider below entropy production in the three basic processes of flow: electrical conduction, atomic diffusion, and heat conduction. In heat conduction, temperature is not constant and should be treated as a variable. On the other hand, in atomic diffusion and electrical conduction, we can assume a constant temperature process. Below, we first consider electrical conduction, and we note that entropy production is equal to Joule heating.

### 9.3.1 Electrical Conduction

We consider one-dimensional conduction of a pure metal at constant temperature and constant volume. In this case, we have $dn_i = 0$, and $dV = 0$ in Eq. (9.9), so

$$TdS = dU \tag{9.11}$$

To consider the internal energy change in electrical conduction, we begin with the relation in physical unit conversion that "1 newton-meter = 1 joule = 1 coulomb-volt." We recall that the unit of electric energy is eV, which is coulomb-volt. It means that the internal electrical energy change can be expressed as a product of charge and its electrical potential. In electrical conduction, we define, j = current density = ampere/$cm^2$ = coulomb/$cm^2$-sec, and also we define $\phi$ = voltage.

Figure 9.3 depicts electric conduction in a conductor of constant cross-section of A between $x$ and $x + dx$, and the voltage drop is $\Delta\phi$. Thus, $jAt$ gives the charge or coulomb. If we consider a conduction of constant current density ($j$ = constant) across a cell of volume $V = Adx$, which will cause a voltage drop of $\Delta\phi$, the internal energy change is given below.

$$TdS = dU = jAdt\Delta\phi = jAdt\left[\phi(x) - \phi(x+\Delta x)\right]$$

$$= -jVdt\left[\frac{\phi(x+\Delta x) - \phi(x)}{\Delta x}\right] = jVdt\left[-\frac{d\phi}{dx}\right]$$

where $V = Adx$. So we obtain

$$\frac{TdS}{Vdt} = j\left[-\frac{d\phi}{dx}\right] = jE = j^2\rho \tag{9.12}$$

where $E = -d\phi/dx = \rho j$ is electric field, and $\rho$ is resistivity. It is Onsager's equation.

Thus, in this simple case, the rate of entropy production is a product of $jE$, where $j$ is the conjugate electric current density or flux (coulomb/$cm^2$-sec), and $E$ is the conjugate driving force (which is negative gradient of electric potential or the electric field). Also $j^2\rho$ is called "Joule heating" per unit volume per unit time. So entropy production is Joule heating. Its unit is energy/$cm^3$-sec.

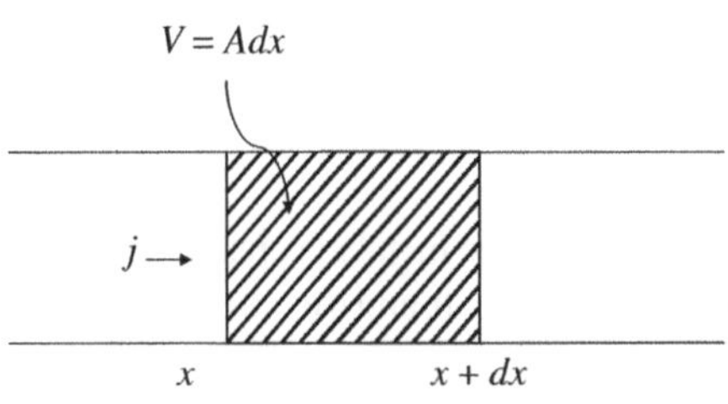


**Figure 9.3** Schematic diagram depicting electric conduction in a conductor of constant cross-section of A between $x$ and $x + dx$, and the voltage drop is $\Delta\phi$.

In Eq. (9.12), we note that Joule heating alone cannot be in a steady state. While the applied current can be steady or the rate of charges being transported is constant, the entropy production increases, so the temperature without heat dissipation will go up. To reach a steady state, the system needs an outgoing heat flux of $J_Q = j^2\rho$, which simultaneously will be accompanied by the outgoing entropy fluxes of $J_Q/T$.

#### 9.3.1.1 Joule Heating

We have shown in the above that entropy production in electric conduction is Joule heat. Usually, the power of Joule heating is written as

$$P = IV_d = I^2R = j^2\rho V \tag{9.13}$$

where $V_d$ (= $IR$) is voltage drop, $I$ is applied current, and $I/A = j$, and $A$ is the cross-sectional area of the sample, $R$ is resistance of the sample, and $R = \rho A/l$, and $l$ is the length of the sample, so the volume $V = Al$. Thus, $I^2R$ is Joule heating per unit time (power = energy/time) of the entire sample, and $j^2\rho$ is Joule heating per unit volume per unit time of the sample. Thus, a low-power device means a low entropy production device.

Because Si devices operate electrically, Joule heating is a built-in cause of reliability. To calculate Joule heating, we consider flip-chip technology, in which typically every two solder joints is connected by an Al interconnect on the chip side, shown in Figure 9.4a. The current density in the Al is about $10^6$ A/cm$^2$ and the current density in the solder joint is about $10^4$ A/cm$^2$. This is because the cross-section of the Al interconnect line is about two orders of magnitude smaller than that of the solder joint. To calculate the Joule heating in them, if we take $j = 10^6$ A/cm$^2$ and $\rho = 10^{-6}$ Ω-cm for Al, we obtain Joule heating of $\rho j^2 = 10^6$ joule/cm$^3$sec. For the solder joint, if we take $j = 10^4$ A/cm$^2$ and $\rho = 10^{-5}$ Ω-cm for SnAgCu solder, we obtain Joule heating of $\rho j^2 = 10^3$ joule/cm$^3$sec, which is much less than that of Joule heating in Al. Figure 9.4b shows the simulation of current density distribution.

The Al interconnect is much hotter than the solder joint, in turn the chip side will be hotter than the substrate side, and consequently there will be a temperature gradient across the solder joint. If the temperature gradient is large enough, thermomigration can occur in the solder joint.

While Joule heating is a kind of waste heat, it cannot be used to do work; however, it generates heat and will increase the temperature of the conductor. Consequently, it will increase the resistance of the conductor and in turn will cause more Joule heating. An increase of temperature of the conductor will lead to thermal expansion of the conductor, and this is the basic reason

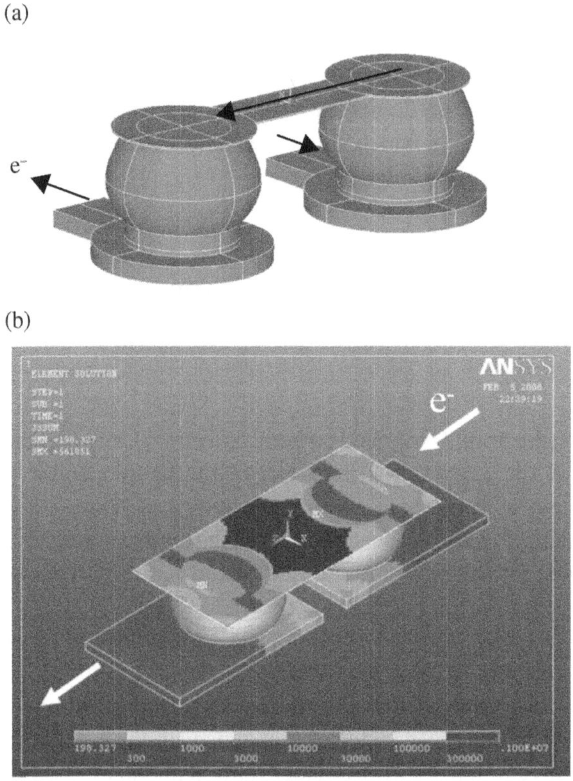


**Figure 9.4** (a) Schematic diagram of flip chip technology, in which typically every two solder joints is connected by an Al interconnect on the chip side. (b) The simulation of current density distribution in the pair of solder joints. The Al interconnect is hotter than the solder joints.

of thermal stress in devices when various materials of different thermal expansion coefficients are integrated. Stress and stress gradient may lead to fracture and creep. When the thermal stress is cyclic because semiconductor devices are being turned on and off frequently, fatigue can happen because of the accumulation of plastic strain energy in the system. Furthermore, the higher temperature caused by Joule heating will increase atomic diffusion, in turn the rate of phase change, interdiffusion, and interfacial reactions in the devices will increase. Thus, Joule heating is a major concern of device reliability, especially in 3D IC packaging where heat dissipation is difficult.

It is worth mentioning the application of Joule heating in fuse to prevent overheating in household appliances. Thin solder strips have been used as fuse. When the appliance draws a current close to 10 Amp, the fuse will melt and the circuit will open. If we assume the fuse has a cross-section of $2 \times 0.1$ mm, the current density in the fuse will be about $5 \times 10^4$ A/cm$^2$. It will lead to melting of the solder fuse. Indeed, when such a high current density was applied to flip-chip solder joints, melting of solder bump was observed. Knowing the heat capacity of the solder, we can calculate the temperature increase due to Joule heating, provided that we know the rate of heat dissipation. Because Si chip itself is a very good heat conductor, this is the reason why Al and Cu interconnects on Si can take a current density close to $10^6$ A/cm$^2$, which is much higher than that in a household wire or an extension cord.

### 9.3.2 Atomic Diffusion

We consider a binary system in isobaric condition,

$$TdS = dH - \sum_{i=1}^{2} \mu_i dn_i \tag{9.14}$$

If we take the binary system to be an element and its isotope or an ideal solution, then $dH = 0$. We have

$$\frac{TdS}{dt} = -\sum_{i=1}^{2} \mu_i \left( \frac{\partial n_i}{\partial t} \right)$$

If we divide the above equation by volume "$V$" and take $n_i/V = C_i$ and for simplicity we recall the continuity equation in one dimension, we have

$$\frac{\partial C_i}{\partial t} = -\nabla \cdot J_i = -\frac{\partial J_i}{\partial x}$$

We use the well-known differentiation of a product of $ydx = d(xy) - xdy$, and we have

$$\frac{TdS}{Vdt} = \sum_{i=1}^{2} \mu_i \frac{\partial J_i}{dx} = -\frac{\partial}{\partial x} \left( -\sum_{i=1}^{2} \mu_i J_i \right) + \sum_{i=1}^{2} J_i \left( -\frac{\partial \mu_i}{\partial x} \right) \tag{9.15}$$

In a steady-state process, the sum of the two terms in the right-hand side of the equation will be zero and they compensate each other. As shown in Eq. (9.10), the first term on the right-hand side of Eq. (9.15) is the divergence term and the second term is the rate of entropy production, which is equal to the product of flux times its driving force.

### 9.3.3 Heat Conduction

We consider one-dimensional heat flux in a pure metal caused by a temperature gradient along x-axis. Let the cell be a thin layer having a cross-sectional area A and a width of $dx$ between $x$ and $x + dx$, shown in Figure 9.5.

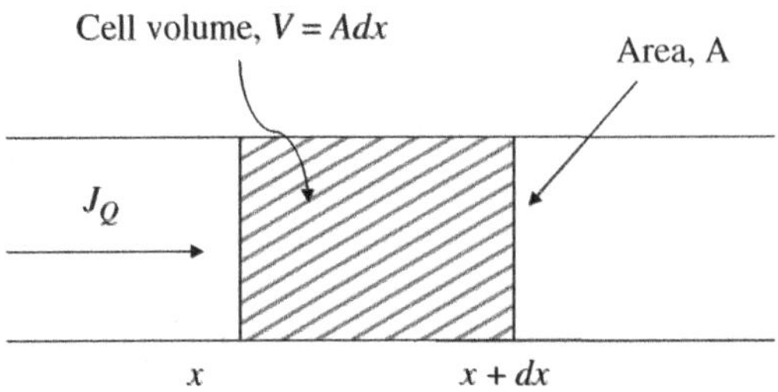


**Figure 9.5** Schematic diagram depicting one-dimensional heat conduction. Let the cell be a thin layer having a cross-sectional area A and a length of $dx$ between $x$ and $x + dx$.

The volume of the cell is $V = Adx$. Assume the process to be isobaric and no atomic flow. Then the change of entropy in the cell is determined by the change of enthalpy, and in turn, this change of enthalpy will be determined by the difference between incoming and outgoing heat fluxes. From Eq. (9.7), we have

$$\begin{aligned} TdS = dH &= J_Q(x)\cdot A\cdot dt - J_Q(x+\Delta x)\cdot A\cdot dt \\ &= -V\frac{J_Q(x+\Delta x)-J_Q(x)}{\Delta x}\cdot dt = -V\frac{\partial J_Q}{\partial x}dt \end{aligned} \tag{9.16}$$

Then we divide both sides by $TVdt$ and again use the well-known differentiation of a product [$ydx = d(yx) - xdy$], we have

$$\frac{\partial S}{Vdt} = -\frac{1}{T}\frac{\partial J_Q}{\partial x} = -\frac{\partial}{\partial x}\left(\frac{J_Q}{T}\right) + J_Q\frac{\partial}{\partial x}\left(\frac{1}{T}\right)$$

In the above, temperature is treated as a variable. Again, according to Eq. (9.10), the ratio of $J_Q/T$ in the first team in the right-hand side of the above equation is an entropy flux, and $\frac{\partial}{\partial x}\left(\frac{J_Q}{T}\right)$ is a divergence of this flux, meaning the rate of entropy change per cell volume caused by the difference of incoming and outgoing entropy fluxes. The second term in the right-hand side of the above equation is the rate of entropy production per unit volume per unit time within the cell. This second term is a product of heat flux and the gradient of inverse temperature, which can be interpreted as the thermodynamic force (or the conjugate force to be discussed in section below) driving the heat flux. Because

$$\frac{\partial}{\partial x}\left(\frac{1}{T}\right) = -\frac{1}{T^2}\left(\frac{\partial T}{\partial x}\right)$$

Thus

$$\frac{TdS}{Vdt} = -T\frac{\partial}{\partial x}\left(\frac{J_Q}{T}\right) - J_Q\frac{1}{T}\left(\frac{\partial T}{dx}\right). \quad (9.17)$$

In a steady-state process, the two terms on the right-hand side of the equation compensate each other. The first term is due to entropy divergence, and the second term is due to entropy production. We note that the dimension of the above equation is correct. It is energy/cm$^3$-sec, and we recall that the unit of heat flux $J_Q$ is energy/cm$^2$-sec.

### 9.3.4 Conjugate Forces When Temperature Is a Variable

What has been shown by the three examples above is that in irreversible processes, $TdS/Vdt$ is equal to the product of flux "$J$" times its driving force "$X$."

From Eq. (9.17), on entropy production, we have

$$\frac{TdS}{Vdt} = J_Q X_Q = -J_Q\left(\frac{1}{T}\frac{dT}{dx}\right) \quad (9.18)$$

So the conjugate force to $J_Q$ is $X_Q = -1/T(dT/dx)$. Thus, we write

$$J_Q = L_{QQ}X_Q = L_{QQ}\left(-\frac{1}{T}\frac{dT}{dx}\right) = -\kappa\frac{dT}{dx} \quad (9.19)$$

we obtain $L_{QQ} = T\kappa$, or the coefficient $L_{QQ}$ is equal to temperature times thermal conductivity.

For the conjugate forces in atomic diffusion and in electrical conduction, we have assumed implicitly that they occur under constant temperature. When we have temperature as a variable, we shall have the following conjugate forces for the three kinds of flows: heat flow, mass flow, and charge flow as below,

$$X_Q = T\frac{d}{dx}\left(\frac{1}{T}\right) \quad (9.20a)$$

$$X_{Mi} = -T\frac{d}{dx}\left(\frac{\mu_i}{T}\right) \quad (9.20b)$$

$$X_E = -T\frac{d}{dx}\left(\frac{\phi}{T}\right) \quad (9.20c)$$

The unit of force in Eqs. (9.20a) and (9.20c) is different from that in Eq. (9.20b). The latter has the unit of a gradient of potential energy, but the former is not. This is because the dimension of the conjugate force is determined by the dimension of the corresponding flux. In heat conduction, the flux has the unit of energy/cm$^2$-sec, so the conjugate force that accompanies the heat flux is $Td(1/T)/dx) = -(1/T)(dT/dx)$, which has the unit of cm$^{-1}$. In atomic diffusion, the flux has the unit of # of particles/cm$^2$-sec, so the conjugate force has the unit of energy/cm. In electrical conduction, the flux has the unit of # of charge particles/cm$^2$-sec, or e/cm$^2$-sec, so the conjugate force has the unit of V/cm and their product is eV/cm$^3$-sec

## 9.4 Cross-Effects in Irreversible Processes

Both electromigration and thermomigration are cross-effects in irreversible processes, because they are fluxes of atoms driven, respectively, by electron flow under an electrical potential gradient or by a heat flow under a temperature gradient. On the other hand, stress-migration is creep, and it is a flux of atoms driven by a stress potential gradient. Thus, creep does not occur under a uniform pressure or hydrostatic pressure; hence, it is an irreversible process too. However, stress potential gradient is part of the chemical potential gradient that drives atomic diffusion, so creep is a primary flow, not a cross-effect. The fundamentals of electromigration, thermomigration, and stress-migration will be presented in later chapters.

Electromigration in Al and Cu interconnects has been the most persistent reliability, problem in very-large-scale integration of circuits in Si technology. Under high current density, above $10^5$ amp/cm$^2$, at the device operation temperature around 100 °C, voids and hillocks have been found to form, respectively, at the cathode and the anode of the Al and Cu interconnects. The study of electromigration in Al short stripes showed that there is a critical length, below which there is no electromigration, as shown in Figure 9.6a and b, in the upper right corner of the figure. Blech and Herring proposed that this is due to the effect of back stress in the Al stripes. The gradient of back stress potential has induced an atomic flux to counter the atomic flux of electromigration. When these fluxes balance each other, there is no electromigration. Below we shall consider the interaction between electromigration and stress-migration by using irreversible processes. It is a cross-effect between atomic flow and charge flow.

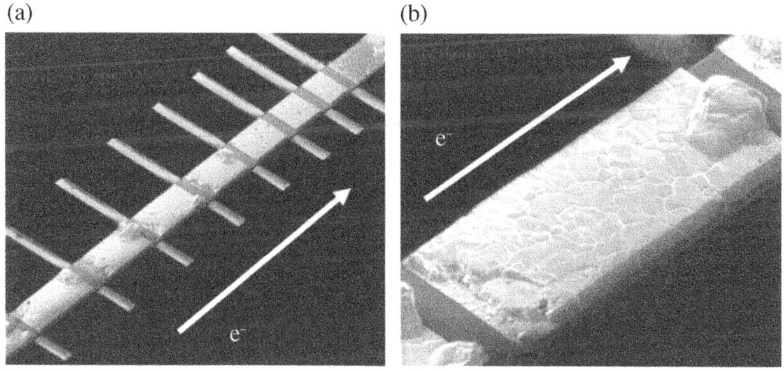


**Figure 9.6** (a) SEM images of a set Al short stripes in electromigration. There is a critical length, below which there is no electromigration, as shown in the upper right corner. (b) SEM image of the enlargement of a stripe in the lower left corner in (a), where depletion at the cathode and hillock formation at the anode are shown.

## 9.5 Cross-Effect Between Atomic Diffusion and Electrical Conduction

Figure 9.7 depicts a schematic diagram of a short Al strip patterned on a baseline of TiN. Since TiN is a poor conductor, electrons will make a detour from the TiN to the Al to reduce the resistance. Under a high current density of electrons going from left to right, electromigration occurs in the Al strips and transports Al atoms from the cathode to the anode, leading to depletion or void formation at the cathode and pile-up or hillock formation at the anode, as shown in Figure 9.7. Hence, the damage of electromigration can be recognized directly using the short strip test samples. The depletion rate at the cathode can be measured, and the drift velocity of electromigration can be deduced. Furthermore, it was found that the longer the strip, the longer the depletion at the cathode side in electromigration, as shown in Figure 9.6a. But below a "critical length," there was no observable depletion.

The dependence of depletion on strip length was explained by the effect of back stress. [5] In essence, when electromigration transports Al atoms in

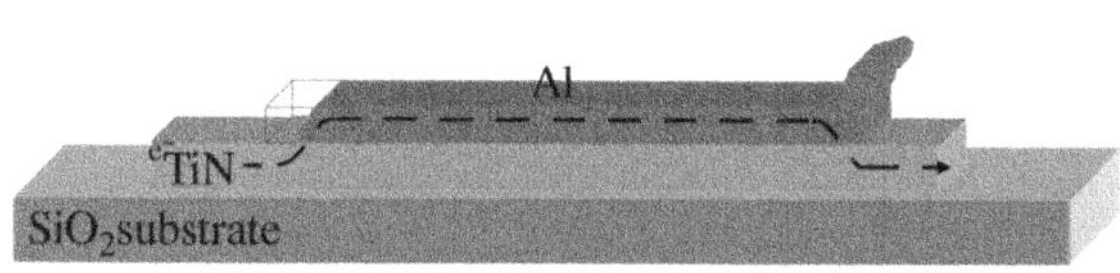


**Figure 9.7** Schematic diagram of a short Al strip patterned on a baseline of TiN.

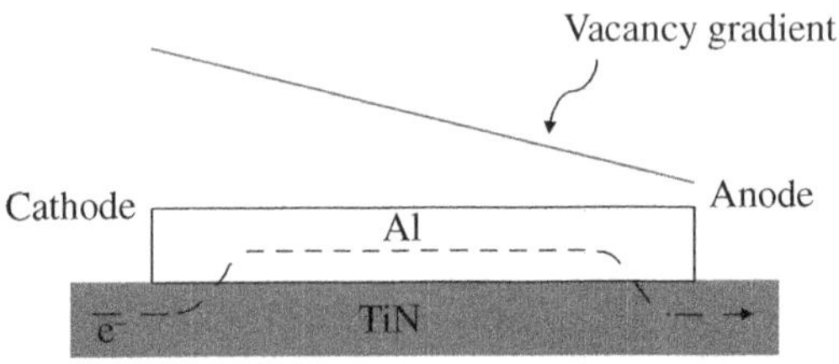


**Figure 9.8** Schematic diagram to depict a vacancy concentration gradient decreasing from the cathode to the anode in the Al strip, due to the back stress in the Al stripe.

a stripe from the cathode to the anode, the latter will be in compression and the former in tension. On the basis of Nabarro-Herring model of equilibrium vacancy concentration in a stressed solid, to be discussed in Chapter 12, the tensile region has more vacancy and the compressive region has less vacancy than the equilibrium vacancy in the unstressed region, so there is a vacancy concentration gradient decreasing from the cathode to the anode in the Al strip, as shown in Figure 9.8. The vacancy gradient induces a flux of Al atoms diffusing from the anode to the cathode, and it opposes the Al flux driven by electromigration from the cathode to the anode. The vacancy concentration gradient depends on the length of the stripe: the shorter the stripe, the larger the slope or the greater the gradient. At a certain short length defined as the "critical length," the vacancy gradient generated by the back stress is large enough to counterbalance electromigration, so there is no depletion at the cathode and no extrusion at the anode occur. [6–11]

To analyze this effect of back stress, we shall use irreversible processes by combining electrical and mechanical forces on atomic diffusion. We assume that these processes occur at constant temperature, so temperature is not a variable. The electrical force proposed by Huntington and Grone is taken below, see Chapter 10. [6]

$$F_{em} = Z^* eE \tag{9.21}$$

where $Z^*$ is the effective charge number, $e$ is electrical charge, and $E$ is electrical field.

The mechanical force is taken as the gradient of chemical potential in a stressed solid, see Chapter 12. Because we consider pure Al interconnect, there is no concentration gradient in Al, so the chemical potential due to concentration change is zero. Instead we have stress potential gradient to drive atomic flux. [5]

$$F_{me} = -\nabla\mu = -\frac{d\sigma\Omega}{dx} \tag{9.22}$$

where $\sigma$ is hydrostatic stress (no shear stress) in the Al, and $\Omega$ is atomic volume, and $\sigma\Omega$ is defined as stress potential. In essence, it is a creep process, in which pressure is not constant but temperature is constant. Thus, we have a pair of phenomenological equations for atomic flux and electron flux below [8],

$$J_{em} = -C\frac{D}{kT}\frac{d\sigma\Omega}{dx} + C\frac{D}{kT}Z^{*}eE \tag{9.23a}$$

$$j = -L_{21}\frac{d\sigma\Omega}{dx} + n\mu_e eE \tag{9.23b}$$

where $J_{em}$ is atomic flux in units of atoms/cm$^2$-sec, and $j$ is electron flux in units of coulomb/cm$^2$-sec. $C$ is the concentration of atoms per unit volume, and $n$ is the concentration of conduction electrons per unit volume. $D/kT$ is atomic mobility, and $\mu_e$ is electron mobility. The unit of $D/kT$ and the unit of $\mu_e$ are basically the same, and they are cm$^2$/eV-sec and cm$^2$/V-sec, respectively, where V is electric potential. $L_{21}$ is the phenomenological coefficient of irreversible processes, and it contains the deformation potential.

In Eq. (9.23a), the first term in the right-hand side is creep, and it is the primary flux, and the second term is due to electromigration and it is the cross-effect of electrical force on atomic diffusion. In Eq. (9.23b), the last term is the primary flux in electric conduction and the first term is the cross-effect of mechanical force on electrical conduction.

### 9.5.1 Electromigration and Stress-Migration in Al Strips

In Eq. (9.23a), if we take $J_{em} = 0$, i.e. there is no net electromigration flux or no damage. The expression for the "critical length" is obtained as

$$\Delta x = \frac{\Delta\sigma\Omega}{Z^{*}eE} \tag{9.24}$$

Because the resistance of the conductor can be taken to be constant at constant temperature, we have instead the "critical product" or "threshold product" of "$j\Delta x$" by moving the current density from the right-hand side to the left-hand side of the equation, and we recall that $E = \rho j$.

$$j\Delta x = \frac{\Delta\sigma\Omega}{Z^{*}e\rho} \tag{9.25}$$

In the above equation, all the parameters on the right-hand side are given for a conductor. Thus, we can calculate the critical product. For Al and Cu interconnects, we take $j = 10^6$ A/cm$^2$ and $\Delta x = 10\,\mu$m, we have a typical value of critical product to be about 1000 A/cm.

The critical length can be measured experimentally by extending the time of electromigration to a sufficient long period until the mass transport in the stripe ceases, as shown in Figure 9.6. We can use Eq. (9.24) to calculate the critical length $\Delta x$ of Al stripe. If we assume that the extrusion at the anode is accompanied by a certain amount of plastic deformation, the change in elastic stress can be taken to be the value corresponding to the elastic limit. For Al, we have $\sigma_{Al} = -1.2 \times 10^9$ dyn/cm$^2$, $\Omega_{Al} = 16 \times 10^{-24}$ cm$^3$, $e = 1.6 \times 10^{-19}$ C, and $E = j\rho$, where $j = 3.7 \times 10^5$ A/cm$^2$ and $\rho_{Al} = 4.15 \times 10^{-6}$ Ω-cm at 350 °C. By substituting these values into Eq. (9.19), we obtain

$$\Delta x_{Al} = -\frac{78\mu m}{Z^*}$$

By taking $Z^* = -26$ for bulk Al, we have the critical length to be 3 μm, which is of the right order of magnitude but shorter than the experimental finding of 10–20 μm. Because the Al strips are polycrystalline thin films, grain boundary diffusion has played a dominant role in electromigration, so the $Z^*$ for atoms diffusing in grain boundaries should be different from that in the bulk or lattice diffusion. It seems that for polycrystalline Al thin films, $Z^*$ is below 10.

In Eq. (9.23b), if we take the coefficient $L_{21} = ne\mu_e N^*$ where $N^*$ is a parameter to be discussed below, and if we consider a short stripe deposited on an insulating substrate and take $j = 0$, we have

$$N^* = -\frac{1}{\Omega}\left|\frac{d\varphi}{d\sigma}\right| \text{at } j = 0 \tag{9.26}$$

where $d\phi/d\sigma$ at $j = 0$ is deformation potential, which is defined as the electrical potential per unit stress difference at zero current. By using Onsager's reciprocity relation, $L_{12} = L_{21}$, we obtain the expression of

$$\frac{d\phi}{d\sigma} = -\frac{Z^* D\rho e}{kT} \tag{9.27}$$

The dimension of $d\phi/d\sigma$ and $N^*$ are cm$^3$/C and C$^{-1}$, respectively.

To calculate the deformation potential as shown in Eq. (9.27), we take $Z^* = -26$ for Al at $T = 500$ °C, $kT = 0.067$ eV, lattice diffusivity of Al at 500 °C to be about $2 \times 10^{-10}$ cm$^2$/sec, and resistivity to be about $4.83 \times 10^{-6}$ Ω-cm, we obtain

$$\frac{d\phi}{d\rho}_{j=0} = 3.7x10^{-13} cm^3/C$$

However, we expect the order of magnitude of deformation potential to be close to Ω/e, which is about $10^{-4}$ cm$^3$/C or $10^{-10}$ V/Nm$^{-2}$. It seems that what we have calculated on the basis of Eq. (9.27) is too small. This is because the deformation process involved in Eq. (9.7) is by creep, which depends on a long-range atomic diffusion, and it is a very slow deformation process. On the other hand, the mechanical deformation in the usual sense does not involve thermally activated atomic diffusion, rather than atomic displacement under an applied mechanical stress proceeds by the acoustic mode of atomic motion, i.e. at the speed of sound, which is about $10^5$ cm/sec or equal to $a_0\upsilon$, where $a_0 = 3 \times 10^{-8}$ cm is the interatomic distance and $\upsilon = 10^{13}$ $sec^{-1}$ is atomic vibration frequency. Thus, we take $D = (a_0)^2\upsilon$ in Eq. (9.27), we have $\frac{d\varphi}{d\rho}_{j=0} = 0.2x10^{-4} cm^3/C$, which is of the expected order of magnitude. Hence, the deformation potential due to creep is indeed very small.

## 9.6 Irreversible Processes in Thermomigration

When a temperature gradient is applied to a homogeneous alloy, the alloy becomes inhomogeneous by phase separation. This is known as Soret effect. One of the component in the alloy will diffuse against its concentration gradient and leads to phase separation. Finally, it establishes a concentration gradient and reaches a steady state. Thus, in thermomigration we shall consider the interaction between atomic flow and heat flow, and the corresponding conjugate forces are $X_M$ and $X_Q$ as shown in Eqs. (9.28a) and (9.28b), respectively. $J_M$ is atomic flow and $J_Q$ is heat flow.

$$J_M = L_{MM}X_M + L_{MQ}X_Q$$
$$= C\frac{D}{kT}\left[-T\frac{d}{dx}\left(\frac{\mu}{T}\right)\right] + C\frac{D}{kT}\frac{Q^*}{T}\left(-\frac{dT}{dx}\right) \tag{9.28a}$$

$$J_Q = L_{QM}X_M + L_{QQ}X_Q = L_{QM}\left[-T\frac{d}{dx}\left(\frac{\mu}{T}\right)\right] + \kappa\frac{dT}{dx} \tag{9.28b}$$

where $Q^*$ is heat of transport, which is positive when the flux of atoms is diffusing from the hot end to the cold end because $dT/dx$ is negative, and it is negative when atoms are driven from the cold end to the hot end because $dT/dx$ is positive. The last term in the equation of $J_Q$ is Fourier's law and $\kappa$ is thermal conductivity.

In the equation of $J_M$, if we let $J_M = 0$, we obtain

$$-T\left(\frac{1}{T}\frac{d\mu}{dx} - \frac{\mu}{T^2}\frac{dT}{dx}\right) = \frac{Q^*}{T}\frac{dT}{dx}$$

Rearrange the terms. We have

$$\frac{d\mu}{dT} = \frac{\mu - Q^*}{T}$$

It shows that when $\mu - Q^* = 0$, $\Delta\mu = 0$.

### 9.6.1 Thermomigration in Unpowered Composite Solder Joints

Figure 9.9a, b, and c show, respectively, the schematic diagram of a flip chip on a substrate, the cross section of a composite flip-chip solder joint between the chip and the substrate, and a SEM image of the cross-section of a flip-chip solder joint. In Figure 9.9a, the small squares on the substrate are electrical contact pads. The composite solder joint was composed of 97Pb3Sn on the chip side and eutectic 37Pb63Sn on the substrate side. The solder bump has a height of 105 μm. The contact opening on the chip side has a diameter of 90 μm. The tri-layer thin films of under-bump-metallization (UBM) on the chip side were Al (~0.3 μm)/Ni(V) (~0.3 μm)/Cu (~0.7 μm). On the substrate side, the bond-pad metal layers were Ni (5 μm) coated with a thin film of Au (0.05 μm).

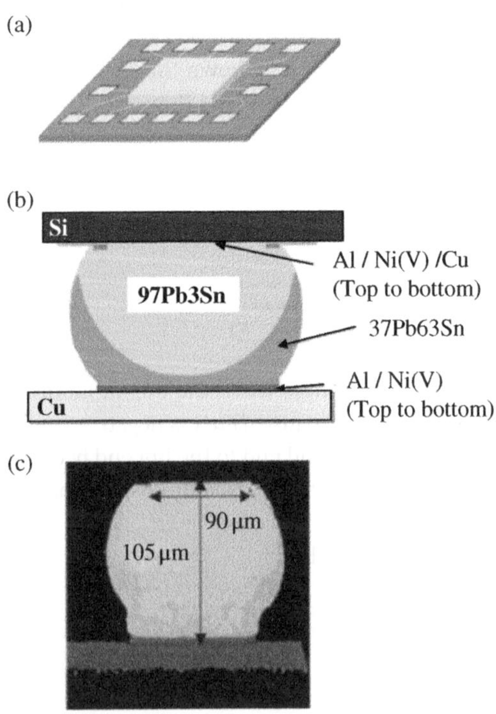


**Figure 9.9** (a) Schematic diagram of a flip chip on a substrate, (b) the cross section of a composite flip-chip solder joint between the chip and the substrate, and (c) SEM image of a composite flip-chip solder joint.

As a control experiment, a constant temperature annealing of the flip-chip composite samples was performed in an oven at 150 °C and atmospheric pressure for a period of one, two weeks, and one month. The microstructures of the cross section of the composite solder joint were examined under optical microscope and scanning electron microscope, as shown in Figure 9.10. Composition was analyzed using energy dispersive X-ray (EDX) and electron probe microanalysis (EPMA). Neither mixing nor homogenization between the high-Pb and the eutectic was observed after the one-month annealing, and the image was nearly the same as that shown in Figure 9.10a. The reason of no mixing is given below. According to the eutectic phase diagram of Sn-Pb, at 150 °C, the chemical potential of the high-Pb phase of 97Pb3Sn and the eutectic phase of 37Pb63Sn is about the same. Thus, there is no driving force for intermixing.

To detect thermomigration by using the temperature gradient induced from Joule heating, a set of 24 bumps on the peripheral of the Si chip was tested. Figure 9.11a depicts a row of 24 solder bumps from right to left at the peripheral of a chip and each bump has the original microstructure as shown in Figure 9.9c before electromigration stressing to be discussed below. We recall that the darker region in the bottom area of each bump is the eutectic SnPb and the brighter region in the top part is 97Pb3Sn.

Electromigration was conducted through only four pairs of bumps in the row of 24 bumps on the peripheral of the chip. They were the pairs of No. 6/7, 10/11, 14/15, and 18/19 as numbered in Figure 9.11a. The arrows indicated the electron path. The electron current went from one of the contact pads to the bottom of one of the bump, up the bump to the Al thin film interconnect on the Si stripe, then went to the top of the next bump, down the bump, and to the other contact pad on the substrate. It is worth noting that we can pass current through just one pair of bumps or several pairs in the row to conduct electromigration. The Joule heating from the Al thin film line on the stripe is the heat source. Due to the excellent thermal conduction of Si, the neighboring unpowered solder joints should have experienced a thermal gradient similar to the pairs stressed by current.

Cross-sectioning was performed after the bump-pair 10 and 11 failed after five hours of current stressing at $1.6 \times 10^4$ A/cm$^2$ at 150 °C. To study thermomigration, the unpowered neighboring bumps were examined. The effect of thermomigration is clearly visible across the entire row of the unpowered solder joints, as shown in Figure 9.11b. Because in all of them Sn has migrated to the Si side, the hot end, and Pb has migrated to the substrate side, the cold end. The redistribution of Sn and Pb or the redistribution of the eutectic phase and the high-Pb phase was caused by temperature gradient across the solder joints because no current was applied to them. For the

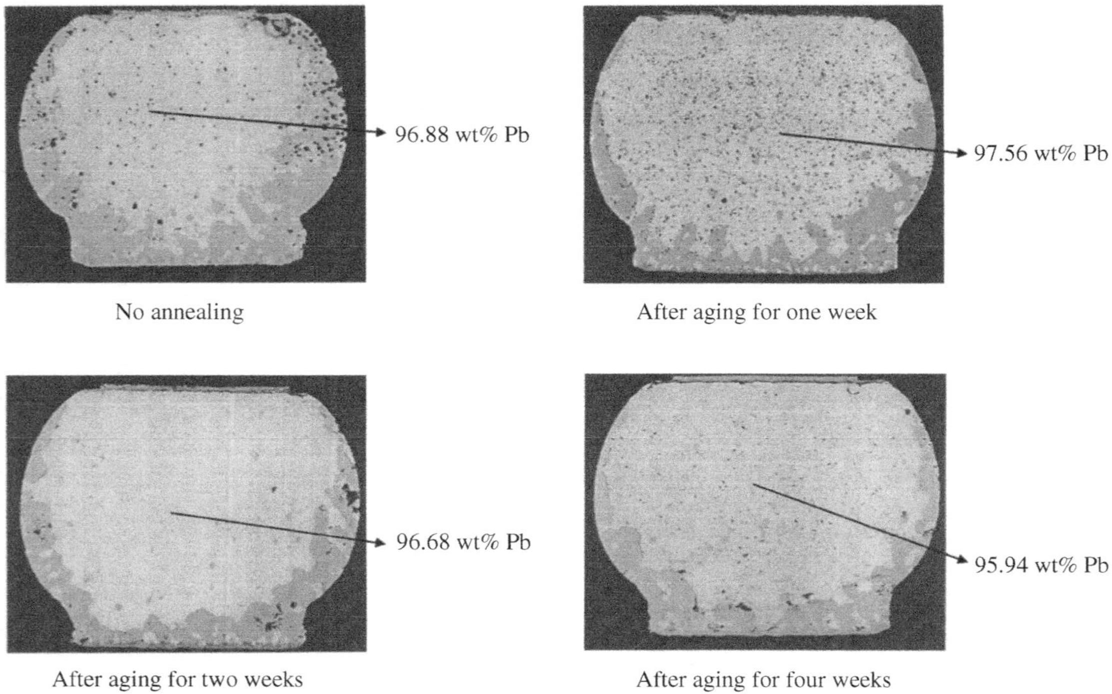


**Figure 9.10** SEM cross-sectional images of composite solder bump before and after annealing at constant temperature of 150 °C for one and two weeks, and one month.

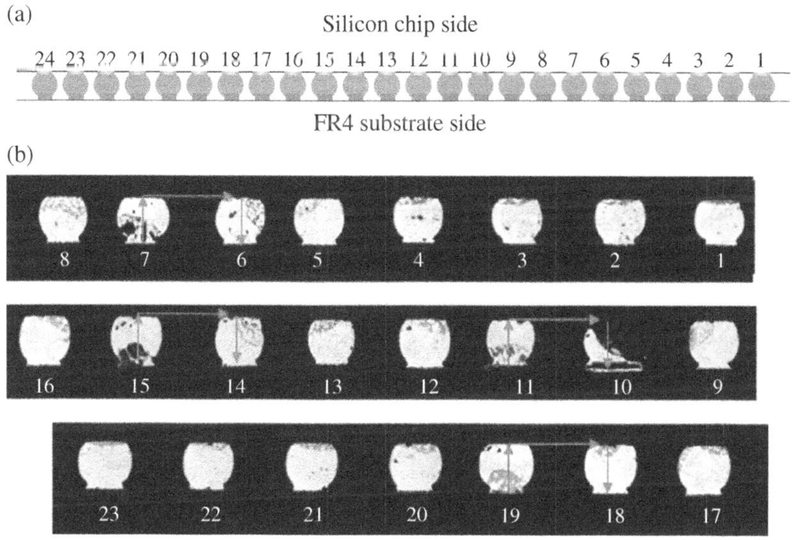


**Figure 9.11** (a) Schematic diagram depicts a row of 24 solder bumps from right to left at the peripheral of a chip. (b) SEM images of the row of solder bump after the pairs of 6/7, 10/11, 14/15, and 18/19 were stressed by at $1.6 \times 10^4$ A/cm$^2$ at 150 °C. The other unpowered bumps showed redistribution of Sn and Pb due to thermomigration.

unpowered bumps that were the nearest neighbor of the powered bumps, the Sn redistribution is also tilted toward the powered bumps. For example, the powered bump 10 is to the left of the unpowered bump 9, the Sn-rich region in bump 9 is tilted to the left and a void is observed. Then, the powered bump 15 is to the right of the unpowered bump 16, the Sn-rich region in bump 16 is tilted to the right. In those bumps further away from the powered bumps, for example from bump 1 to bump 4 and bump 21–23, Sn accumulated rather uniformly on the Si side.

## 9.7 Cross-Effect Between Heat Conduction and Electrical Conduction

Cross-effect between heat flow and electrical flow is the thermal-electrical effect, especially the Seebeck effect and the Peltier effect are well known. The Seebeck effect is the generation of an electrical flow or electrical potential by a temperature gradient. It is the basis of applications of thermal couple to measure temperature of materials. The Peltier effect is the generation of a heat flow by an electrical potential gradient. It is the theoretical basis of solid-state cooling devices.

The interaction between heat conduction and electrical conduction can be represented by a pair of equations of irreversible processes below. Because temperature is not constant, it becomes a variable in the forces in these two equations. $J_Q$ and $J_E$ are heat flow and charge flow, respectively.

$$J_Q = L_{QQ}X_Q + L_{QE}X_E = L_{QQ}T\frac{d}{dx}\left(\frac{1}{T}\right) - L_{QE}T\frac{d}{dx}\left(\frac{\varphi}{T}\right)$$

$$J_E = L_{EQ}X_Q + L_{EE}X_E = L_{EQ}T\frac{d}{dx}\left(\frac{1}{T}\right) - L_{EE}T\frac{d}{dx}\left(\frac{\varphi}{T}\right) \tag{9.29}$$

where $L_{QQ} = T\kappa$, $L_{EE} = ne\mu_e$, and $L_{QE}$ and $L_{EQ}$ are coefficients of cross-effects. $L_{QE}$ is for heat flow induced by an electrical field, and $L_{EQ}$ is for the electrical flow induced by a temperature gradient. On the basis of Onsager's reciprocity relationship, $L_{QE} = L_{EQ}$. We shall analyze the above two equations for the understanding of thermal-electric effects. We can rewrite the pair of equation in Eq. (9.29) in terms of $dT/dx$ and $d\phi/dx$ as below.

$$J_Q = \left(-L_{QQ} + \phi L_{QE}\right)\frac{1}{T}\left(\frac{dT}{dx}\right) - L_{QE}\frac{d\phi}{dx} \tag{9.30}$$

$$J_E = \left(-L_{EQ} + \phi L_{EE}\right)\frac{1}{T}\left(\frac{dT}{dx}\right) - L_{EE}\frac{d\phi}{dx}$$

### 9.7.1 Seebeck Effect

Figure 9.12a depicts a single metal wire of a given length, with its two ends kept at two temperatures, $T_1$ and $T_2$, where $T_1 > T_2$. Thus, there is a temperature gradient, and we expect it to drive an electrical flow of charge carriers. Since the wire is open ended, no electric flow will occur. Thus, in the second equation in Eq. (9.30), we have $J_E = 0$ and

$$0 = \left(-L_{EQ} + \varphi L_{EE}\right)\frac{1}{T}\left(\frac{dT}{dx}\right) - L_{EE}\frac{d\varphi}{dx}$$

Thus, we have

$$\frac{\Delta\varphi}{\Delta T} = \frac{\left(-L_{EQ} + \varphi L_{EE}\right)}{TL_{EE}} \tag{9.31}$$

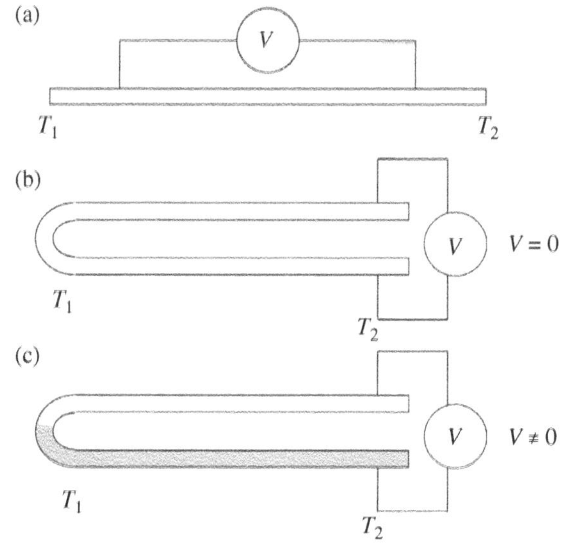


**Figure 9.12** (a) Schematic diagram depicts a single metal wire of a given length, with its two ends kept at different temperatures, $T_1$ and $T_2$, where $T_1 > T_2$. (b) Schematic diagram of a bent wire having two branches The bent end is placed at $T_1$ and the other two ends at $T_2$. The potential difference across the two ends at $T_2$ is zero, $V = 0$. (c). Schematic diagram of two kinds of metallic wires, A and B, joined at one end. The joined end is kept at $T_1$, and the unjoined ends at $T_2$. There is a potential difference between the two open ends at $T_2$. The joined end is the probe in a thermo-couple.

This is called Thomson effect. It means that there will be an electrical potential difference between the ends due to the temperature gradient. In the above, we obtain $\Delta\phi/\Delta T$ by assuming $J_E = 0$.

However, if we applied an electric current through the wire in Figure 9.12a, the applied potential will increase in one direction but will decrease if it is applied in the opposite direction. In addition to the conventional Joule heating when electrical conduction occurs at constant temperature, there will be an addition or reduction in Joule heating due to the Thomson effect.

Now, we double the length of the wire and bend it at the middle into two branches and place the bent end at $T_1$ and the other two ends at $T_2$, as shown in Figure 9.12b. While there will be an electric potential difference between the end at $T_1$ and the two ends at $T_2$, there will be no net potential difference between the two ends kept at $T_2$ because the potential changes in both branches are the same.

However, if we replace one branch of the bent wire by a different metal, or if we take two kinds of metallic wires, A and B, and join them at the end kept at $T_1$, as shown in Figure 9.12c, we have a thermal couple. If we place the joined end at a high temperature and the unjoined ends of the couple at a reference temperature of 0 °C or room temperature, we have a thermal couple, and we can measure a potential difference $\Delta\phi$ between the two ends kept at the reference temperature. This is because the potential changes in the two wires are not the same, so a potential difference occurs. If we have calibrated the couple at different temperatures, we obtain the Seebeck coefficient,

$$\frac{\Delta\phi}{\Delta T} = \varepsilon_{AB} = \varepsilon_A - \varepsilon_B$$

where $\varepsilon_{AB}$ is the combination of thermal electric properties of A wire and B wire in the couple as shown in Eq. (9.30). For convenience, thermal electric properties of many pairs of individual material of $\varepsilon_A$ and $\varepsilon_B$, have been measured. So, we can choose a pair to serve as thermal couple to measure temperature in various temperature ranges.

### 9.7.2 Peltier Effect

Peltier effect is opposite to Seebeck effect. If we keep the two ends of the sample as shown in Figure 9.12c at constant temperature, and if we applied an electric field, we can transfer heat from one end to the other end. If we let $J_Q = 0$ in the first equation in Eq. (9.30), we have

$$0 = \left(-L_{QQ} + \phi L_{QE}\right)\frac{1}{T}\frac{dT}{dx} - L_{QE}\frac{d\phi}{dx}$$

Thus, we have

$$\frac{\Delta T}{\Delta\phi} = \frac{TL_{QE}}{\left(-L_{QQ} + \phi L_{QE}\right)} \tag{9.32}$$

In the above equation, $\Delta T$ is proportional to $\Delta\phi$ and $L_{QE}$. The larger the $\Delta T$, the better the cooling effect. However, due to the existence of $\Delta T$, heat will be transferred from the hot end to the cold end and will reduce the cooling effect. In order to decrease the heat transfer, we will need to reduce heat conductivity in the sample. Yet, for most conductors, heat conductivity is linearly proportional to electric conductivity. In the applications of Peltier effect, the challenge is to find a conductor, which has good electrical conduction, but poor heat conduction.

From the view point of reliability, the cross-effect between atomic diffusion and electrical conduction or between atomic diffusion and heat conduction is much more important than the cross-effect between heat conduction and electrical conduction. Therefore, we will not discuss more about the thermal-electric effect, which actually has been covered in many books and review articles.

## Problems

**1** What is the physical meaning of entropy in thermodynamics?

**2** Derive the equation of entropy production in electrical conduction.

**3** In Figure 9.1(a), how can we define pressure in the closed container? Then try to derive the ideal gas law of $pV = RT$.

**4** Consider electrical conduction of a Cu wire, which is 10 cm long and has a cross-section of 100 $\mu m^2$. When we apply a current density of $10^5$ A/cm$^2$, what is the Joule heating?

## References


1 Prigogine, I. (1967). *Introduction to Thermodynamics of Irreversible Processes*, 3e. New York: Wiley-Interscience.
2 Ragone, D.V. (1995). *Thermodynamic of Materials*, Chapter 8 on "Nonequilibrium Thermodynamics", vol. Volume II. New York: Wiley.
3 Balluffi, R.W., Allen, S.M., and Carter, W.C. (2005). *Kinetics of Materials*, Chapter 2 on "Irreversible Thermodynamics: Coupled Forces and Fluxes". New York: Wiley-Interscience.
4 Tu, K.N. (2011). *Electronic Thin Film Reliability*, Chapter 10 on "Irreversible Processes in Interconnect and Packaging Technology". Cambridge, UK: Cambridge University Press.
5 Blech, I.A. and Herring, C. (1976). Stress generation by electromigration. *Applied Physics Letters* 29: 131–133.
6 Huntington, H.B. and Grone, A.R. (1961). Current-induced marker motion in gold wires. *Journal of Physics and Chemistry of Solids* 20: 76.
7 Kirchheim, R. (1992). Stress and electromigration in Al-lines of integrated circuits. *Acta Metallurgica et Materialia* 40: 309–323.
8 Tu, K.N. (1992). Electromigration in stressed thin films. *Physics Review* B45: 1409–1413.
9 Korhonen, M.A., Borgesen, P., Tu, K.N., and Li, C.-Y. (1993). Stress evolution due to electromigration in confined metal Lines. *Journal of Applied Physics* 73: 3790–3799.
10 Clement, J.J. and Thompson, C.V. (1995). Modeling electromigration-induced stress evolution in confined metal lines. *Journal of Applied Physics* 78: 900.
11 Wang, P.C., Cargill, G.S. III, Noyan, I.C., and Hu, C.K. (1998). Electromigration-induced stress in aluminum conductor lines measured by x-ray microdiffration. *Applied Physics Letters* 72: 1296.

# 10

# Electromigration

## 10.1 Introduction

In a Si chip of the size of our finger nail, there are billions of transistors, and the total length of Cu interconnects is over 10 kilometers in order to connect all the transistors to form the very-large-scale integration of circuits. So far, electromigration has been the most serious and persistent reliability problem in the interconnect technology. This is because typically a current density of $10^5$ to $10^6$ A/cm$^2$ is conducted by the Cu line. Under such a high current density, atomic diffusion and rearrangement are enhanced, leading to void formation (open) in the cathode and extrusion formation (short) in the anode of the interconnect. As the trend of device miniaturization demands smaller and smaller components, the cross-section of nanoscale interconnect keeps shrinking and the current density keeps increasing, so does the probability of circuit failure induced by electromigration. This is why electromigration remains the most important reliability subject, which has attracted much attention.

For comparison, no electromigration occurs in an ordinary extension cord used at home and in laboratories. The electric current density in the cord is low, about $10^2$ A/cm$^2$, and also the ambient temperature is too low for atomic diffusion to occur in Cu wire in a cord coated with rubber. However, because Si itself is an excellent thermal conductor, the Joule heating can be conducted away, so the current density in the Cu interconnect can go up to $10^5$ A/cm$^2$.

The free electron model of conductivity of normal metals assumes that the valence electrons are free to move in the metal, unconstrained by the perfect lattice of atoms except for scattering due to phonon vibration and structural defects such as vacancies, impurities, dislocations, and grain boundaries. The scattering is the cause of electric resistance.

*Electronic Packaging Science and Technology*, First Edition. King-Ning Tu, Chih Chen, and Hung-Ming Chen.

When an atom is out of its equilibrium position, for example a diffusing atom at the activated state, it possesses a very large scattering cross-section, and so a very large resistance. Nevertheless, when the electric current density is low, the scattering or the momentum exchange between the electrons and the diffusing atoms does not enhance the displacement of the latter, so it has very little effect on atomic diffusion. However, the scattering by electrons in a high current density, above $10^4$ A/cm$^2$, can enhance atom diffusion along the direction of electron flow. The enhanced atomic displacement and the accumulated effect of mass transport under the influence of an electric field (mainly due to a high density of electric current rather than a high electric voltage) is called electromigration.

Furthermore, in a device having a very dense integration of circuits, such as in a 3D IC device, the effect of heat management or heat removal on electromigration is the most serious yield and reliability issue, and it is becoming the limiting factor in VLSI in the near-future microelectronic technology. [1] Heat affects electromigration, and there is a positive feedback between them to be discussed Chapter 13. Typically, a server is cooled in an air-conditioned room in order to maintain the device working temperature at about 100 °C, but in a mobile device, it is hard to do so.

## 10.2 To Compare the Parameters in Atomic Diffusion and Electric Conduction

Because electromigration is the interaction between atomic diffusion and electron flow, we need to consider both atomic flux and electron flux, as well as their coupling due to the cross-effect in irreversible processes. It is helpful to make a direct comparison of the parameters used to define these two kinds of flux. In Table 10.1, we list them side-by-side for comparison.

In atomic diffusion, we have defined "driving force" as negative chemical potential energy gradient, so the chemical force acting on the diffusing atom is given as $F = -d\mu/dx$, where $\mu$ is chemical potential energy. In electrical conduction, $E = -d\psi/dx = -dV/dx$, and $E$ is electric field, rather than electric force. Electric force is given as $eE$. It means that if we place a charge of "$e$" in the electric field of $E$, the charge will feel a force of $eE$ acting on it. We recall that $\psi$ or $V$ is defined as electric potential, which is not electric potential energy. Electric potential energy is defined as $e\psi$ or $eV$. Again we note that the unit of thermal energy is given as $kT$ and the unit of electric energy as $eV$. In atomic diffusion, the activation energy can be given as kcal/mole or eV/atom.

**Table 10.1** Comparison between atomic flux and electron flux.

| **Atomic flux** | **Electron flux** |
|---|---|
| Chemical potential: μ | Electric potential: Φ<br>Applied voltage: $V$ |
| Chemical force: $F = -\frac{\partial \mu}{\partial x}$ | Electric field: $E = -\frac{\partial \phi}{\partial x} = -\frac{\partial V}{\partial x}$ |
| Mobility: $M = \frac{D}{kT}$ | Electron mobility: $\mu = \frac{e\tau}{m}$ |
| Drift velocity: $v = MF$ | Drift velocity: $v = \mu E$ |
| Atomic flux: $J = C_v = CMF$ | Electron flux: $j = nev = ne\mu E = \frac{E}{\rho}$<br>(electron current density) |
| Viscosity (friction coefficient): $1/M$ | Resistivity: $\rho = \frac{1}{ne\mu} = \frac{m}{ne^2\tau}$ |
| Divergency: $\nabla \cdot j = \frac{\partial J}{\partial x} + \frac{\partial J}{\partial y} + \frac{\partial J}{\partial z} = -\frac{\partial C}{\partial t}$ | Divergency (Gauss' theorem):<br>$\nabla \cdot j = -\frac{\partial (ne)}{\partial t}$ |

To compare mobility of a charge and mobility of an atom, their unit should be the same. Charge carrier mobility $\mu_e$ is given by $v/E$, where $v$ is velocity and $E$ is electric filed, so its unit is cm$^2$/V-sec. Also, charge carrier mobility $\mu_e = e\tau/m^*$, where $\tau$ is scattering time and $m^*$ is electron mass. Using Newton's law that $F = ma$, we have the unit of mass as force/acceleration, which is equal to energy-sec$^2$/cm$^2$ or eVsec$^2$/cm$^2$. Again we obtain the unit of charge carrier mobility ($= e\tau/m^*$) to be cm$^2$/V-sec. On the other hand, atomic mobility is $D/kT$, where $D$ is diffusivity, which has the unit of cm$^2$/sec and $kT$ is thermal energy, so the unit of atomic mobility is cm$^2$/J-sec or cm$^2$/eV-sec. In comparison between charge mobility and atom mobility, the charge "$e$" has been cancelled in carrier mobility.

On electric current and electric current density, we note that electric current is defined as the total number of electrons (or charges) passing through the cross-section of area A per unit time of the conductor. Thus, $I/A = j$, where $j$ is defined as electric current density. The unit of $I$ is amp or coulomb/sec, and the unit of $j$ is A/cm$^2$ or coulomb/cm$^2$sec. Basically, coulomb is the unit of static charge, and amp is the unit of moving charge, so coulomb/sec = amp.

For comparison, in atomic diffusion, the atomic flux is defined as $J$ = number of atoms/cm$^2$sec. In electric conduction, $j$ or current density is defined as $j$ = number of charge/cm$^2$sec = coulomb/cm$^2$sec = A/cm$^2$.

## 10.3 Basic of Electromigration

The driving force and kinetics of electromigration were analyzed by Huntington. [2] Figure 10.1a and b are schematic diagrams depicting the diffusion of an atom in a face-centered cubic lattice. It goes across the activated state and exchanges position with the neighboring vacancy. Under the scattering or collision between a high current density of electrons and the diffusing atom, the latter is enhanced in its rate of exchanging position with the vacancy and in going to a new equilibrium position. We note that the diffusing atom has a high collision cross-section at the activated state because it is away from its equilibrium position and it has pushed its nearest neighboring atoms apart, so the collision cross-section is greatly increased by about a factor of 10. The force of the collision is defined as "electron wind force." Many studies on electromigration have followed Huntington's model. [3–7]

The electron wind force of electromigration is $F = Z^*eE = Z^*e\rho j$. We recall that when there is an electric field of $E$, if we place a charge "$e$" in the field, the charge will experience a driving force of $eE$ to move in the field direction because the field is a vector. Now if we place an atom in the electric field, the atom will experience a driving force of $Z^*eE$ and we may regard $Z^*e$ as the effective number of charges of the atom in electromigration. Typically, for metallic conductors such as Al and Cu, their $Z^*$ is about one order of magnitude larger than the number of valence electrons of the metal atom. Therefore, $Z^*$ for Al is about 30, and $Z^*$ for Cu is about 10. Physically, we expect so because the cross-section of collision of a diffusing atom at the activated state is about 10 times larger than that of an atom in the equilibrium state.

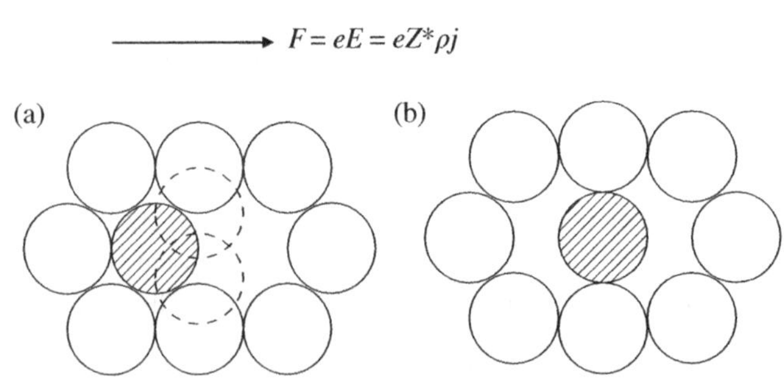


**Figure 10.1** Schematic diagram depicting atomic diffusion in a face-centered-cubic structure to go across the activated state. (a) The atom on the left-hand side of the vacancy attempts to exchange position with the vacancy. It is being helped by the flow electrons from left to right. (b) The jumping atom is at the activated state, and it has a much larger cross-section of scattering than an atom at an equilibrium state.

No doubt, the effect of electron wind force on the atomic diffusion covers the entire period from the very beginning of its motion to the end of the exchange of position between the atom and the vacancy, not just when the atom is at the activated state.

### 10.3.1 Electron Wind Force

In electromigration, the force acting on a diffusing atom is taken to be

$$F_{em} = Z^* eE = \left(Z^*_{el} + Z^*_{wd}\right) eE \tag{10.1}$$

where $Z^*_{el}$ is the nominal valence of the diffusing atom when the dynamic screening effect is ignored, and it is responsible for the electric field effect, and $Z^*_{el}eE$ is called the direct force, and it is acting in the direction opposing electron flow. The magnitude of $Z^*_{el}$ is taken to be the nominal number of valence electrons of the metal atom. $Z^*_{wd}$ is an assumed charge number representing the effect of momentum exchange between electrons and the diffusing atom, and $Z^*_{wd}eE$ is called the electron wind force, and it is acting in the same direction as electron flow. Generally, $Z^*_{wd}$ is found to be of the order of 10 for a good conductor, so electron wind force is much greater than the direct force for electromigration in metals. Hence, the enhanced flux of atomic diffusion is in the same direction as electron flow in electromigration. For this reason, in electromigration, we prefer to use the direction of electron flow.

To estimate the electron wind force, the ballistic approach to the scattering process was developed by Huntington and Grone. [2] The model postulates a transition probability of free electron per unit time from one free electron state to another free electron state due to the scattering of the electron by the diffusing atom. The force, i.e. the momentum transfer per unit time, is calculated by summing over the initial and final states of the scattered electrons. The step-by-step derivation of the model has been presented elsewhere. [2] A simple derivation is given below.

During the elastic scattering of electrons by a diffusing atom, we assume the system momentum is conserved. The average change in electron momentum in the transport direction equals to $m_e\langle v\rangle$, rather than $2m_e\langle v\rangle$, where $m_e$ is the electron mass and $\langle v\rangle$ is the mean velocity of electron in the direction of electron flow. This is because the atom moves. The force on the moving atom induced by the scattering is

$$F_{Wd} = \frac{m_e < v >}{\tau_{col}} \tag{10.2}$$

where $\tau_{col}$ is the mean time interval between two successive collisions. The net momentum lost per second per unit volume of electrons to the diffusing atom is then $nm_e\langle v\rangle/\tau_{col}$, and the force on a single diffusing atom is

$$F_{wd} = \frac{nm_e < v >}{\tau_{col} N_d}$$

where $n$ is the electron density and $N_d$ is the density of diffusing atoms. The electron density can be given as $j = -ne\langle v\rangle$. Substituting $\langle v\rangle$ into the last equation, we obtain

$$F_{wd} = -\frac{m_e j}{e\tau_{col} N_d} = -\frac{m_e}{ne^2\tau_{col}}\frac{neE}{\rho N_d} = -\left[\frac{\rho_d}{N_d}\right]\left[\frac{n}{\rho}\right]eE$$

where $\rho = E/j$ is the resistivity of a conductor, and $\rho_d = m/ne^2\tau_{col}$ is the metal resistivity due to the diffusing atoms, and $E$ is the applied electric field.

Besides the electron wind force, the electric field will produce a direct force on the diffusing atoms to be given by

$$F_{direct} = Z^*_{el} eE$$

where $Z^*_{el}$ can be regarded as the nominal valence of the metal atom when the dynamical scattering effect around the atom is ignored. So the total force will be

$$F_{EM} = \left\{Z^*_{el} - Z\left[\frac{\rho_d}{N_d}\right]\left[\frac{N}{\rho}\right]\right\}eE$$

where $N$ is the atomic density of the conductor and $n = NZ$. The last equation can be rewritten as

$$F_{EM} = Z^* eE$$

And $Z^* = \left\{Z^*_{el} - Z\left[\frac{\rho_d}{N_d}\right]\left[\frac{N}{\rho}\right]\right\}$

where $Z^*$ is defined as the effective charge number of the atom in electromigration.

The above model shows that in essence the effective charge number can be given in terms of the ratio of specific resistivity of a diffusing atom and a normal lattice atom,

$$Z^*_{wd} = Z\frac{\rho_d / N_d}{\rho / N}$$

where $\rho = m_0/ne^2\tau$ and $\rho_d = m^*/ne^2\tau_d$ are the resistivity of the equilibrium lattice atoms and the diffusing atoms, respectively; $m_0$ and $m^*$ are the free electron mass and effective electron mass, respectively, and they are equal; and $\tau$ and $\tau_d$ are the relaxation times of a lattice atom and a diffusing atom, respectively. In a FCC lattice, there are 12 equivalent jump paths along the <110> directions. For a given electron current direction, the average resistivity of a diffusing atom must be corrected by a factor of 1/2. By rewriting the above equation, we have

$$Z^* = -Z\left[\frac{1}{2}\frac{\frac{\rho_d}{N_d}}{\frac{\rho}{N}} - 1\right] \tag{10.3}$$

where we have taken $Z_{el}$ as $Z$, the nominal valence of the metal atom. Conceptually, it means to calculate $Z^*$, we need to know the specific resistivity ratio of a diffusing atom to a lattice atom.

### 10.3.2 Calculation of the Effective Charge Number

The specific resistivity of an atom in a metal is assumed to be proportional to the elastic cross-section of scattering, which in turn is assumed to be proportional to the average square displacement from equilibrium, or $<x^2>$. The cross-section of a normal lattice atom can be estimated from Einstein model of atomic vibration in which the energy of each mode is

$$\frac{1}{2}m\omega^2 <x^2> = \frac{1}{2}kT$$

where the product of $m\omega^2$ is the force constant of the vibration, and $m$ and $\omega$ are atomic mass and angular vibrational frequency, respectively.

To obtain the cross-section of scattering of a diffusing atoms, $< x_d^2>$, we assume that the atom and its surrounding atoms, as shown in Figure 10.1 have acquired the motion activation energy of diffusion, $\Delta H_m$, which is independent of temperature.

$$\frac{1}{2}m\omega^2 <x_d^2> = \Delta H_m$$

Then, the ratio of the last two equations gives the ratio of cross-section of scattering,

$$\frac{<x_d^2>}{<x^2>} = \frac{2\Delta H_m}{kT}$$

Substituting the last equation into the equation of $Z^*$, we have

$$Z^* = -Z\left[\frac{\Delta H_m}{kT} - 1\right] \tag{10.4}$$

Now the value of $Z^*$ can be calculated at a given temperature. The calculated values of $Z^*$ agree quite well with those measured for Au, Ag, Cu, Al, and Pb. Also, at 480 and 640 °C, the measured and calculated $Z^*$ for Al (taking $\Delta H_m = 0.62\,\text{eV/atom}$) are about −30 to −26, respectively. The temperature dependence of $Z^*$ calculated for Au is also found to agree well with the measured value by Huntington and Grone. [2]

### 10.3.3 Atomic Flux Divergence Induced Electromigration Damage

In electric conduction, Kirchhoff's law states that the sum of all currents in and out of a point must be zero, thus the in-flux equals to the out-flux, so there is no electric flux divergence at the point, except if the point is a source or sink of charges. But atomic flux can have flux divergence at a physical point, for example at a triple point (or a triple line) where three grain boundaries meet in a polycrystalline microstructure. If we assume the diffusivity and the effective grain boundary width of diffusion of the three are the same, there will be atomic flux divergence at the triple point where the in-flux comes from one grain boundary, but the out-flux goes out along the other two grain boundaries, or vice versa. The net effect of flux divergence after a long time event is the accumulation or depletion of vacancies (or atoms). This has been found to occur in Al interconnects because electromigration in Al interconnect at the device operation temperature of 100 °C is controlled by grain boundary diffusion. The accumulation of vacancies can lead to void formation at the site of divergence, and it can become an open in the interconnect microstructure.

We note that if there is no atomic flux divergence, or if the atomic flux is in a steady state in an interconnect, in turn the opposite vacancy flux is in a steady state, there will be no electromigration damage while there is electromigration. We can assume that the vacancy concentration is at equilibrium everywhere in the sample, or the sources and sinks of vacancies are fully operative in the sample, so lattice shift occurs completely to accommodate the absorption and emission of vacancies. According to the classical model of interdiffusion by Darken, there will be no void formation when equilibrium vacancy is assumed. This is because void nucleation needs super-saturation of vacancies. Below, we shall discuss briefly the flux divergence sites.

The most common site of vacancy absorption and emission is the free surface of an interconnect. However, when the surface is oxidized and if the oxide is stable and protective, as the Al oxide on Al interconnects (or Sn oxide on Sn plated surfaces), the Al/oxide interface is no longer an effective source and sink of vacancies, so flux divergence can occur. For Cu interconnects, the free surface after chemical-mechanical planarization is the most important path of flux divergence, where electromigration damage occurs.

The next common site of flux divergence is the interphase interface, such as the interface between a W via and an Al line. This is because solubility and diffusivity change dramatically across the interface.

Then, grain boundaries, especially the triple points, are sites of flux divergence. No doubt, we must consider dislocations, especially the kink site on an edge dislocation line. As in dislocation climb, the kink sites can absorb and emit atoms and vacancies. Consequently, lattice planes can be created or destroyed. When the created or destroyed lattice plane can migrate in the lattice as needed, it leads to lattice shift and the event has no effect on microstructure damage or no void formation. The lattice plane migration or lattice shift is manifested by marker motion in Darken's analysis of interdiffusion. But, if the ends of a lattice plane are pinned by a protective oxide interface, in turn the plane cannot migrate, flux divergence will occur. In interdiffusion, it leads to Frenkel or Kirkendall void formation. In electromigration in Al, this is the reason of back-stress formation as well as void and hillock formation. We recall that in Figure 10.5, the back-stress effect and void and hillock formation in Al short stripes are shown.

Furthermore, in void formation, after nucleation, the growth of a void requires a gradient of vacancy concentration for the supply of ensuing vacancies. In hillock growth, it requires a normal stress gradient. Since we mentioned back-stress, we should point out that according to Nabarro and Herring model of stress potential in a stressed solid, the compressive region will be able to absorb some vacancies to reduce the compressive stress and the tensile region will be able to absorb some excess atoms to reduce the tensile stress. It may accommodate a small degree of flux divergence, but not a lot.

In multilayered Cu interconnects, due to the dual-damascene process, there are interfaces between the chemical-mechanically polished top surface of Cu and the sputtered $SiO_xN_{1-x}$. Unlike the interface between Al and $Al_2O_3$, the sputtered $SiO_xN_{1-x}$ is not a protective oxide on Cu, thus their interface is a good source and sink for vacancies. So the diffusion of Cu on the interface is rapid, similar to that on a free surface of Cu. During electromigration, void formation and migration occur on the interface. It dominates circuit failure in multilayered Cu interconnects. We recall that in Figure 1.7, the electromigration induced damages in multilayered Cu interconnects are shown. From the point of view of Joule heating, the thinning of the cathode will increase the current density, in turn Joule heating, and in turn temperature increase. This becomes a positive feedback, which will enhance a run-away electromigration failure. Furthermore, when the void formation migrates to the bottom side of a via, a rapid failure occurs. This is because it only takes a thin pancake-type of void at the bottom of the via to cause an electric open.

### 10.3.4 Back Stress in Electromigration

The model of Huntington and Grone, especially the flux of electromigration, has been taken to be the theoretical basis in nearly all experimental studies of electromigration.

$$J_{EM} = C < v > = CMF = C\frac{D}{kT}Z^{*}eE = \frac{1}{\Omega}\frac{D}{kT}Z^{*}e\rho j \quad (10.5)$$

It indicates that if we measure the drift velocity using short stripes and know the diffusivity $D$, we will be able to calculate $Z^*$.

On the measurement of $Z^*$, an effective method was developed by Blech in using short stripes of Al on TiN. [8, 9] Blech patterned a set of short stripes of Al of different length on a TiN thin line, and one of them is shown schematically in Figure 10.6. Because Al is a better conductor than TiN, the electrons will detour from the latter and enter into the former. Electromigration damage occurs in these Al short stripes at both the cathode and the anode ends. There is the depletion or void formation at the cathode ends and the growth of hillock at the anode ends, as depicted in Figure 10.6. What is the most informative of the Blech stripes is that we can observe the damage due to electromigration directly, so we can measure the kinetics of electromigration by measuring the rate of depletion at the cathode end, in turn we can calculate the value of $Z^*$.

In the early stage of study of electromigration in Al interconnects, there were two key challenges. One was how to measure $Z^*$ experimentally in order to verify if it is in agreement with the theoretical model of electron wind force. The other was how to reduce electromigration damage in Al interconnect technology. However, when it was realized that electromigration in Al interconnects lines was controlled by grain boundary diffusion at the device operation temperature, the way to reduce electromigration in Al was found by adding a small amount, about 1% of Cu, to retard self-grain-boundary diffusivity in Al. [3] We cannot add more Cu because it will increase the resistivity of the Al line.

**Table 10.2** Diffusivity at 100 °C.

| | Melting point (°K) | Temperature ratio 373 °K/Tm | Diffusivities at 100 °C [373 °K] ($cm^2$/sec) |
|---|---|---|---|
| Cu | 1356 | 0.275 | Surface $D_s = 10^{-12}$ |
| Al | 933 | 0.4 | Grain boundary $D_{gb} = 6 \times 10^{-11}$ |
| Pb | 600 | 0.62 | Lattice $D_l = 6 \times 10^{-13}$ |
| Eutectic SnPb | 456 | 0.82 | Lattice $D_l = 2 \times 10^{-9}$ to $2 \times 10^{-10}$ |

To finish this section on the basic model of electromigration, we note that lattice diffusion is assumed as shown in Figure 10.1. However, electromigration in Al interconnect around the device operation temperature of 100 °C is found to occur by grain-boundary diffusion. In other words, when we discuss the electromigration flux in Eq. (10.5), we will need to consider not only the thermodynamic driving force but also the kinetics of diffusion. Actually, when we review the experimental electromigration damages in Al, Cu, and solder interconnects in microelectronic devices, we find that they are dominated by grain boundary diffusion in Al, surface diffusion on Cu, and lattice diffusion in Pb-free solder, at the device operation temperature of 100 °C. Different kinetic paths will lead to different failure sites and modes. Table 10.2 lists the corresponding diffusivity of Al, Cu, and Sn near 100 °C.

## 10.4 Current Crowding and Electromigration in 3-Dimensional Circuits

Current crowding occurs when the flow of electrons turns, which is common in a 3D structure. Even in 2D IC devices, the multilevel metal interconnect structure is 3D. For example, in a multilevel interconnects, via is used to connect wires in two levels. Whether it is left-to-right or up-to-down, all electrons take the shortest path in order to reduce resistance. Figure 10.2 shows an example of current crowding, where a Ni silicide line is formed under a high current density in the heavily doped Si channels. In Figure 10.2a, a straight silicide line is formed in the Si channel, because the cathode and the anode are aligned along a straight line. In Figure 10.2b, the silicide line turns because the electrodes are at right angle to each other, making a 90-degree turn. What is unique in Figure 10.2b is that the silicide

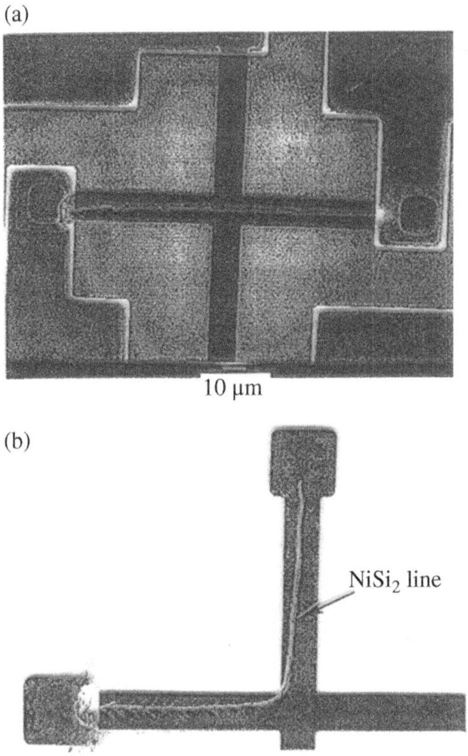


**Figure 10.2** (a) SEM image of a straight Ni silicide line formed between two Ni contact electrodes arranged in a line. (b) SEM image of current crowding of a Ni silicide line formed between two Ni contact electrodes making a 90-degree turn.

line was found to form along the inner corner of the channel, rather than in the middle of the channel. Because the formation was driven by electric current, it shows the nature of current crowding.

However, the effect of current crowding on electromigration-induced damage is unusual; the damage tends to occur not in the region of high current density but in the region of low current density, which is against the expectation on the basis of electron wind force. This is because in the low current region, the electron wind force is low so there should be very little electromigration. [10, 11]

Using simulation of current distribution in a conductor, we can show easily that the current crowding phenomenon is a function of geometry and resistance of the conductor. Besides turns, when a conductor changes its thickness or width or resistance, current crowding occurs too. Furthermore, when a current goes across a contact interface, for example the Al/TiN interface near the cathode end, as well as near the anode end in Figure 10.3a, the degree of current crowding depends not only on the resistance of Al and

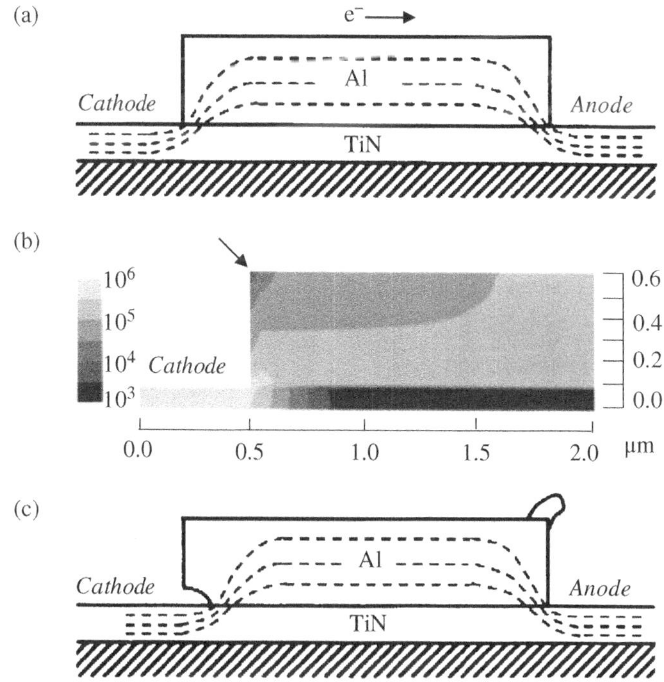


**Figure 10.3** (a) Schematic diagram of the cross-section of an Al short stripe on TiN. Current occurs as electrons enter the cathode end and exit the anode end. (b) The two-dimensional current density distribution in the cross-section of half an Al short stripe on TiN. Current crowding occurs at the end of the contact interface between the Al and TiN. However, the upper corner of the short stripe is the low current-density region, as indicated by a short arrow. (c) If vacancies follow the current and go to the low end of the cathode, we expect a pancake-type of void to form along the interface between the Al and TiN. A pancake-type void cannot deplete the entire cathode. On the other hand, if the growth of the void occurs at the upper end of the cathode and is fed by a flux of vacancies, taking a detour and being driven to the upper end of the cathode, the void can grow downward to deplete the entire cathode.

TiN but also on the contact resistance of the Al/TiN interface. The larger the contact resistance, the lesser the current crowding.

In flip-chip solder joint technology, the solder bump has a diameter of about 100 μm, and it is connected to an Al or Cu line of about 1 μm in thickness, as shown in Figure 10.4a. If we assume the width of the line and the diameter of the bump are the same, there is a change of current density by two orders of magnitude when the same current is passing from the line to the bump or vice versa. A very large current crowding occurs in the transition

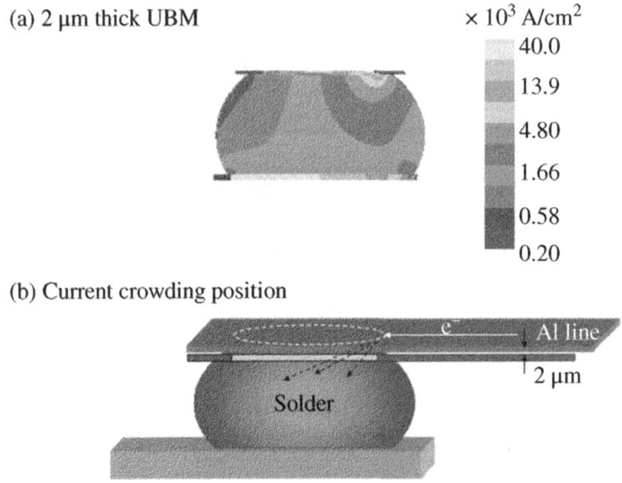


**Figure 10.4** The lower schematic diagram (b), depicting a joint of an Al line of 2 μm in thickness and 100 μm in width on a solder bump with a contact opening of 100 μm in diameter. The contact opening is defined by the dielectric layer between the Al and the solder bump. The upper simulation diagram (a) is a two-dimensional current distribution simulation of the cross-section of the joint. Current crowding occurs in the entrance at the upper right corner as the electrons flow from the Al line into the solder bump.

region between them. The 2D simulation of current distribution in the upper part of Figure 10.4b shows that the current density in the transition region is about a factor of 10 higher than that within the solder joint. The transition region has been the most common failure site of electromigration in flip-chip solder joints.

Today, the simulation of electric current distribution in 2D and 3D structures can be obtained readily because commercial programs are available. In a simulation, we can see easily where are the high current density regions, as well as the low current density regions. Under electromigration, if a void forms in a high current density region, it is expected, but if the void forms in a low current density region, it needs explanation.

### 10.4.1 Void Formation in the Low Current Density Region

Okabayashi et al. [12] prepared in situ TEM samples, as depicted in Figure 10.5a, for direct observation of void formation and hillock formation in electromigration. The width of the Al line was designed so that it could be imaged clearly in TEM from the side view, rather than from the top view.

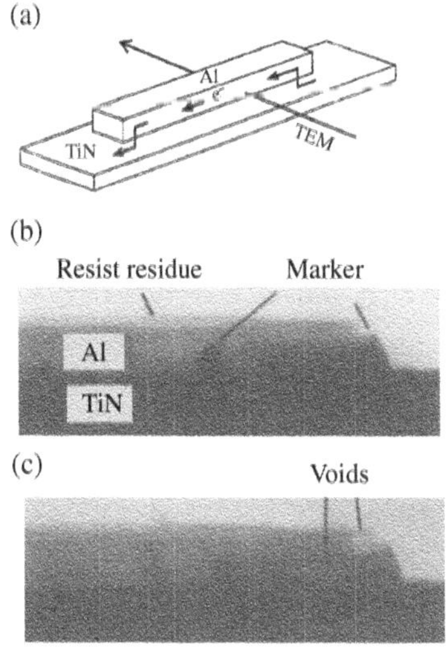


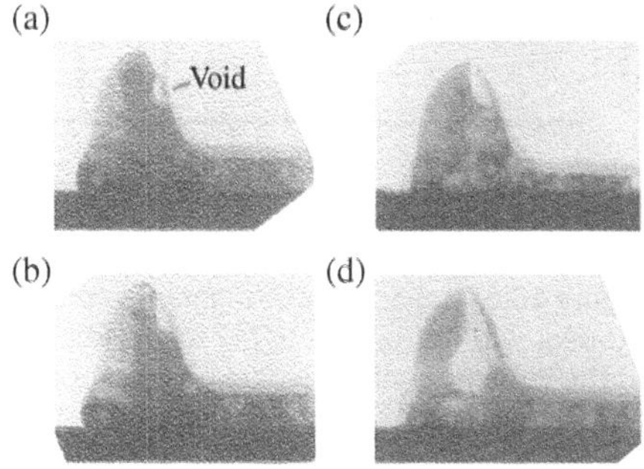


**Figure 10.5** (a) Schematic diagram of an in situ TEM samples, where the electron beam is directed to go across the sides of the sample during electromigration for direct observation of void formation and hillock formation in electromigration. (b) After 14 seconds of electromigration, a void was seen at the upper left corner of the stripe, i.e. the upper corner of the cathode end. We note that the upper corner has a very low current density. (c) Furthermore, after a long time of electromigration, the polarity of the applied current was reversed, and a void was formed at the upper tip of the hillock (which was the anode before the change of polarity). Again, the current density at the upper tip of the hillock should be very low, yet the void there kept growing with electromigration.

In other words, the electron beam in the TEM is normal to the applied electron current of 50 mA, from the side view, as shown in Figure 10.5a. After 14 seconds of electromigration, a void was seen at the upper left corner of the stripe, i.e. the upper corner of the cathode end. We note that the upper corner has a very low current density as given by the simulation of current distribution in the short stripe, as shown in Figure 10.3b. Furthermore, after a long time of electromigration, the polarity of the applied current was reversed, and a void was formed at the upper tip of the hillock (which was the anode before the change of polarity), see Figure 10.5c. Again, the current density at the upper tip of the hillock should be very low, yet the void there kept growing with electromigration. These results show that electromigration induced void formation, which occurs in the low current density region when current crowding occurs.

Shingubara et al. [13] patterned U-shape baseline of TiN, on which Al short stripes were deposited. Some of the short stripes were made to stick

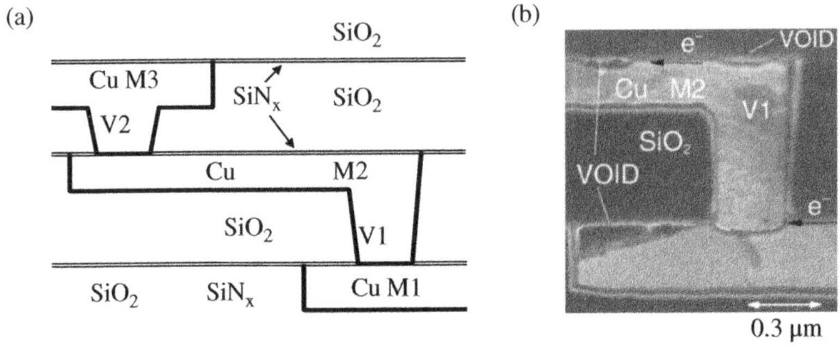


**Figure 10.6** (a) Schematic diagram depicts a Cu interconnect conductor with a 90-degree turn from M1 to M2 to M3. (b) SEM image of a triangular void formation in the left corner of M1 where no current flowed through.

out from the U-shape baseline of TiN, so that the Al short stripe has an overhang out of the baseline. The overhang was about 20 μm long. It is clear that there will be no current in the overhang during electromigration when current is applied to the TiN baseline. However, voids were observed experimentally to form in the overhang in electromigration.

Hu et al. [7] conducted electromigration in a three-level damascene interconnect structure of Cu with two Cu vias (V1 and V2) connecting three levels of lines (M1, M2, and M3). In electromigration test at a current density of $2.5 \times 10^6$ A/cm$^2$ at 295 °C over 100 hours, a large triangular shape of void was found to form to the left of V1 via, see Figure 10.6. However, the current density in the region of triangular void formation is low. Also some long and shallow voids were found to form on the surface of the line (M2) above the V1 via, which is in agreement with the fact that electromigration in Cu occurs by surface diffusion.

It is worth noting that the mechanism of formation of the long void on the surface of M2 line is different from that in the formation of the triangular void to the left of V1 via. The latter forms in the low current density region, while the former forms in the high current density region driven by electron wind force. What is the electromigration driving force to drive vacancies to go to the low current density region to form the triangular void will be analyzed below.

The observations presented in the above showed that some void formations induced by electromigration occur in low current density region in Al and Cu interconnects. This is quite unexpected on the basis of our understanding of the electron wind force of electromigration, which means that the lower the current density, the lower the driving force. Actually in the low current density region, the electron wind force is so weak that there should be no electromigration. We further note that all the phenomena of void formation in the low current density region occur near a turn of electron flow.

The two-dimensional current density distribution in the cross-section of the Al short stripe is shown in Figure 10.3b. Indeed, current crowding is found to occur at the end of the two lower corners of the contact interface between the Al and TiN, where electrons flow in and flow out of the stripe. However, both the two upper corners of the short stripe are regions of very low current density. On the basis of electron wind force, we expect that vacancies will be driven to the current crowding region, i.e. the cathode end of the contact interface between the Al and TiN where the electron exits the Al and enters the TiN. Because the electron wind force will keep driving more and more vacancy to the exit region, thus when the vacancy concentration in the exit region reaches the super-saturation needed for nucleation, a void would have nucleated. The lateral growth of the void would push the exit of electrons inward from the lower end because the void would have now occupied the end. The lateral growth of the void would be fed by the ensuing flux of vacancies supplied by electromigration, so we expect that the void would have grown in a pancake-type shape along the Al/TiN interface, see Figure 10.3c.

What is wrong about the above expectation is that it will not lead to the depletion of the entire cathode end of the Al short stripe. It only leads to the formation of a pancake type of void along the Al/TiN interface. However, this is not what has been observed. Actually the void growth extends to the upper corner of the cathode. [12] Experimentally, the entire cathode of Al short stripe is depleted completely in electromigration, as depicted in Figure 9.7. In order to deplete the entire cathode, the vacancies must go to the low current density region at the upper left corner, as depicted in Figure 10.3b. Therefore, the void must start or nucleate from the upper end of the left corner of the stripe, and the growth of the void should propagate downward and toward the anode. Consequently, it enables us to measure the depletion rate at the cathode end.

We should consider and examine whether or not the void nucleation at the upper left corner can be explained by stress-migration or by thermomigration. If we assume that more and more vacancies are being driven to the high current density region (we could assume it to be in tension), a concentration gradient of vacancy is created between the high and low current density regions. The latter can be assumed to have no stress. The gradient will drive vacancies from the high to the low current density region. Then the question is will it lead to void formation in the low current density region. Because a vacancy concentration gradient has been assumed, the vacancy concentration in the high current density region is always higher than that in the low current density region. Yet, the nucleation of a void requires super-saturation of vacancies, and because the vacancy concentration in the high current density region is higher, it is unreasonable to assume that void will be nucleated in the low current density region rather than in the high current density region.

For thermomigration, it will be helpful if we review briefly what happens when a Cu kettle is used to boil water. The outer temperature of the kettle will be over 600 °C and the insider about 100 °C. If the thickness of wall of kettle is 1 mm, we have a temperature gradient of 5000 °C/cm, which is very large, and we expect thermomigration to occur across the wall. The temperature gradient will drive Cu atoms to diffusion, from hot to cold, from the outside to the inside. If it would happen, we should expect the inside diameter of the kettle to enlarge with time because more and more atoms diffuse to it. Somehow, if we examine the kettles at home after use of years, they do not become larger at all. The reason is that at the hot end, there are more vacancies, and the diffusion of vacancies tends to balance the thermomigration of atoms. Thus, thermomigration in a pure metal tends to be negligible.

### 10.4.2 Current Density Gradient Force in Electromigration

Generally speaking, electromigration occurs due to electron wind force; however, the force cannot explain void formation in the low current density region. Therefore, a new force of electromigration, the current density gradient force, which is normal to the electron wind force, has been proposed. [10, 14, 15] The new force will divert vacancies to diffuse to the low current density region before they reach the high current density region driven by the electron wind force. Because vacancies go to the low current density region, the concentration of vacancy can reach super-saturation and void can nucleate there.

We consider the electromigration in a conductor, which has a 90-degree turn as depicted in Figure 10.7a. The current density gradient force is due to the existence of a gradient of electric potential in current crowding from the turn. Figure 10.7b shows the simulation of electrical potential gradient in the region of the turn.

Before the turn, the conductor is straight, so the current density is uniform and at the same time there is a uniform vacancy flux moves from the anode side to the cathode side or an atomic flux moving in the opposite direction. Approaching the turn, current crowding occurs; the current density is higher in the inner side of the turn and lower in the outer side of the

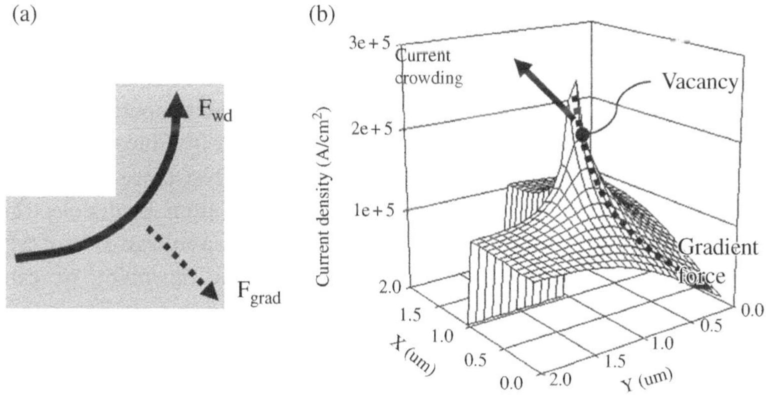


**Figure 10.7** (a) The schematic diagram on the left side depicts a conductor with a 90-degree turn. (b) The diagram on the right shows the simulation of electrical potential gradient in the 90-degree turn of the conductor. Current crowding occurs in the inner turn. The electric potential energy gradient force is depicted to go down from the inner turn to the lower right corner.

turn. For those vacancies, which follow the high current density path in the inner side, its potential energy increases and its equilibrium concentration should decrease with respect to those in the uniform region. However, the low current density region in the outer side can have a higher equilibrium vacancy concentration than that in the uniform region. The electric potential gradient will push the excess vacancies in the high current density region to go to the low current density region. In other words, the electric potential energy gradient force drives vacancies to go from the high current density region to the low current density region.

The diffusion of vacancies is uphill, going from a low vacancy concentration region to a high vacancy concentration region, i.e. against the vacancy concentration gradient. However, the uphill flux of vacancy is actually driven by the downhill electric potential energy gradient. Quantitatively, an analysis is given below.

When an electric potential is applied to the conductor, the chemical potential energy of every atom and vacancy is increased. Because of resistance, the increase in potential energy of vacancies (or solute atoms) is much larger than that of the matrix atoms. Furthermore, the increase is proportional to current density. In the high current density region, the vacancies will have a higher potential energy, so the concentration of vacancies in the high current density region will be reduced. It means that the formation energy of a vacancy in the high current density region is higher, so the concentration will be lower, similar to that in a mechanically compressed solid.

We consider a single crystal Al stripe and assume that a vacancy in the Al crystalline lattice has a specific resistivity of $\rho_v = R_v$. The specific resistivity may depend on current density due to Joule heating since resistivity depends on temperature. However, for simplicity, we ignore the temperature effect here and assume that it is independent of current density. Since vacancy is a lattice defect, we can regard its specific resistivity (resistance of a single vacancy) as the excess resistance over that of a lattice atom. Under electromigration in a current density of $j_e$, or in a current of $I$, a voltage drop of $\Delta V = IR_v$ occurs across the vacancy. From the energetic viewpoint, we can regard that the vacancy has a potential of ΔV above the surrounding lattice atoms. Knowing the charge of the vacancy, we have the potential energy of the vacancy in the current density of $j_e$ to be $Z^{**}e\Delta V$, where $Z^{**}$ is the effective charge number of the vacancy and $e$ is the charge of an electron. The potential energy of a vacancy in the current density $j_e$ or a current of $I$ is given as below.

$$P_v = Z^{**}e\Delta V \tag{10.6}$$

If we assume the equilibrium vacancy concentration in the single crystal without any electrical current ($j_e = 0$) to be $C_v$,

$$C_v = C_0 \exp\left(-\Delta G_f / kT\right) \tag{10.7}$$

where $C_0$ is the atomic concentration of the crystal and $\Delta G_f$ is the formation energy of a vacancy in the crystal. When we apply a current density of $j_e$ to the crystal, the vacancy concentration will be reduced to

$$C_{ve} = C_0 \exp\ \left(\Delta G_f + Z^{**}e\Delta V\right) / kT \tag{10.8}$$

Under a uniform high current density, the equilibrium vacancy concentration in the crystal decreases. In other words, the electric current dislikes any excess high resistive obstacles (or defects) and prefers to get rid of them until the equilibrium is reached. When there is a current density gradient as in current crowding, a driving force exists to do so,

$$F = -dP_v / dx \tag{10.9}$$

where $P_v$ has been defined in Eq. (10.6) as the electric potential energy of a vacancy in current density $j_e$. This force drives the excess vacancies to diffuse in the direction normal to the direction of current flow. Consequently, a component of the vacancy flux is moving in the direction normal to the electric current flow, as depicted in Figure 10.7b,

$$J_{cc} = C_{ve}\left(D_v / kT\right)\left(-dP_v / dx\right) \tag{10.10}$$

where $D_v/kT$ is the mobility, and $D_v$ is the diffusivity of vacancies in the crystal. Since a constant flux of vacancies keeps coming from the anode to the cathode due to electromigration, the total flux of vacancies moving toward the cathode is given by the vector sum of the two components below,

$$J_{sum} = J_{em} + J_{cc} = C_{ve}\left(D_v / kt\right)\left(-Z^* eE - dP_v / dx\right) \tag{10.11}$$

where the first term is due to electromigration driven by the current density (the electron wind force) and the second term is due to current crowding driven by the current density gradient. In the first term, $Z^*$ is the effective charge number of the diffusing Al atom, and $E = j_e\rho$ (where $\rho$ is the resistivity of the Al). We note here that we assume the vacancy flux is opposite but equal to the Al flux. Also it is important to note that the sum in the bracket is a vector sum; the first term is directed along the current direction and the second term is directed normal to the current direction. In other words, the vacancies are driven by two forces in the current crowding region. They are depicted in Figure 10.7a at the upper left corner. Because the current turns continuously in the current crowding region, the direction of $J_{sum}$ changes with position.

How large is the gradient force? If we apply a current density of $10^5$ A/cm$^2$ through the stripe and assume that the current density will drop to zero across the thickness of the stripe of 1 μm, the gradient can be as high as $10^9$ A/cm$^3$. The gradient force is of the same magnitude as the electron wind force.

However, the electric potential gradient across an atom is quite small so the current density gradient force on a lattice atom is negligible. But because the electrical resistance of a vacancy is about one hundred times larger than that of an atom, the electric potential gradient force on a vacancy is significant.

From the kinetic point of view, the diffusion of vacancies driven by the current density gradient force occurs mainly by lattice diffusion. While it is possible to consider grain boundary diffusion if a grain boundary happens to exist in the region of current crowding, it is hard to imagine how it can occur by surface diffusion because there is no surface diffusion path. Nevertheless, vacancy diffusion and interfacial diffusion can occur in Cu at the device operation temperature of 100 °C.

In a short summary, we postulate that defects such as vacancies and solute atoms have a higher potential in the high current density region than that in the low current density region. The potential gradient in the current crowding area provides a driving force to push these defects from the high current density region to the low current density region. As a consequence, the voids tend to form in the low current density region rather than in the high current density region. In other words, electromigration failure tends to occur in a low current density region in 3-D interconnects because of current crowding.

### 10.4.3 Current Crowding Induced Pancake-Type Void Formation in Flip-Chip Solder Joints

We recall that the unit of atomic flux is number of atoms per unit area per unit time, so a change of the cross-sectional area will lead to atomic flux change. When a conduction path changes its cross-section such as an Al or a Cu thin film line connecting a solder joint, as shown in Figure 10.4, atomic flux divergence occurs across the contacts because of the large change in cross-section and also because of the change in atomic diffusivity between the line and the solder bump. Furthermore, the current density changes too due to current crowding, as shown in Figure 10.4a.

In flip-chip solder joint technology, the solder bump has a diameter of about 100 μm, and it is connected to an Al or Cu line of 1 μm or so in thickness, if we assume the width of the line and the diameter of the bump are the same, there is a change of current density by two orders of magnitude when the same current is passing through them. A very large current crowding occurs in the line-to-bump contact or in the transition region. Figure 10.4a has shown a simulation of current density change from a line to a flip-chip solder joint or vice versa. The phenomenon is similar to a water-fall. Combining atomic flux divergence and current crowding, the contact has been the most important failure site of electromigration in flip-chip solder joints.

In Figure 10.8, the upper part shows SEM images of a daisy chain of flip-chip solder joint after electromigration. The white arrows show the current path. On No. 1, 3, and 5 solder joints, where the electrons enter the joint

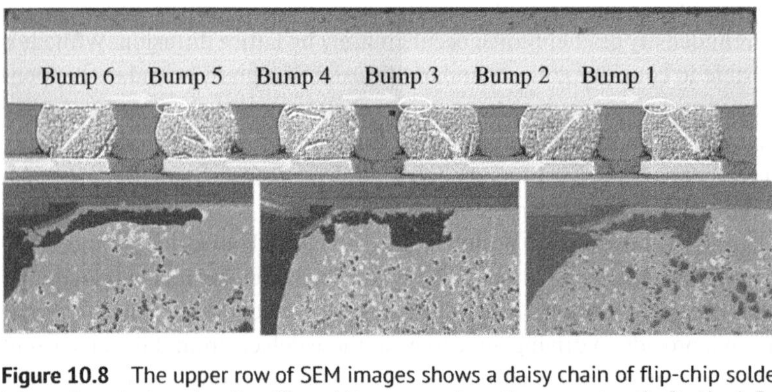


**Figure 10.8** The upper row of SEM images shows a daisy chain of flip-chip solder joints after electromigration. The white arrows show the current path. On No. 1, 3, and 5 solder joints, where the electrons enter the solder joint from the upper left corner, a pancake-type of void was formed. The lower row of enlarged SEM images shows a pancake-type of void in No. 1, 3, and 5 solder joints.

from the upper left corner, a pancake-type of void was formed. The lower part of Figure 10.8 shows enlarged images of a pancake-type of void in No. 1, 3, and 5 solder joints. However, in No. 2, 4, and 6 solder joints, where the electrons exit the joint from the lower right corner, current crowding is much less, and no void forms.

The simulation of current crowding, which is similar to a water-fall as shown in Figure 10.4, is based on a direct contact of Al to solder joint, without UBM or with a very thin UBM less than 1 μm. On the other hand, a thick UBM decreases current crowding. Simulation shows that when a thick UBM is used, about 10 μm of Cu, current density distribution will spread over the entire contact, resulting in little current crowding.

## 10.5 Joule Heating and Heat Dissipation

Heat generation is mainly due to Joule heating, which is entropy production on the basis of irreversible processes, as presented in Chapter 9. For electric conduction, the equation below shows that the entropy production is the product of the conjugated flux of $j$ (current density = coulomb/cm$^2$-sec) and the conjugated driving force of $E$ (electric field $E = j\rho$, where $\rho$ is resistivity).

$$\frac{TdS}{Vdt} = jE = j^2\rho \tag{10.12}$$

where $T$ is the temperature, $V$ is the volume of sample, $dS/dt$ is the entropy production rate, and $j^2\rho$ is the Joule heating per unit volume per unit time, in units of Watt/cm$^3$.

The generated Joule heating will be absorbed and conducted through the device structure. The heat generation and heat conduction in a conductor is fully coupled and governed by the following 1-dimensional Fourier's equation of heat transfer in solids [16],

$$\frac{\partial T}{\partial t} = \frac{\kappa}{\rho C_p}\frac{\partial^2 T}{\partial x^2} + \frac{q}{\rho C_p} \tag{10.13}$$

where $\rho$ is the materials density of conductor, $C_p$ is the heat capacitance, $\kappa$ is the thermal conductivity, and $q$ is the power density (Watt/cm$^3$) induced by Joule heating, and $q = jE$. The first term on the right-hand side of the above equation represents divergence of heat fluxes in a unit volume, or the net change of heat in the volume due to heat flux in and out from the neighboring volumes. The second term represents heat generation within the unit volume due to Joule heating. It is worth mentioning that the unit of ($\kappa/\rho C_p$) is the same as atomic diffusivity, cm$^2$/sec. Clearly, if we ignore the second term, it is the same as Fick's second law of diffusion. Here, we recall Eq. (9.10) in Chapter 9 for comparison. Thus, the difference between atomic diffusion and heat conduction is that there is always a temperature change in the volume under heat conduction consideration. Because entropy is positive, heat is generated in the volume so the temperature of the volume increases.

Dissipation of Joule heating to the ambient occurs in complex ways, depending on the device structure and materials, as well as the ambient. To simplify the picture, we consider a steady-state model in which the heat is dissipated in terms of heat flux to the ambient in two ways. The first is through Si to the surrounding air, and the second through the substrate to the surrounding air. Typically, heat flux according to Fourier's law is given as $J_{heat} = -\kappa\,(dT/dx)$, which is a vector. Here, we assume experimentally it is governed by the following equation,

$$J_{heat} = \left(h_{Si} + h_{sub}\right)\left(T - T_{ext}\right) \tag{10.14}$$

where $J_{heat}$ is the dissipation heat flux (joule/cm$^2$-sec or Watt/cm$^2$), $h$ is the experimental thermal transfer coefficient (W/(cm$^2$·K)). $T_{ext}$ is the ambient temperature and $T$ is the temperature of the conductor under consideration. For Si, we may assume $h_{Si}$ to be 900 W/(cm$^2$·K), and for the substrate, we may assume $h_{sub}$ to be 5 W/(cm$^2$·K). With these assumed values of thermal transfer coefficient, it is clear that most heat is dissipated through the Si side in the model. Models of heat transfer simulation are available.

To increase the area of the Si surface in order to enhance the heat dissipation, a fin structure can be attached to the Si surface or the substrate surface by using the thermal-interface-material (TIM), which should have an excellent thermal transfer coefficient. For the reason, the search for TIM has been quite active.

### 10.5.1 Joule Heating and Electromigration

How to couple Joule heating and electromigration is difficult. The thermal-electric effect deals with heat flow and charge flow. Electromigration relates charge flow and atomic flow. Thermomigration relates heat flow to atomic flow. If we add Joule heating to electromigration, we have three kinds of flow together. We have a $3 \times 3$ matrix equation. Below, we just consider the physical picture of the coupling.

About the effect of Joule heating on electromigration, Joule heating will increase the temperature of the sample, but at the same time, dissipation of heat will create a temperature gradient. Both of them affect atomic diffusion. First, we consider temperature gradient on atomic motion, this is thermomigration. We combine thermomigration and electromigration by adding them together. Second, we consider the effect of a uniform temperature increase due to Joule heating on electromigration. It means the rate of electromigration will be faster at a high temperature due to a faster atomic diffusion.

About the effect of electromigration on Joule heating, when electromigration leads to a change of resistance or a change of current density of the sample, Joule heating will increase. To increase current density means to reduce the cross-section of the sample, assuming that the applied current remains unchanged. In Cu interconnects, electromigration occurs by surface diffusion on the Cu surface near the cathode end, and it tends to reduce the thickness, as well as the cross-section of the sample. When the cathode end is becoming thinner and thinner, the current density should increase, so Joule heating, as well as the temperature, will increase. In turn, it will increase the rate of electromigration [17]. This positive feedback will accelerate failure quickly. [18]

### 10.5.2 Joule Heating on Mean-Time-to-Failure in Electromigration

Product assurance in microelectronics industry requires statistical reliability tests and data. Besides physical analysis of the failure mode and understanding of the failure mechanism, statistical analysis of failure distribution is needed in order to predict and to project the lifetime of a product. In Chapter 13, a unified discussion of mean-time-to-failure (MTTF) of electromigration, thermomigration, and stress-migration will be given on the basis on entropy production. It verifies that in Black's equation on MTTF for electromigration, where the current density dependent is $(T/j)^2$ not $(1/j)^2$, nor $(1/j)$. To calculate the parameters in Black's equation, the effect of Joule heating must be taken into consideration.

## Problems

**1** Ag is a better conductor than Cu and Al, why don't we use Ag as interconnects?

**2** It is well known that we can add 1 atomic % of Cu into Al interconnects to greatly improve its resistance to electromigration. What element can we add to Cu interconnects to improve its resistance to electromigration? Explain the criteria of your selection.

**3** Electric current crowding occurs in multilayered interconnect structure when there is a turn in the path of the conduction. Besides turns, what other structure feature can lead to current crowding?

**4** What is the current density in an extension cord used at home for a 100-watt table lamp? Assume the Cu wire in the cord has a diameter of 0.1 mm and is 10 ft long, and the applied voltage is 110 volt.

**5** Calculate the cross-section of scattering of an Au atom at 800 °C based on the Einstein model of atomic vibration.

**6** In an FCC metal, show that the displacement of an atom from its equilibrium position to the activated position is $(2^{1/2}a)/4$, where $a$ is the lattice parameter. Take $< x_d^2 > = \left(2^{1/2} a/4\right)^2$ and show that $(1/2)\, m\omega^2 a^2/8$ is a good approximation of $\Delta H_m$.

**7** Calculate the electric force of $eE$ at a current density of $10^5$ amp/cm$^2$, and the chemical potential $\sigma\Omega$ at the elastic limit for Au. Then calculate the critical length of Au.

**8** In conducting electromigration in an Au wire, because of uneven heat conduction, there is a temperature gradient in the wire. Take the forces as $X_1 = -dT/dx$ and $X_2 = -d\psi/dx$, then write down the pair of equations similar to those in Eq. 13.23 and explain the phenomenological coefficients in the equations.

**9** A drift velocity of 2 m/sec was measured in a Blech test structure of Cu at 350 °C under a current density of $1 \times 10^6$ amp/cm$^2$. Calculate the effective charge number $Z^*$ of Cu by assuming surface diffusion and also grain boundary diffusion of Cu.

## References


1 Chen, K.N. and Tu, K.N. (2015). Materials challenges in three-dimensional integrated circuits. *MRS Bulletin* 40 (3): 219–222.

2 Huntington, H.B. and Grone, A.R. (1961). Current-induced marker motion in gold wires. *Journal of Physics and Chemistry of Solids* 20: 76.

3 Ames, I., d'Heurle, F.M., and Horstman, R. (1970). Reduction of electromigration in aluminum films by copper doping. *IBM Journal of Research and Development* 4: 461.

4 d'Heurle, F.M. and Ho, P.S. (1978). *Thin Films: Interdiffusion and Reactions* (eds. J.M. Poate, K.N. Tu and J.W. Mayer), 243. New York: Wiley-Interscience.

5 Ho, P.S. and Kwok, T. (1989). Electromigration in metals. *Reports on Progress in Physics* 52: 301.

6 Sorbello, R.S. (1997). *Solid State Physics*, vol. 51 (eds. H. Ehrenreich and F. Spaepen), 159–231. New York: Academic Press.

7 Hu, C.K., Gignac, L., Malhotra, S.G. et al. (2001). Mechanisms for very long electromigration lifetime in dual-damascence Cu interconnections. *Applied Physics Letters* 78: 904.

8 Blech, I.A. and Herring, C. (1976). Stress generation by electromigration. *Applied Physics Letters* 29: 131–133.

9 Blech, I.A. (1976). Electromigration in thin aluminum films on titanium nitride. *Journal of Applied Physics* 47: 1203–1208.

10 Tu, K.N., Yeh, C.C., Liu, C.Y., and Chen, C. (2000). Effect of current crowding on vacancy diffusion and void formation in electromigration. *Applied Physics Letters* 76: 988–990.

11 Yeh, E.C.C., Choi, W.J., Tu, K.N. et al. (2002). Current-crowding-induced electromigration failure in flip chip solder joints. *Applied Physics Letters* 80 (4): 580–582.

12 Okabayashi, H., Kitamura, H., Komatsu, M., and Mori, H. (1996). *In-situ* side-view observation of electromigration in layered Al lines by ultrahigh voltage transmission electron microscopy. AIP Conference Proceedings, vol. 373, 214.

13 Shingubara, S., Osaka, T., Abdeslam, S. et al. (1998). Void formation mechanism at no current stressed area. AIP Conference Proceedings, vol. 418, 159.

14 Tu, K.N. (2003). Recent advances on electromigration in very-large-scale-integration of interconnects. *Journal of Applied Physics* 94: 5451–5473.

15 Tu, K.N., Liu, Y., and Li, M. (2017). Effect of Joule heating and current crowding on electromigration in mobile technology. *Applied Physics Reviews* 4: 011101.

16 Liu, Y., Li, M., Kim, D.W. et al. (2015). Synergistic effect of electromigration and Joule heating on system level weak-link failure in 2.5D integrated circuits. *Journal of Applied Physics* 118: 135304.

17 Li, M., Kim, D.W., Gu, S. et al. (2016). Joule heating induced thermomigration failure in un-powered microbumps due to thermal crosstalk in 2.5D IC technology. *Journal of Applied Physics* 120: 075105.

18 Carslaw, H.S. and Jaeger, J.C. *Conduction of Heat in Solids*, 2e. New York: Oxford Science Publications.

# 11

# Thermomigration

## 11.1 Introduction

In Chapter 9, we have mentioned that when an inhomogeneous binary solid solution or alloy is annealed at constant temperature and constant pressure, it will become homogeneous. On the other hand, when a homogeneous binary alloy is annealed at constant pressure but under a temperature gradient, the alloy will become inhomogeneous. This de-alloying phenomenon is called the Soret effect [1].

On the basis of the Soret effect, we can appreciate it why the study of thermomigration has to wait until the reliability of C-4 solder joint becomes an issue [2–7]. This is because solder is an alloy or a two-phase eutectic alloy. On the other hand, while electromigration is significant in Al and Cu interconnects, we have had very little discussion of thermomigration in them.

In 3D IC packaging, because of dense packing, not only Joule heating is serious but also heat dissipation is poor. To enhance heat dissipation, we need to have a high temperature gradient. [8] Unfortunately, a high temperature gradient can lead to fast thermomigration. This is especially serious in a small structure. For example, we consider a micro-bump of solder joint of 10 μm in diameter. If there is a 1°C temperature difference across the bump, the temperature gradient will be 1000 °C/cm, which can cause thermomigration in the micro-bump.

## 11.2 Driving Force of Thermomigration

In thermoelectric effect, a temperature gradient can move electrons. Similarly, a temperature gradient can drive atoms. In essence, the electrons in the hot end have higher energy in scattering or stronger interaction with diffusing

*Electronic Packaging Science and Technology*, First Edition. King-Ning Tu, Chih Chen, and Hung-Ming Chen.

atoms, hence atoms move down the temperature gradient. On the driving force of atomic diffusion, it is driven by chemical potential gradient as below.

$$J = C < v >= CMF = C\frac{D}{kT}\left(-\frac{\partial \mu}{\partial x}\right)$$

where $\mu$ is chemical potential energy. Now we consider temperature gradient as the driving force, we rewrite

$$J = C\frac{D}{kT}\frac{Q^*}{T}\left(-\frac{\partial T}{\partial x}\right) \tag{11.1}$$

where $Q^*$ is defined as heat of transport in thermomigration. Comparing the above two equations, we see that $Q^*$ has the same dimension as $\mu$, so it is the heat energy per atom.

The definition of $Q^*$ is that it is the difference between the heat carried by the moving atoms and the heat of the atom at the initial state (the hot end or the cold end, depending on the sign of $Q^*$).

To define the sign of $Q^*$, we examine Eq. (11.1). We shall consider $J$ in the equation between two points: point 1 at $(x_1, T_1)$ and point 2 at $(x_2, T_2)$ in a Cartesian coordinate, as shown in Figure 11.1, and we assume $T_1 > T_2$, and atomic flux moves from hot to cold, i.e. from point 1 to point 2. Then, $\Delta T/\Delta x$ is negative, so $Q^*$ is positive. Recall that this is also the reason why all the flux equation as shown in Chapter 9 have a negative sign. Thus for an element which moves from the hot end to the cold end, $Q^*$ is positive. For an element moving from cold to hot, its $Q^*$ is negative.

## 11.3 Analysis of Heat of Transport, $Q^*$

The heat flow, $J_Q$, and atomic flow, $J_M$, in thermomigration can be expressed by temperature gradient and chemical potential gradient, respectively, below.

$$J_Q = -L_{QQ}\frac{1}{T}\frac{dT}{dx} - L_{QM}T\frac{d}{dx}\left(\frac{\mu}{T}\right)$$

$$J_M = -L_{MQ}\frac{1}{T}\frac{dT}{dx} - L_{MM}T\frac{d}{dx}\left(\frac{\mu}{T}\right)$$

When an alloy is held in a temperature gradient until a concentration gradient is established to balance the former, it comes to a steady state that the mass flow $J_M$ will be zero. Taking $J_M = 0$, we have from the last equation,

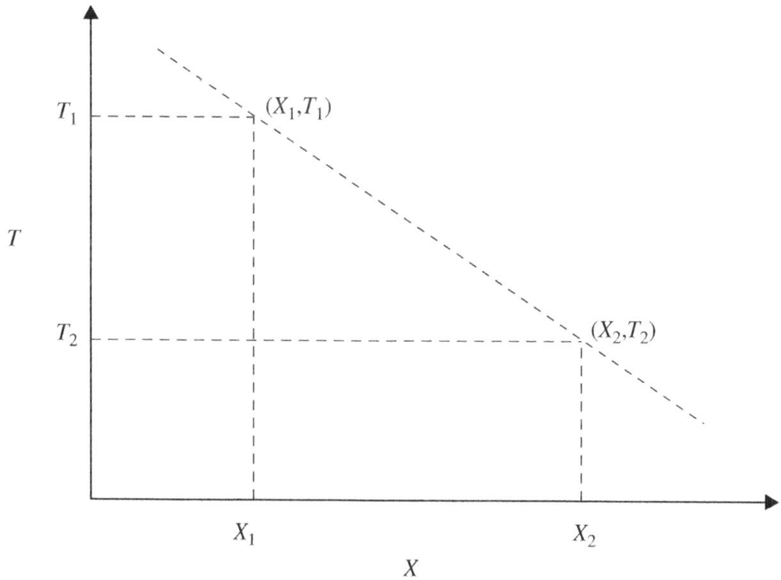


**Figure 11.1** We shall consider $J$ in the equation between two points; point 1 at $(x_1, T_1)$ and point 2 at $(x_2, T_2)$ in a Cartesian coordinate, and we assume $T_1 > T_2$, and atomic flux moves from hot to cold, i.e. from point 1 to point 2. Then, $\Delta T/\Delta x$ is negative, so $Q^*$ is positive.

$$L_{MQ}\frac{1}{T}\frac{dT}{dx} = -L_{MM}T\frac{d}{dx}\left(\frac{\mu}{T}\right)$$

By eliminating $T/dx$, we have

$$d\left(\frac{\mu}{T}\right) = -\frac{L_{MQ}}{L_{MM}}\frac{dT}{T^2}$$

By differentiation, we have

$$d\left(\frac{\mu}{T}\right) = \frac{1}{T}d\mu - \mu\frac{1}{T^2}dT$$

Using the thermodynamic relations, $\mu = H - TS$ and

$$d\mu = -SdT + Vdp$$

Then, we have,

$$d\left(\frac{\mu}{T}\right) = \frac{Vdp}{T} - H\frac{dT}{T^2} = -\frac{L_{MQ}}{L_{MM}}\frac{dT}{T^2}$$

Thus,

$$\frac{Vdp}{T}=\left(H-\frac{L_{MQ}}{L_{MM}}\right)\frac{dT}{T^2}$$

To gain an understanding of the meaning of $L_{MQ}/L_{MM}$, we consider the ratio of heat flow to mass flow under isothermal condition, i.e. when $dT/dx = 0$. We have

$$\frac{J_Q}{J_M}=\frac{L_{QM}}{L_{MM}}=\frac{L_{MQ}}{L_{MM}}$$

where we note that by using Onsager's relation that $L_{QM} = L_{MQ}$. The term $L_{MQ}/L_{MM}$ represent the energy flow associated with a mass flow. Defining $Q' = L_{MQ}/L_{MM}$, we have

$$\frac{Vdp}{T}=\left(H-Q\right)\frac{dT}{T^2}=Q*\frac{dT}{T^2}$$

where we define the heat of transport, $Q^* = H - Q'$. It represents the difference between the energy associated with the atoms in the flows ($Q'$) and the enthalpy of atoms ($H$) in the reservoir from which the flow starts. If $H$ is bigger than $Q'$, or atoms are going from hot to cold, $Q^*$ is positive.

In iron–carbon alloys, Shewmon showed that under a temperature gradient, carbon atoms move from the cold end to the hot end and a steady state was established [1]. The value of $Q^*$ for carbon in iron is about −24 kcal/mol near 700 C.

Here, we must caution that we typically deal with an alloy which has two elements. Often, we find one of them is moving from hot to cold and the other is from cold to hot. Then, we might say the $Q^*$ is positive for the first one and yet $Q^*$ is negative for the second one. This could be wrong.

Consider a two-phase mixture of almost pure components; hereafter, the indexes 1 and 2 correspond to phases, as well as to species. The shape of the sample is assumed to be unchanged, so we have the constraint of constant volume in every part of the sample. It means that in the laboratory reference frame, the sum of volume fluxes of the two species should be zero.

$$\Omega_1 J_1 At=\Omega_2 J_2 At$$

or

$$\Omega_1 J_1=\Omega_2 J_2$$

Where $J_1$ and $J_2$ are fluxes of atoms per unit area per unit time, and $\Omega_1$ and $\Omega_2$ are atomic volumes, respectively. A is the cross-section of the sample, and $t$ is the reaction time. Under the assumption of constant volume, in the two-phase system, $J_2$ is in the opposite direction of $J_1$.

Thus, when we found that $J_1$ moves from hot to cold, its $Q^*$ is positive, and $J_2$ moves from cold to hot, its $Q^*$ is negative. Actually, $J_2$ is passive and is driven by $J_1$, rather than by the temperature gradient. Therefore, in the two-phase case, we should only consider that dominant flux in thermomigration, or we rather take the one which goes from hot to cold.

## 11.4 Thermomigration Due to Heat Transfer Between Neighboring Pairs of Powered and Unpowered Solder Joints

In Chapter 9, Section 9.6.1, we have discussed thermomigration in unpowered composite solder joint. It was found that the effect of thermomigration can be observed directly in a composite solder joint because the optical or SEM images of high-Pb and eutectic SnPb solders can be distinguished clearly. From Figures 9.9–9.11, the phenomenon of thermomigration in a composite solder joints have been explained, we will not repeat them here.

Furthermore, in Figure 9.4, we have shown the calculation of Joule heating in flip-chip technology, where typically every two solder joints are connected by an Al interconnect on the chip side, shown in Figure 9.4a. The Al interconnect is much hotter than the solder joint, in turn the chip side will be hotter than the substrate side, consequently there will be a temperature gradient across the solder joint. If the temperature gradient is large enough, thermomigration can occur in the solder joint. Figure 9.4b shows the simulation of current density distribution.

Figure 11.2 shows the temperature distribution on the cross-section of a solder joint when the applied current is 0.5 amp, measured by infrared sensor. The temperature is not uniform; the chip side is hotter than the substrate side because the Joule heating due to the short Al interconnect is greater. The temperature difference across the solder bump of 100 μm in height is about 5 °C, so the temperature gradient is about 500 °C/cm. This temperature gradient is large enough that thermomigration can occur in the solder bump. However, when the temperature gradient can be transferred laterally to the neighboring and unpowered solder joints, where thermomigration can occur.

To conduct thermomigration, the test sample was prepared as shown in the schematic diagram in Figure 11.3a. In a row of four composite solder joints, we passed current to conduct electromigration in the pair on the left-hand side. Due to Joule heating of the Al interconnect on the Si chip side at the top, the chip side is hotter than the substrate side at the bottom. Because Si is a good thermal conductor, this temperature difference extends to the pair of joints unpowered on the right-hand side. Owing to

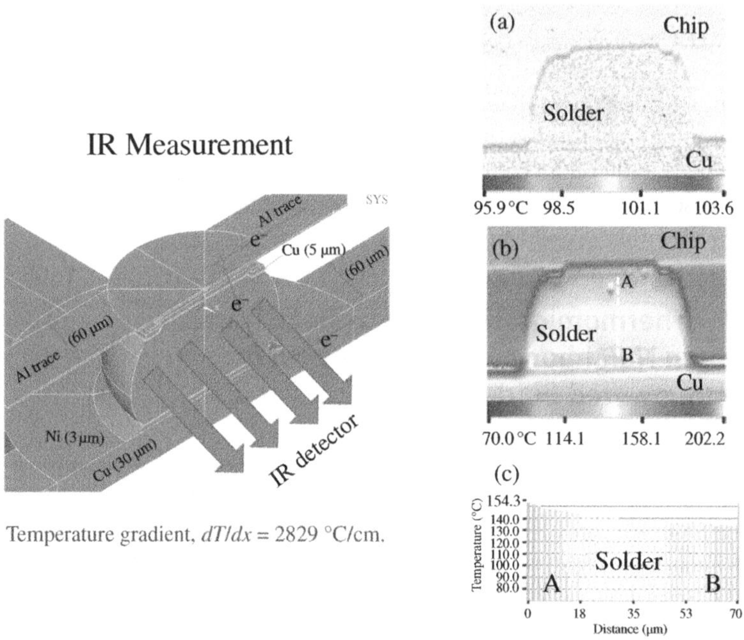


**Figure 11.2** The temperature distribution on the cross-section of a solder joint measured by infrared sensor. (a) Temperature map before applying current. The solder joint was placed on a hot plate maintained at 100 °C. (b) Temperature map when the applied current is 0.55 amp. (c) The temperature profile along the white dashed line AB in (b).

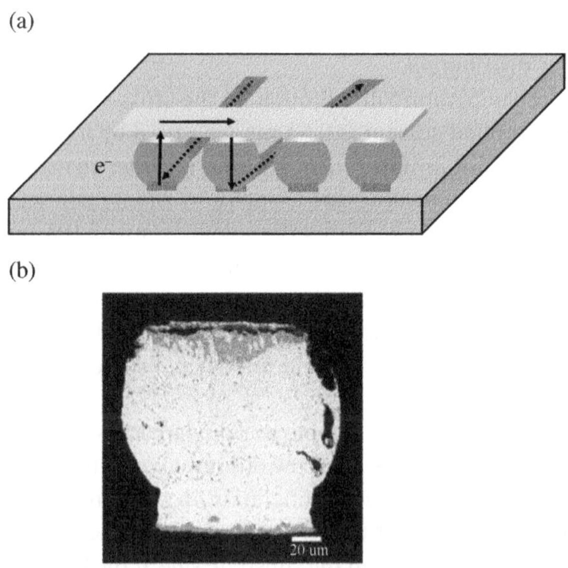


**Figure 11.3** (a) Schematic diagram of a row of four composite solders, in which the two on the left-hand side were tested by electromigration, but the two on the right-hand side were not. (b) Thermomigration was found to occur in the two unpowered composite solder joints on the right. The eutectic solder has migrated to the top. Also a pancake-type void has been formed on the top side.

the fact that there was no electric current passing through the pair on the right, only thermomigration occurs. The image in Figure 11.3b shows that the eutectic phase has moved to the top side. Also a pancake-type void is there because Pb is the dominant diffusing species going from hot to cold, so vacancies move up. The Sn atoms actually diffuse against the temperature gradient by going up. It is obvious that the pair of solder joints on the left should have experienced a combined effect of electromigration and thermomigration.

## Problems

**11.1** In Al and Cu interconnects, why is thermomigration unimportant and seldom mentioned? Yet in flip-chip solder joints, thermomigration becomes important?

**11.2** The heat of transport for most metals is negative, indicating that thermomigration of atoms is going from the hot end to the cold end. Why?

**11.3** If we anneal a diffusion couple of $Sn_{40}Pb_{60}$ and $Sn_{60}Pb_{40}$ at 150 °C, what will happen? Then if we anneal the SnPb couple at 200 °C, what will happen?

**11.4** It is known that in solder joint of eutectic SnPb, we observe electromigration when the applied current density is 10.4 amp/cm$^2$ at 150 °C. How large is the driving force in terms of $Z^*eE$? If we need force of the same magnitude in thermomigration, what will the temperature gradient be?

**11.5** To boil water in a copper kettle, we assume that the inside and outside temperatures are 100 and 600 °C, respectively, and the thickness of the wall of the copper kettle is 1 mm. Will thermomigration occur? Will the size of kettle become bigger and bigger with time?

## References


1 Shewmon, P. (1989). *Diffusion in Solids*, 2e. Warrendale, PA: TMS.

2 Roush, W. and Jaspal, J. (1982). Thermomigration in Pb-In solder. *IEEE Proceedings*, CH1781, 342–345.

3 Ye, H., Basaran, C., and Hopkins, D.C. (2003). Thermomigration in Pb-Sn solder joints under joule heating during electric current stressing. *Applied Physics Letters* 82: 1045–1047.

**4** Chuang, Y.C. and Liu, C.Y. (2006). Thermomigration in eutectic SnPb alloy. *Applied Physics Letters* 88: 174105.

**5** Huang, A., Gusak, A.M., Tu, K.N., and Lai, Y.-S. (2006). Thermomigration in SnPb composite flip chip solder joints. *Applied Physics Letters* 88: 141911.

**6** Hsiao, H.-Y. and Chen, C. (2009). Thermomigration in Pb-free SnAg solder joint under alternating current stressing. *Applied Physics Letters* 94: 092107.

**7** Hsiao, H.-Y., Chang, Y.-W., Ouyang, F. et al. (2012). Thermomigration in solder joints. *Materials Science and Engineering Research* 73: 85–100.

**8** Arpact, V.S. (1966). *Conduction Heat Transfer*. Reading, MA: Addition-Wiley.

# 12

# Stress-Migration

## 12.1 Introduction

Conceptually, there is a fundamental difference between stress-migration and electromigration or thermomigration. The latter two are cross-effects on the basis of irreversible processes, as discussed in the last three chapters. This is because the atomic flow in electromigration or thermomigration is accompanied by electron flow or heat flow, respectively. Yet, there is no "stress flow" accompanying the atomic flow in stress-migration, especially if we assume elastic stress. Stress-migration is a primary flow of atoms, driven by stress potential gradient, which is also a chemical potential gradient [1–3]. Often, stress-migration is called a steady-state diffusional creep.

Stress potential is defined as $\sigma\Omega$, where $\sigma$ is stress and $\Omega$ is atomic volume. Thus, the driving force of stress-migration is given as

$$F = -\frac{d\sigma\Omega}{dx} \tag{12.1}$$

And the atomic flux in stress-migration is given as

$$J = CMF = \frac{1}{\Omega}\frac{D}{kT}\left(-\frac{d\sigma\Omega}{dx}\right) = \frac{D}{kT}\left(-\frac{d\sigma}{dx}\right)$$

where in a pure metal, we have $C\Omega = 1$.

A common example of steady-state diffusional creep is the sagging of Pb pipes by their own weight under gravity in certain very old buildings. Room temperature is a relatively high temperature for Pb, whose melting point is 327 °C; therefore, atomic diffusion is sufficient for creep to occur over years. A modern application of creep is the use of pure and well-annealed O-ring

*Electronic Packaging Science and Technology*, First Edition. King-Ning Tu, Chih Chen, and Hung-Ming Chen.

of Cu as pressure seal in ultrahigh vacuum systems, where the compressive stress under tightened screws will allow the diffusion or redistribution of a few Cu atoms to fill in any tiny gaps in the seal. In packaging technology, an example of stress-migration is the growth of Sn whisker under a compressive stress gradient, to be discussed in this chapter.

## 12.2 Chemical Potential in a Stressed Solid

Under a steady state of stress (or a constant load), even it is within elastic limit, the solid will respond by a slow deformation or relaxation due to diffusional creep. Since elastic strain energy is small as compared to chemical energy, so we shall deal with creep in pure metals only, in the absence of chemical effects. We have the change of Helmholtz free energy $F$ as

$$dF = -SdT - pdV$$

If the change occurs at constant temperature as in room temperature creep, we have

$$p = -\frac{\partial F}{\partial V}$$

The above equation can be interpreted to mean that stress (pressure) is an energy density (i.e. energy per unit volume). For a given volume, the energy change equals the energy density times the given volume. So, for an atomic volume, $\Omega$, we have

$$p\Omega = -\frac{\partial F}{\partial V}\Omega = -\frac{\partial F}{\partial\left(\frac{V}{\Omega}\right)} = -\frac{\partial F}{\partial N}$$

where $N$ is the number of atoms in volume $V$. We note that the last term is chemical potential by definition, where the negative sign is used to indicate that a decrease in volume by pressure results in an increase in energy because pressure is a compressive stress, which is negative. Since chemical potential is defined by Helmholtz (or Gibbs) free energy per atom, the chemical potential change in a stress solid is

$$\mu = \pm\sigma\Omega \tag{12.2}$$

where the positive or negative sign refers, respectively, to external tensile or compressive stress.

To gain a quantitative feeling of the magnitude of $\sigma\Omega$, we consider a piece of Al stressed at the elastic limit (i.e. strain is 0.2%). Young's modulus for Al is $Y = 6 \times 10^{11}$ dyne/cm$^2$, so the stress is

$$\sigma = Y\varepsilon = 1.2\times10^9\ \text{dyne}/\text{cm}^2 = 1.2\times10^9\ \text{erg}/\text{cm}^3$$

Since Al has a face-centered cubic lattice with a lattice constant of 0.405 nm, there are four atoms in a unit cell of $(0.405)^3$ or $0.602\times19^{23}$ atoms/cm$^3$. Thus

$$\sigma\Omega = \frac{1.2x10^9 erg}{0.602x10^{23} atom} = 2x10^{-14} erg/atom = 0.0125 eV/atom$$

To compare this value to the value of strain energy per atom below, we have

$$E_{elastic} = \int \sigma d\varepsilon = \frac{1}{2}Y\varepsilon^2 = \frac{1}{2}x1.2x10^9 erg/cm^3 x\frac{2}{1000}$$
$$= 1.2x10^6 erg/cm^3 = 10^{-5} eV/atom$$

which is much smaller than the stress potential energy. The strain energy is the increase in energy per atom due to the strain. The stress potential energy is the energy changes in adding or removing one atom from the stressed solid.

In thermally activated processes such as diffusion, the chemical potential enters as an exponential factor. We consider the case of Al stressed to its elastic limit at 400 °C, we have $kT = 0.058$ eV, and

$$\exp-\left(\frac{\sigma\Omega}{kT}\right) = \exp-\left(\frac{0.0125}{0.058}\right) = 1.23$$

Usually creep occurs at a much lower stress (or $\sigma\Omega << kT$), so we can linearize the exponential term by taking,

$$\exp\left(-\frac{\sigma\Omega}{kT}\right) = 1 - \frac{\sigma\Omega}{kT}$$

Now, we consider the deposition of an Al thin film on a thick quartz substrate at room temperature, as shown in Figure 12.1a. Then, we arise the temperature to 400 °C. Owing to the expansion of Al, the sample will bend downward and the Al is under compression, as shown in Figure 12.1b. If we keep the sample at 400 °C for a while, creep will occur to relax the bending, as shown in Figure 12.1c. Then we lower the temperature to 100 °C, the Al film is under tension, as shown in Figure 12.1d. Because the creep rate at 100 °C is slow, we can measure the bending curvature to determine the tensile stress in the Al film, to be discussed below.

The thermal expansion co-efficient for Al and quartz in the above temperature range are about $\alpha = 25\times10^{-6}$/C and $\alpha = 0.5\times10^{-6}$/C, respectively. Thus, the thermal strain is

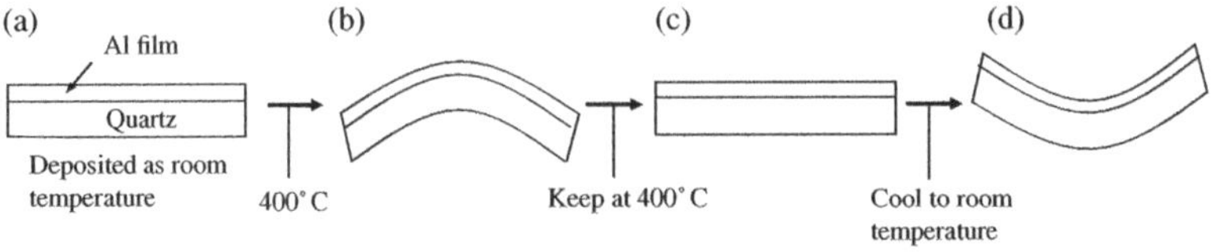


**Figure 12.1** (a) The deposition of an Al thin film on a thick quartz substrate at room temperature. (b) We rise the temperature to 400 °C. Owing to the expansion of Al, the sample will bend downward and the Al is under compression. (c) If we keep the sample at 400 °C for a while, creep will occur to relax the bending. (d) Then we lower the temperature to 100 °C, the Al film is under tension.

$$\varepsilon = \Delta\alpha\Delta T = 25\times10^{-6}\times300 = 0.75\%$$

Then, the thermal stress is

$$\sigma = Y\varepsilon = 4.5\times10^{9}\,\text{dyne}/\text{cm}^{2} \text{ and } \sigma\Omega = 0.045\,\text{eV}$$

## 12.3 Stoney's Equation of Biaxial Stress in Thin Films

To measure the stress in the Al thin film on quartz as mentioned above, we note that the stress is biaxial [4]. As shown in Figure 12.2a, the stresses act along the two principal axes in the plane of the film, but there is no stress in the direction normal to the free surface of the film. But there is strain in the normal direction.

To express the biaxial stress, we start with a three-dimensional isotropic cubic structure as shown in Figure 12.2b. The linear dimensions in the *x*, *y*, and

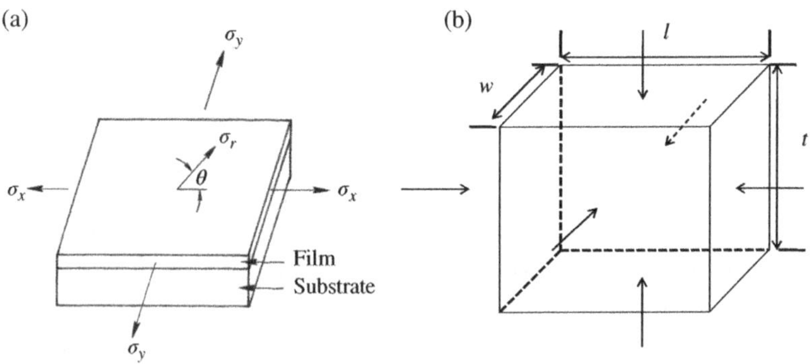


**Figure 12.2** (a) It shows that the stresses act along the two principal axes in the plane of the film, but there is no stress in the direction normal to the free surface of the film. But there is strain in the normal direction. (b) To express the biaxial stress, we start with a three-dimensional isotropic cubic structure. The linear dimensions in the *x*, *y*, and *z* axes are *l*, w, and t, respectively.

$z$ axes are $l$, $w$, and $t$, respectively. We begin by considering the bulk structure under a hydrostatic compression. We shall apply the compression in the $x$, $y$, and $z$ directions in sequence. First, we apply the pressure $p$ in the x-direction, so

$$p = -Y\frac{\Delta l_1}{l}$$

Thus, the strain in the x-direction is

$$\frac{\Delta l_1}{l} = -\frac{p}{Y}$$

Second, we apply the compression in the y-direction, and we have

$$\frac{\Delta w}{w} = -\frac{p}{Y}$$

Owing to Poisson's effect, we have tensile strain in the x-direction given by

$$\frac{\Delta l_2}{l} = +v\frac{p}{Y} = -v\frac{\Delta w}{w}$$

Third, we apply the compression in the z-direction, and we have tensile strain in the x-direction given by

$$\frac{\Delta l_3}{l} = +v\frac{p}{Y}$$

The total strain in the x-direction is given by

$$\frac{\Delta l}{l} = \frac{\Delta l_1}{l} + \frac{\Delta l_2}{l} + \frac{\Delta l_3}{l} = -\frac{p}{Y}(1-2v)$$

Or

$$\varepsilon_x = -\left(\frac{\sigma_x}{Y} - v\frac{\sigma_y}{Y} - v\frac{\sigma_z}{Y}\right) = -\frac{1}{Y}\left[\sigma_x - v\left(\sigma_y + \sigma_z\right)\right]$$

Now we change the stress to tension and we have the following equations.

$$\varepsilon_x = \frac{1}{Y}\left[\sigma_x - v\left(\sigma_y + \sigma_z\right)\right]$$

$$\varepsilon_y = \frac{1}{Y}\left[\sigma_y - v\left(\sigma_x + \sigma_z\right)\right]$$

$$\varepsilon_z = \frac{1}{Y}\left[\sigma_z - v\left(\sigma_x + \sigma_{zy}\right)\right]$$

In thin film biaxial stress state, we assume there is tensile stress within the plane of the film ($x$ and $y$), but no stress in the z-direction ($\sigma_z = 0$). Therefore,

$$\varepsilon_x = \frac{1}{Y}\left(\sigma_x - \nu\sigma_y\right)$$

$$\varepsilon_y = \frac{1}{Y}\left(\sigma_y - \nu\sigma_x\right) \tag{12.3}$$

$$\varepsilon_z = -\frac{\nu}{Y}\left(\sigma_x + \sigma_y\right)$$

From these equations, we have

$$\varepsilon_x + \varepsilon_y = \frac{1-\nu}{Y}\left(\sigma_x + \sigma_y\right)$$

$$\varepsilon_z = -\frac{\nu}{1-\nu}\left(\varepsilon_x + \varepsilon_y\right)$$

In two-dimensional isotropic systems where $\varepsilon_x = \varepsilon_y$, we obtain

$$\varepsilon_z = -\frac{2\nu}{1-\nu}\varepsilon_x$$

$$\sigma_x = \left(\frac{Y}{1-\nu}\right)\varepsilon_x \tag{12.4}$$

Later, we shall apply the relations above to obtain Stoney's equation of the stress of a thin film on a substrate. Now, we begin the thin film stress analysis by assuming that the film thickness $t_f$ is much smaller than that of the substrate thickness of $t_s$, so the neutral plane where there is no stress can be taken to be at the middle of the substrate.

In Figure 12.3a, we have enlarged one end of the substrate to show the neutral plane, the stress distribution in the film, as well as in the substrate, and the corresponding forces and moments. At equilibrium, the moment produced by the stress in the film must be equal to the moment produced by the stress in the substrate, see Figure 12.3b. Because we have assumed that the film thickness is thin, the stress $\sigma_f$ is uniform across the film thickness. The moment $M_f$ (force times the perpendicular distance) due to the force in the film with respect to the neutral plane is

$$M_f = \sigma_f W t_f \frac{t_S}{2} \tag{12.5}$$

where $W$ is the width of film normal to $t_f$. Next, to calculate the moment of the substrate, we first obtain the geometrical relation that

$$\frac{d}{r} = \frac{\Delta d}{t_S/2}$$

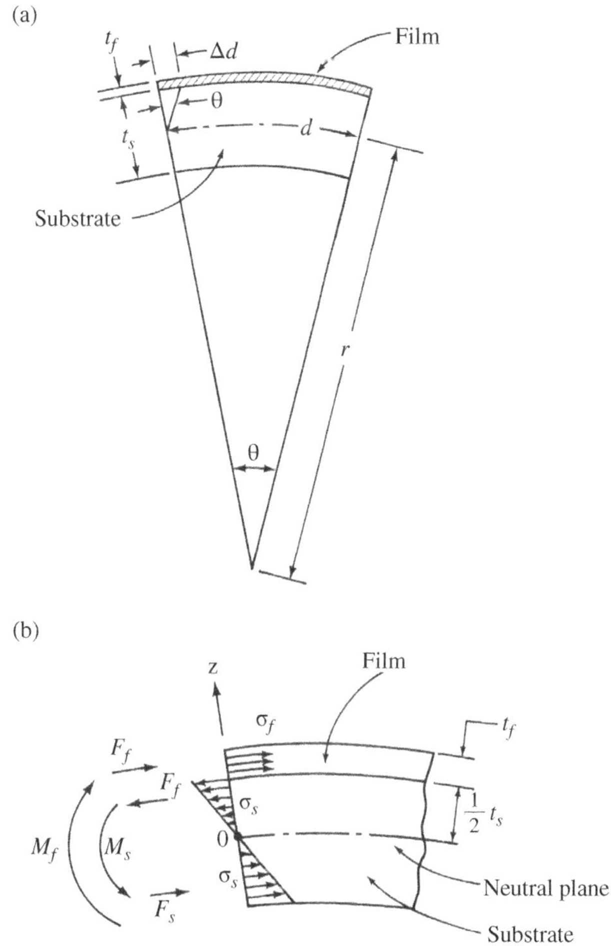


**Figure 12.3** (a) We have enlarged one end of the substrate to show the neutral plane, the stress distribution in the film, as well as in the substrate, and the corresponding forces and moments. (b) At equilibrium, the moment produced by the stress in the film must be equal to that produced by the stress in the substrate.

So,

$$\frac{1}{r} = \frac{\Delta d}{dt_S/2} = \frac{\varepsilon_{\max}}{t_S/2} \tag{12.6}$$

where $r$ is the radius of curvature of the substrate measured from the neutral plane, $d$ is an arbitrary length of the substrate measured at the neutral plane, and $\Delta d/d = \varepsilon_{max}$ is the strain measured at the outer surface of the substrate. Within the substrate, the elastic strain is zero at the neutral plane, yet it increases linearly with distance $z$, which is measured from the neutral plane (i.e. it obeys Hooke's law and increases linearly with stress), so that

$$\frac{\varepsilon_S(z)}{z} = \frac{\varepsilon_{\max}}{t_S/2} = \frac{1}{r}$$

where $\varepsilon_S(z)$ is the strain in a plane which is parallel to the neutral plane and is at a distance of $z$ from the neutral plane. Then, by assuming a state of biaxial stress in the substrate, we have from Eq. (12.4) that

$$\sigma_S(z) = \left(\frac{Y}{1-\nu}\right)_s \varepsilon_S(z) = \left(\frac{Y}{1-\nu}\right)_S \frac{z}{r} \tag{12.7}$$

Therefore, the moment produced by the stress in the substrate is

$$M_S = w\int_{-t/2}^{t_S/2} z\sigma(z)dz = w\int_{-t_S/2}^{t_S/2}\left(\frac{Y}{1-\nu}\right)_S \frac{z^2}{r}dz = \left(\frac{Y}{1-\nu}\right)_s \frac{Wt_S^3}{12r} \tag{12.8}$$

By equating $M_S$ to $M_f$, we obtain Stoney's equation below,

$$\sigma_f = \left(\frac{Y}{1-\nu}\right)_S \frac{t_S^2}{6rt_f} \tag{12.9}$$

where the subscripts $f$ and $s$ refer to film and substrate, respectively. The above equation shows that by measuring the curvature and the thickness of the film and the substrate, and knowing the Young's modulus and Poisson's ratio of the substrate, the biaxial stress in the thin film can be determined. The curvature can be measured by bending beam or laser interference.

## 12.4 Diffusional Creep

Creep is a time-dependent mechanical or deformation behavior of solids. The deformation involves atomic diffusion when the applied load or stress is steady. In Figure 12.4, we depict a hexagonal grain in a polycrystalline pure metal under a shear stress "S," and it can be regarded as under a combination of tensile and compressive stresses, or vice versa. This can be verified by using vector sum because stress (force per unit area) is a vector. The effect of the shear stress or the combined tensile and compressive stresses is to deform the grain from its original shape delineated by the solid lines to that delineated by the broken lines.

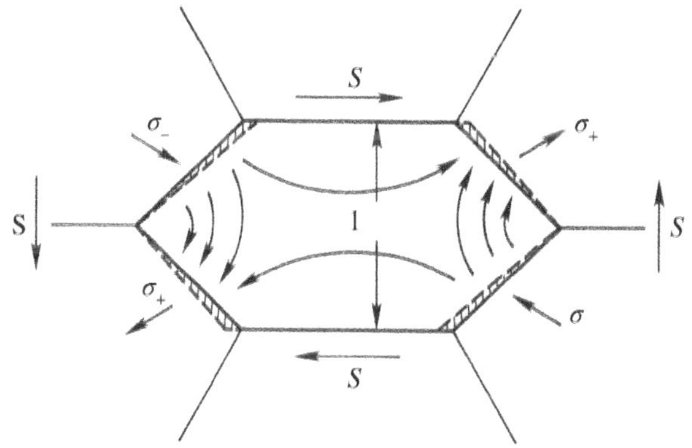


**Figure 12.4** A hexagonal grain in a polycrystalline solid is under a shear stress, "S," which can be decomposed into a pair of tensile and compressive stresses.

If the stress persists, the solid can be deformed slowly by transporting the part of atoms in the shaded areas from the compressive region to the tensile region in order to release the stress. The transport can occur by atomic diffusion as indicated by the curved arrows. To analyze this problem, we shall follow the Nabarro-Herring creep model. In this model, it is assumed that surface and grain boundaries are effective sources and sinks of vacancies, which mediate the mass transport by atomic diffusion or vacancy diffusion in the lattice.

In the tensile region, the chemical potential, which deviates from the equilibrium value, $\mu_0$, due to the stress is given by

$$\mu_1 - \mu_0 = \sigma\Omega$$

Similarly, in the compressive region,

$$\mu_2 - \mu_0 = -\sigma\Omega$$

Thus, the chemical potential difference in going from the compressive region to the tensile region is

$$\Delta\mu = \mu_2 - \mu_1 = -2\sigma\Omega$$

This potential difference will drive atoms to diffuse from the compressive regions to the tensile regions. The force acting on the diffusing atoms is given below,

$$F = -\frac{\Delta\mu}{\Delta x} = \frac{2\sigma\Omega}{l}$$

where $l$ is the grain size. The flux of the diffusing atoms can be given as

$$J = CMF = C\frac{D}{kT}\left(\frac{2\sigma\Omega}{l}\right) = \frac{2\sigma D}{kTl} \tag{12.10}$$

In the above, we have taken $C\Omega = 1$ for a pure metal. The number of atoms transported by a flux in a period "$t$" and through an area "$A$" is $N' = JAt$, then the volume accumulated by the atomic flux is $\Omega N' = \Omega JAt$. The strain is,

$$\varepsilon = \frac{\Delta l}{l} = \frac{\frac{\Omega N}{A}}{l} = \frac{\Omega Jt}{l} \tag{12.11}$$

So, the strain rate or the creep rate is

$$\frac{d\varepsilon}{dt} = \frac{\Omega J}{l} = \frac{2\sigma\Omega D}{kTl^2} \tag{12.12}$$

This is the well-known Nabarro-Herring creep equation. It shows an inverse dependence on $l^2$.

We note that the above derivation is obtained by considering the flux of atom in the lattice. However, if we assume lattice diffusion occurs by vacancy mechanism, we should be able to obtain the same equation by considering the inverse flux of vacancies. To do so, we need to deal with the change of concentration of vacancy in the tensile region and the compressive region.

The formation energy of a vacancy in a face-centered-cubic (FCC) metal is to break all the 12 nearest-neighbor bonds of the atom and remove it and place it on the metal surface. Under tension, the breaking of the bonds will be easier because the tensile stress has stretched the bonds slightly, but under compression, it will be harder because the compressive stress has shortened the bonds slightly. Thus, we can express the change of concentration of vacancy in a stressed solid as,

$$C_v^{\pm} = C\exp\left[\left(-\Delta G_f \pm \sigma\Omega\right)/kT\right] \tag{12.13}$$

where $C_v^{+} and C_v^{-}$ correspond to the vacancy concentration in the tensile and compressive region, respectively. Assuming $\sigma\Omega << kT$, we have

$$C_v^{\pm} = C_v\left(1 \pm \frac{\sigma\Omega}{kT}\right)$$

where $C_v = C \exp(-\Delta G_f/kT)$ is the concentration of vacancies in the equilibrium state. Then the vacancy concentration difference between the tensile region and the compressive region is

$$\Delta C_v = C_v^+ - C_v^- = C_v \frac{2\sigma\Omega}{kT} \quad (12.14)$$

Thus, the flux of vacancy going from the tensile region to the compressive region is

$$J_v = -D_v \frac{\Delta C_v}{\Delta x} = -\frac{2\sigma\Omega D_v C_v}{kTl} \quad (12.15)$$

where $D_v$ is the diffusivity of vacancy, by taking $DC = D_v C_v$, we obtain

$$J = \frac{2\sigma\Omega DC}{kTl} = \frac{2\sigma D}{kTl} \quad (12.16)$$

which is the same as Eq. (12.10). Thus, whether we consider atomic diffusion or vacancy diffusion, the creep equation is the same. The advantage of considering vacancy diffusion is that when the creep occurs in a microstructure, in which there are interfaces, the vacancy flux can lead to void formation. To calculate the void volume or rate of void growth, it will be easier to consider vacancy diffusion.

In Eq. (12.12), if we plot ln $(T\, d\varepsilon/dt)$ versus $1/kT$, we can determine the activation energy of creep, which has been found to be the same as the activation energy of lattice diffusion in the solid. Indeed, this correlation has been verified in the literature.

## 12.5 Spontaneous Sn Whisker Growth at Room Temperature

### 12.5.1 Morphology

Figure 12.5a is SEM image of a Sn whisker between two Cu bumps having a top layer of Sn, and the whisker is an electrical short. Figure 12.5b shows no Sn whisker when a Ni layer was added between the Cu bump and the Sn layer. The mechanism of spontaneous Sn whisker growth is discussed below.

The lead-frame in electronic packaging, where the Cu legs have a surface coating of Sn about 15 μm thick, enables the spontaneous growth of Sn whiskers at room temperature, as shown in Figure 12.6a and b. The cross-sectional TEM image of Sn whiskers and the corresponding electron diffraction patterns are shown, respectively, in Figure 12.7a and b.

Typically, whiskers occur on the surface of a bilayer of Sn on Cu, because Cu diffuses interstitially in Sn, and Cu and Sn react at room temperature to form $Cu_6Sn_5$. The growth of $Cu_6Sn_5$ along the grain boundaries in Sn

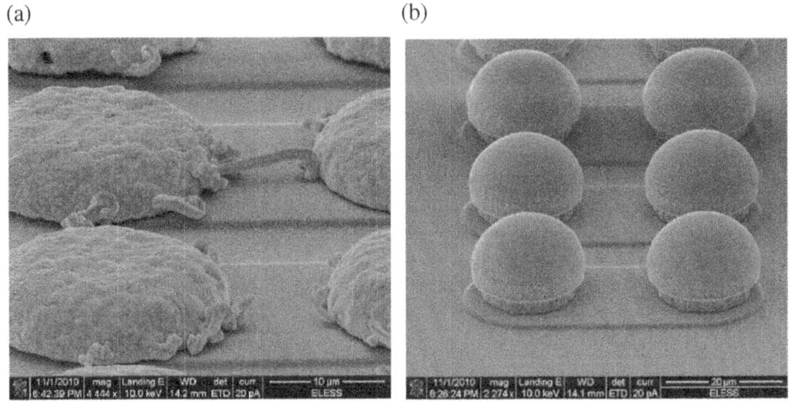


**Figure 12.5** (a) SEM image of a Sn whisker between two Cu bumps having a top layer of Sn, and the whisker is an electrical short. (b) SEM image shows no Sn whisker when a Ni layer was added between the Cu bump and the Sn layer.

**Figure 12.6** (a) SEM image of Sn whiskers on a lead-frame surface. (b) A short whisker of the lead-frame surface.

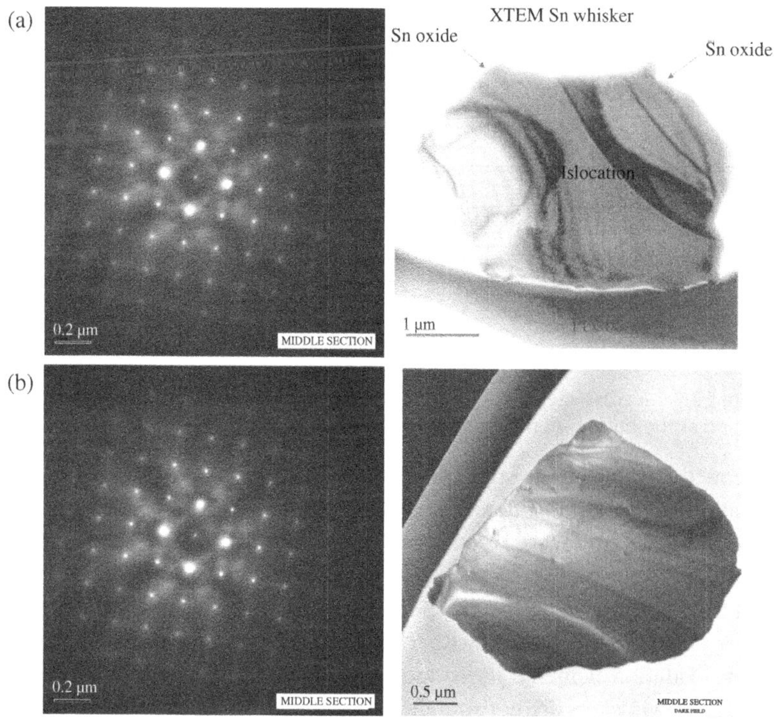


**Figure 12.7** Both (a) and (b) are electron diffraction pattern and cross-sectional image of Sn whiskers. The electron diffraction patterns show the growth orientation of the Sn whisker is (001).

generates a compressive stress in the surrounding region of Sn. The stress is compressive because Cu atoms are added to a fixed volume of Sn. To release the compressive stress, some of the Sn atoms must diffuse away, but they need a stress gradient as the driving force. The growth of the whisker exerts a pulling force at the interface between the whisker and its substrate. Because the whisker is covered completely by Sn oxide, so the pulling can induce a crack in the surface oxide on Sn near the root of the whisker. The crack exposes a free surface of Sn, which is stress-free. Thus, there is a stress gradient to drive the Sn atoms to the root of the whisker, so the whisker can grow.

Figure 12.8 depicts a cross-sectional view of a whisker on a bilayer of Sn/Cu. A crack opening at the bottom of the whisker is indicated by an arrow. The crack exposes a free surface of Sn, which has no normal stress, so a stress gradient exists between the crack surface and the compressive region in Sn. The compressive region is indicated by a broken line square in Figure 12.8. The stress gradient drives Sn atoms to diffuse to the surrounding of the crack

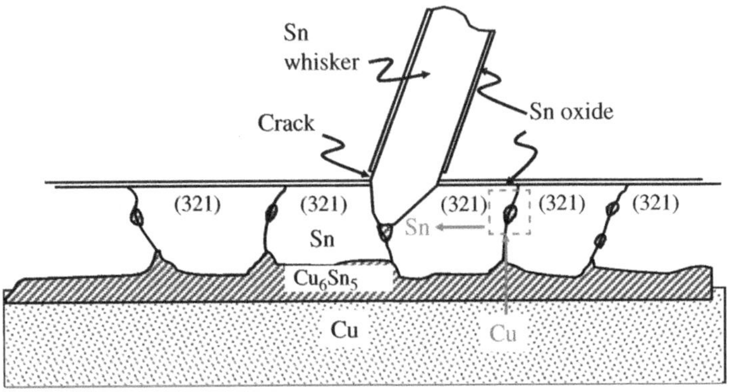


**Figure 12.8** A cross-sectional view of a whisker on a bilayer of Sn/Cu. A crack opening at the bottom of the whisker is indicated by an arrow.

to grow the whisker. The stress gradient is small, measured by synchrotron radiation micro-X-ray diffraction, which will be discussed later. We can regard whisker growth as a room temperature recrystallization, in which the Sn under compressive stress transforms to a stress-free Sn whisker.

In the literature, it has been reported that adding rare earth elements to solder enhances whisker growth. This is because rare earth elements are strong oxidation elements, its oxide cracks the Sn oxide and whisker follows.

Often Sn whiskers are not straight and it bends. This can be explained by the oxide formation at the whisker root. If part of the oxide surrounding the root is not broken, the grow stops on this part. Then the growth on the opposite side will lead to tilting of the whisker. Yet when the whisker has bended to 90 degrees, it cannot bend more, then a straight growth will resume.

Even though we understand the mechanism of whisker growth, how to achieve no whisker growth in a device remains very challenging, because we cannot guarantee that not even one whisker is allowed to grow in a device which requires high reliability such as a satellite. This is why whisker research is still active!

The condition of completely free of Sn whisker growth is difficult to meet, except if we can get rid of Sn-based solder joints or we cover the entire surface of the device by a strong coating. The Ni layer in Figure 12.5b is to block the diffusion of Cu into Sn to prevent the growth of $Cu_6Sn_5$.

In the next section, we shall present a direct measurement of the stress gradient near the root of a whisker, which turns out to be quite small. Combining with the fast kinetics of interstitial diffusion of Cu in Sn, as well as the fast self-diffusion of Sn in Sn at room temperature because of its low melting point, we can understand why Sn whisker growth is very easy and why it is a unique metallurgical phenomenon.

### 12.5.2 Measurement of the Driving Force to Grow a Sn Whisker

Figure 12.9 shows an SEM image of Sn whiskers on a lead-frame surface. The circle indicates the root of the Sn whisker to be studied by synchrotron radiation micro-X-ray diffraction. The image of the tip of the whisker is out of focus because it sticks out. Figure 12.10 shows the area surrounding the Sn whisker (in red color) measured by synchrotron radiation micro-X-ray diffraction [5, 6]. Figure 12.11 shows the measured distribution of compressive stress in the surrounding area around the root of the whisker.

In Figure 12.11, we estimate the stress gradient between the origin ($x = 0$, $y = 0$) at the root of the whisker and the lower left corner point at ($x = -0.5400$, $y = 0.8475$). We found that $\Delta x \approx 10\,\mu\text{m}$ and $\Delta\sigma \approx 4\,\text{MPa}$, and we take the atomic volume of Sn atom to be $27 \times 10^{-24}\,\text{cm}^3$. The driving force is given below,

$$F = -\frac{\Delta\sigma\Omega}{\Delta x} = \frac{4x10^{6}\left(N/m^{2}\right)x27x10^{-24}\left(cm^{3}\right)}{10^{-3}\left(cm\right)}$$
$$= \frac{4x10^{7}\left(dyne/cm^{2}\right)x27x10^{-24}\left(cm^{3}\right)}{10^{-3}cm}$$
$$= \frac{108x10^{-17}erg}{10^{-3}cm} \approx 10^{-12} erg/cm$$

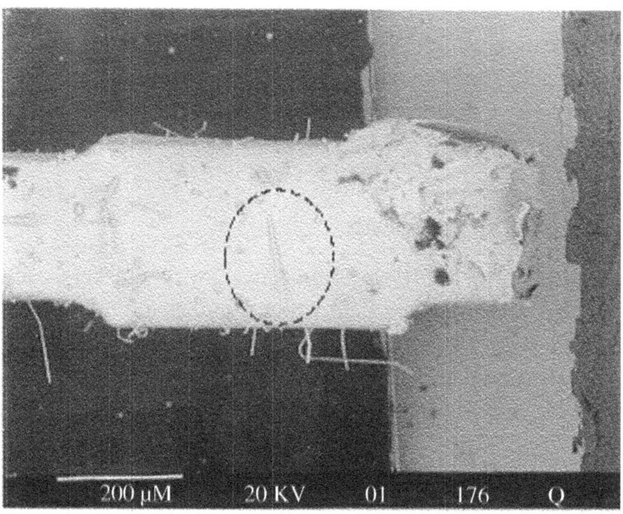


**Figure 12.9** SEM image of a Sn whisker on a leg of lead-frame indicated by a circle. The tip of the whisker of out of focus.

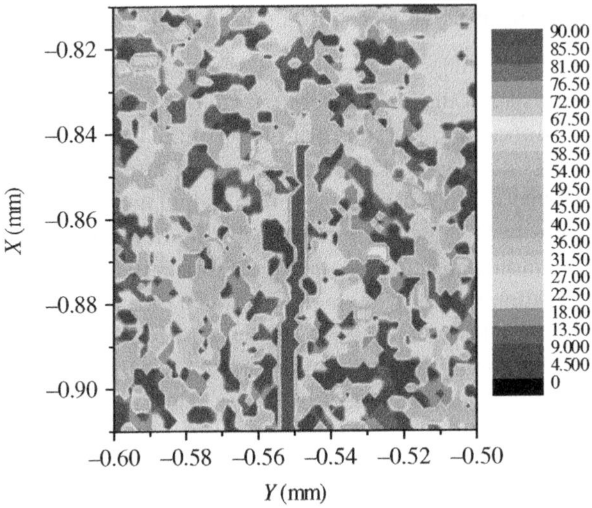


**Figure 12.10** Synchrotron radiation micro-diffraction was used to measure the stress distribution around the root of the whisker as shown in Figure 12.9. The *x* and *y* coordinates enabled us to locate the position precisely.

The work done by this force over a distance of atomic diameter of 0.3 nm is $3 \times 10^{-27}$ joule, which is of the same order of magnitude of those calculated for electromigration and thermomigration in Sn, to be discussed in the next Section. The small driving force needed to grow a whisker is the reason why Sn whisker is so common. Whisker growth has also been found to take place at the anode of a Sn-based solder joint in electromigration, to be discussed in Section 12.5.4. Current crowding occurs at the anode and a whisker was found to be squeezed out.

### 12.5.3 Kinetics of Sn Whisker Growth

To model the Sn whisker growth [7], we assume an array of whiskers, each has a diameter of "2a," and the stress field which supplies atomic flux for the growth has a radius of "b," as shown in Figure 12.12. To evaluate the driving force in the stress field under a cylindrical coordination,

$$F = -\frac{\partial \mu}{\partial r}$$

we consider the early stage of growth, assuming a steady state for simplicity. Thus, we solve in a cylindrical coordinate, two-dimensional, and stead-sate continuity equation of the potential $\sigma\Omega$.

1.5 µm

(Unit:MPa)

| | –0.5400 | –0.5415 | –0.5430 | –0.5445 | –0.5460 | –0.5475 | –0.5490 | –0.5505 | –0.5520 | –0.5535 | –0.5550 |
|---|---|---|---|---|---|---|---|---|---|---|---|
| –0.8340 | –2.82 | –3.21 | –2.26 | 0.93 | 0.93 | –0.23 | –8.17 | 2.22 | 1.49 | 1.6 | –0.03 |
| –0.8355 | –2.26 | –2.64 | –2.64 | –1.04 | 1.37 | 1.37 | –1.31 | 0.87 | 0.87 | 0.87 | –0.7 |
| –0.8370 | –2.53 | –3.21 | –3.21 | –2.64 | –1.04 | 3.61 | 0.75 | 0.87 | 0.7 | 0.7 | –3.19 |
| –0.8385 | –7.37 | –9.62 | –6.57 | –2.64 | 3.61 | 4.52 | 3.61 | 0.29 | –1.31 | 0 | –4.79 |
| –0.8400 | –7.37 | –8.22 | –6.57 | –1.18 | 0.75 | 4.23 | 0.75 | –2.25 | –2.27 | –2.91 | –6.91 |
| –0.8415 | –4.17 | –4.84 | –4.17 | –1.81 | –0.67 | 0.00 | –1.96 | –1.96 | –3.74 | –5.08 | –5.08 |
| –0.8430 | –4.17 | –4.17 | –3.63 | –1.81 | –1.81 | –2.29 | –2.29 | –1.96 | –1.96 | –3.27 | –3.27 |
| –0.8445 | –4.14 | –4.17 | –3.86 | –3.63 | –2.79 | –4.64 | –4.78 | –0.84 | –1.4 | –1.49 | –3.27 |
| –0.8460 | –3.14 | –3.63 | –3.86 | –3.63 | –3.13 | –4.78 | –4.78 | 0.04 | 0.04 | –1.41 | –2.33 |
| –0.8475 | –4.14 | –4.49 | –4.49 | –4.64 | –3.86 | –6.04 | –1.72 | 3.55 | 3.55 | –0.41 | –2.33 |
| –0.8490 | –3.33 | –5.67 | –6.29 | –6.29 | –2.66 | –2.08 | –1.72 | –1.79 | 0 | –1.79 | –3.73 |

Whisker

**Figure 12.11** The measured distribution of compressive stress in the surrounding area of the root of the whisker. In the distribution, we estimate the stress gradient between the origin ($x = 0, y = 0$) at the root of the whisker and the lower left corner point at ($x = -0.5400, y = 0.8475$). We found that $\Delta x \approx 10\,\mu m$ and $\Delta\sigma \approx 4\,MPa$.

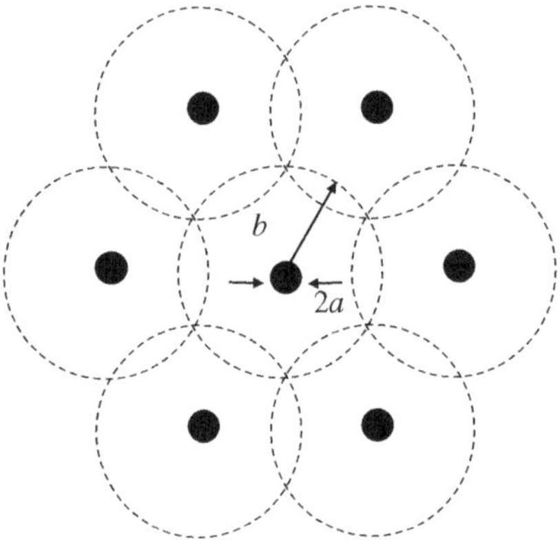


**Figure 12.12** To model the whisker growth, we assume an array of them, each has a diameter of "2a," and the stress field which supplies atomic flux for the growth has a radius of "b."

$$\nabla^2\sigma = \frac{\partial^2\sigma}{\partial r^2} + \frac{1}{r}\frac{\partial\sigma}{\partial r} = 0$$

The reason why we can use the continuity equation to solve "$\sigma$" is that it is a density-like function, which has been discussed in Section 12.2.

The solution has the form of $\sigma = A'' \ln r + B''$. Using the boundary condition at $t$ (time) = 0, which is taken to be the very beginning of whisker growth, we have

$$\sigma = \sigma_0, \text{where}\, r = b.$$

$$\sigma = \sigma, \text{where}\, r = a.$$

we obtain

$$A'' = \frac{\sigma_0}{\ln(b/a)} = \sigma_0 A' B'' = -\frac{\sigma_0 \ln a}{\ln(b/a)}$$

So $\sigma = A'\sigma_0 \ln r + B''$

Then, $F = -\frac{\partial\mu}{\partial r} = -\frac{\partial\sigma\Omega}{\partial r} = -A\frac{\sigma_0\Omega}{r}$

Knowing $F$, the atomic flux at $r = a$ is given,

$$J = CMF = C\frac{D}{kT}\left(-A\frac{\sigma_0\Omega}{a}\right) = -A\frac{\sigma_0 D}{kTa}$$

Hence, the amount of material accumulated at the base of the whisker in a period of $t$ will be equal to the volume of the whisker growth, or $JAt\Omega = h\pi a^2$, where $A$ is the peripheral area of the whisker base, so $A = 2\pi r\delta$, where $\delta$ is the effective thickness of the interface between the whisker and the Sn substrate, and it is about 0.3–0.5 nm.

The growth rate of the whisker will be

$$\frac{dh}{dt} = \frac{JA\Omega}{\pi a^2} = \frac{2\pi a\delta\Omega}{\pi a^2}\left(-A\frac{\sigma_0 D}{kTa}\right) = -\frac{A\,2\sigma_0\Omega\delta D}{kTa^2}$$

where $D$ is the diffusivity along the interface of width $\delta$. The final height of the whisker $h_f$ can be estimated by using the law of conservation of mass, so

$$\pi a^2 h_f = \pi b^2 h\,\varepsilon$$

or

$$h_f = \frac{b_2 h'}{a^2}\sigma_0\frac{(2-3\nu)}{Y}$$

where $h'$ is the thickness of the Sn film and $\varepsilon = \sigma_0(2-3\nu)/Y$ is the strain of the film, and $\nu$ is Poisson's ratio and $Y$ is Young's modulus. The estimate rate and the height of whisker are in reasonable agreement with experimental data.

### 12.5.4 Electromigration-Induced Sn Whisker Growth in Solder Joints

To study electromigration in the anode side of flip-chip solder joints, we need the simulation of current distribution in a pair of solder joints in order to complete the current flow-in and flow-out of them. In Chapter 9, Figure 9.4 depicts a pair of them between a chip of Si on the top side and a polymer substrate on the bottom side. The two solder bumps are connected by a short Al interconnect on the chip side, and they are connected separately on the substrate side each by a Cu bond-pad to the outside. As indicated by the arrows in Figure 9.4a, electron current can be applied from one of the bond-pads to flow through one bump, then the Al interconnect, the other bump, and to get out from the other bond-pad. The dimension and resistance of the Al interconnect, the under-bump-metallization, the solder bump, and the Cu bond-pad are known. We can simulate the current distribution in the flip-chip structure, as shown in Figure 11.3b.

From the current distribution simulation, two features are important. First, the Al interconnect has the highest current density, and second, the current density distribution in the solder bump is nonuniform. The first implies that the Al is the hottest part during device operation because Joule heating is proportional to the square of current density, as in $j^2\rho$. There will

**Figure 12.13** A set of four SEM images of the initiation and growth of a Sn whisker at the anode side of a Pb-free solder joint in electromigration. After 148 hours, a Sn whisker has been pushed out at the upper right corner of the Pb-free solder joint.

be a temperature gradient across the bump because the Joule heating in the Cu bond-pad will be less. The second is that there is current crowding in the solder joint, where electrons enter the solder bump from the Al interconnect, as well as exit the bump into the Al. At the entrance, electromigration will lead to pancake-type void formation as shown in Chapter 10, Figure 10.8, but at the exit, Sn whisker formation can happen, see Figure 12.13.

Figure 12.13a–d is a set of four SEM images of the initiation and the subsequent growth of a Sn whisker at the anode end of the Pb-free solder joint under electromigration [8]. Electrons flew in from the lower left corner and exited from the upper right corner, as indicated by the long arrow. Thus, electromigration is driving Sn atoms to the anode end at the upper right corner. Because the surface is covered by Sn oxide, which is protective, compressive stress is built up at the anode. The compressive stress cracked the surface oxide, see Figure 12.13b, so a small area of free surface which is stress-free is formed. Thus, a stress potential gradient is produced between the stress-free surface and the surrounding anode, which is under compression, and the stress potential gradient enables stress-migration to occur and leads to the Sn whisker growth, see Figure 12.13c and d. The whisker has Sn oxide on the surface, yet the growth of the whisker can stretch and keep the surface oxide near its root to be broken. While electromigration maintains the compressive stress, the broken oxide maintains the stress-free surface. In combination, whisker growth occurs. The broken oxide is a key element in Sn whisker growth.

## 12.6 Comparison of Driving Forces Among Electromigration, Thermomigration, and Stress-Migration

Here, a comparison is made among the driving forces of electromigration, thermomigration, and stress-migration in Sn, especially the growth of Sn whisker under a compressive stress. We first consider electromigration, where the driving force is $F = Z^*eE = Z^*e\rho j$. Furthermore, it is known that electromigration occurs in Sn-based solder joints when the applied current density is above $10^4$ A/cm$^2$ or $10^8$ A/m$^2$ at temperatures above 100 °C. For calculation, we take $Z^* = 10$, $e = 1.6 \times 10^{-19}$ comb, $\rho = 10 \times 10^{-8}$ Ωm, and $j = 1 \times 10^8$ A/m$^2$, we obtain $F = 1.6 \times 10^{-17}$ N. The work done by an atomic jump distance driven by this force will be

$$\Delta w = F \times a = \left(1.6 \times 10^{-17}\,\text{N}\right) \times \left(3 \times 10^{-10}\,\text{m}\right)$$
$$= 4.8 \times 10^{-27}\,\text{Nm} = 4.8 \times 10^{-27}\,\text{J}.$$

Next, we consider thermomigration under a temperature gradient of 1000 °C/cm. It has been found that under such a temperature gradient, thermomigration occurs in solder joint near 100 °C. Across an atom of diameter of $3 \times 10^{-8}$ cm, the temperature difference due to the temperature gradient of 1000 °C/cm is about $3 \times 10^{-5}$ K. The change of thermal energy across an atom is

$$3k\Delta T = 3 \times 1.38 \times 10^{-23}\left(\text{J}/\text{K}\right) \times 3 \times 10^{-5}\,\text{K} = 1.3 \times 10^{-27}\,\text{J}$$

which is of the same order of magnitude as the work done under electromigration given in the above calculation.

Finally, we consider stress-migration by calculating the driving force of Sn whisker growth. On stress-migration, we assume that the force is equal to the stress potential gradient, which is $-d\sigma\Omega/dx$, where $\sigma$ is normal stress (not shear stress) and $\Omega$ is atomic volume. The stress distribution around the root of a Sn whisker has been measured by X-ray diffraction using synchrotron radiation, as shown in Figure 12.11. We estimate the stress gradient between the origin ($x = 0, y = 0$) and the lower left corner point at ($x = -0.5400, y = 0.8475$). We found that $\Delta x \approx 10\,\mu m$ and $\Delta\sigma \approx 4\,MPa$, and we take the atomic volume of Sn atom to be $27 \times 10^{-24}$ cm$^3$. The force is given below,

$$F = -\frac{\Delta\sigma\Omega}{\Delta x} = \frac{4x10^6\left(N/m^2\right)x27x10^{-24}\left(cm^3\right)}{10^{-3}\left(cm\right)}$$
$$= \frac{4x10^7\left(dyne/cm^2\right)x27x10^{-24}\left(cm^3\right)}{10^{-3}cm}$$
$$= \frac{108x10^{-17}erg}{10^{-3}cm} \approx 10^{-12}erg/cm$$

The work done by this force over a distance of atomic diameter of 0.3 nm is $3 \times 10^{-27}$ joule, which is of the same order of magnitude of those calculated in the above for electromigration and thermomigration in Sn.

### 12.6.1 Products of Force

We have three products of force. The first is force times distance = work done. The second is force times mobility = drift velocity, or $<v> = MF$. The third is conjugated force ($X$) times conjugated flux ($J$) = entropy production, or $JX = T\theta$, where $T$ is temperature and $\theta$ is rate of entropy production.

On drift velocity, $<v> = MF$, where $M = D/kT$ in atomic diffusion. Because the atomic diffusivity, $D$, depends on materials and temperature, it was found that in Al interconnects near the microelectronic device operation temperature of 100 °C, electromigration occurs via grain boundary diffusion. However, in Cu interconnects, electromigration occurs via surface diffusion at the same temperature of 100 °C. Furthermore, in solder joints or Sn, electromigration occurs via lattice diffusion. Table 10.2 shows the diffusivity differences of electromigration in Al, Cu, and Sn; mainly, it is due to the difference in melting point of the metals.

Knowing the drift velocity, we can calculate atomic flux, $J = C<v> = CMF$. Then, knowing flux, we can calculate entropy production because we know the driving force already. In comparison, entropy production due to atomic flux is much smaller than that in electron flux, because the latter is much larger than the former. For a direct comparison, we consider below the Joule heating due to electromigration in Al interconnect and joule heating due to electrical conduction in Al interconnect.

In Section 9.3.1.1 on Joule heating, we showed that to calculate the Joule heating of electrical conduction in Al interconnect, if we take $j = 10^6$ A/cm$^2$ and $\rho = 10^{-6}$ Ω-cm for Al, the joule heating is

$$\frac{TdS}{Vdt} = jE = \rho j^2 = 10^6 \frac{joule}{cm^3 \sec}$$

For electromigration in Al interconnect, we have

$$\frac{TdS}{Vdt} = JX = \left( C\frac{D}{kT}F \right) F = C\frac{D}{kT}\left(Z * e\rho j\right)^2$$

We take $C = 10^{23}/\text{cm}^3$, $T = 400$ K, $D = 10^{-16}$ cm$^2$/sec, $Z^*e = 10^{-18}$ coulomb, $\rho = 10^{-6}$ Ω-cm, and $j = 10^6$ A/cm$^2$. Substituting these data into Eq. (13.32), we obtain

$$\frac{TdS}{Vdt} = \frac{10^{-19}\, joule^2}{\left(0.033eV\right) cm^3 \sec} = 20 \frac{joule}{cm^3 \sec}$$

where kT = 0.033 eV at T = 400 K, and 1 joule = $6.24 \times 10^{18}$ eV. Clearly, the Joule heating due to electrical conduction is much larger.

## Problems

**1** Both electromigration and thermomigration are cross-effects in irreversible processes. How come stress-migration is not a cross-effect?

**2** Upon applying an elastic stress to a piece of metal, it will deform. What is the primary flux of flow in the metal?

**3** We have model of Nabarro-Herring creep on the basis of lattice diffusion, and also Coble creep based on grain-boundary diffusion. Can we have creep on the basis of surface diffusion? If so, please explain.

**4** An Al film of 1 μm thick is deposited without thermal stress on a Si wafer of 100 μm thick at temperature of 120 °C. The wafer and film are cooled to 20 °C. Using the values below and assume the Poisson's ratio is 0.272 for Si, calculate the thermal strain and stress of the Al film and also calculate the radius of its curvature. The thermal expansion coefficients of Al and Si are $\alpha = 24.6 \times 10^{-6}$/°C and $2.6 \times 10^{-6}$/°C, respectively. The Young's modulus for Al and Si are $0.7 \times 10^{11}$ N/m$^2$ and $1.9 \times 10^{11}$ N/cm$^2$, respectively.

**5** The stress in films on wafers can be determined from the amount of bow of the wafer. The bow can be measured by a surface profilometer with a scan of length of 2 $l$. Show that Stoney's equation can be expressed as

$$\sigma = \left(\frac{\delta}{3l^3}\right)\left(\frac{Y}{1-\nu}\right)\left(\frac{t_S^2}{t_f}\right)$$

6 where $\delta$ is the maximum bow height of the profile-meter scan and $\rho$ is half the scan length.

## References


1 Mott, N.F. and Jones, H. (1958). *The Theory of the Properties of Metals and Alloys*. New York: Dover.

2 Nix, W.D. (1989). Mechanical properties of thin films. *Metallurgical Transactions A* 20: 2217–2245.

3 Spaepen, F. (2000). Interfaces and stress in thin films,'. *Acta Materialia* 48: 31–42.

4 Murakami, M. and Segmiller, A. (1988). Analytical techniques for thin films. In: *Treatise on Materials Science and Technology*, vol. 27 (eds. K.N. Tu and B. Rosenberg). Boston: Academic Press.

5 Sheng, G.T.T., Hu, C.F., Choi, W.J. et al. (2002). Tin whiskers studied by focused ion beam imaging and transmission electron microscopy. *Journal of Applied Physics* 92: 64–69.

6 Choi, W.J., Lee, T.Y., Tu, K.N. et al. (2003). Tin whisker studied by synchrotron radiation micro-diffraction. *Acta Materialia* 51: 6253–6261.

7 Tu, K.N. (1994). Irreversible processes of spontaneous whisker growth in bimetallic Cu-Sn thin film reactions. *Physics Review* B49: 2030–2034.

8 Ouyang, F.-Y., Chen, K., Tu, K.N., and Lai, Y.-S. (2007). Effect of current crowding on whisker growth at the anode in flip chip solder joints. *Applied Physics Letters* 91: 231919.

# 13

# Failure Analysis

## 13.1 Introduction

When a device is manufactured to provide a specific function in application, it is hoped that the microstructure in the device will not change in its lifetime of use. Unfortunately, this is not true! Failure occurs when an unwanted microstructure change has taken place such as void formation at the cathode in an interconnect in the device, for instance. This is because Si device is a current–voltage device, so we have to apply electric current to operate the device. It is an open system. Under a high current density, electromigration induces microstructure change in the interconnect and can lead to circuit failure due to opening by void formation at the cathode or shorting by whisker extrusion at the anode. Furthermore, the high current density can cause Joule heating and the temperature rise will produce thermal stress between materials having different thermal expansion coefficients in the device. Heat conduction will produce a temperature gradient, which can be large enough to cause thermomigration. The thermal stress and temperature gradient will induce atomic diffusion, phase instability, and microstructure change.

What is unique in these microstructure changes is that they occur in the domain of nonequilibrium thermodynamics, or they are irreversible processes. This is because of the coexistence of electron flow, atomic flow, and heat flow in the open system. Thus, the reliability science on the basis of entropy production is needed, as we have discussed in Chapter 9, in order to provide the basic understanding of microstructure change in irreversible processes that leads to device failure.

*Electronic Packaging Science and Technology*, First Edition. King-Ning Tu, Chih Chen, and Hung-Ming Chen.

Furthermore, product assurance in microelectronics industry requires statistical reliability tests of a new product. In the previous chapters, we have discussed physical analysis of the failure mode and understanding of the failure mechanism. Yet, failure distribution is needed in statistical analysis in order to predict and to project the lifetime of a product. It can provide two important information about the reliability of a product. First is the mean-time-to-failure (MTTF) of the product under a given condition of its use in the field of consumers. For example, we need to know the MTTF of a device to be used under the hood of an automobile, which is a very hush environment of high temperatures. Second is that in the projection of future application of an existing device, the applied current density may have to be increased due to the increase of functionality. Then, we need to know what will be the maximum current density, which can be applied to an existing device so that it can still perform without failure within a required lifetime?

To conduct statistical distribution analysis, there are two important requirements. First, we must have the equipment, which can measure the failure of a large number of devices as a function of time, temperature, and a specific driving force such as current density in electromigration, which can stress the device until it fails. Second, we must have a large set of test samples, which should be very close to the real consumer product. Thus, it is important to use samples provided by an industry laboratory, not arbitrary samples prepared in a campus environment. Figure 13.1 shows a set of

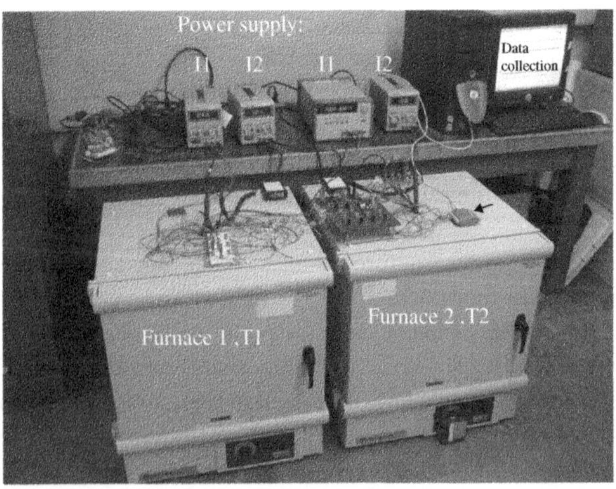


**Figure 13.1** Shows such an equipment for statistical analysis of failure of flip-chip solder joints under electromigration. It consists of two furnaces, four power sources (for safety reasons, the applied current is limited to 2 Amp), a multichannel control unit, and a personal computer for recording.

equipment for statistical analysis of failure of flip-chip solder joints under electromigration. It consists of two furnaces, four power sources (for safety reasons, the applied current is limited to 2 Amp), a multichannel control unit, and a personal computer for recording.

For example, in the study of flip-chip solder joint reliability, we have multiple flip chips on a board for accelerated tests, or we must be able to connect electrically several single flip chip together for the test so that a meaningful size of data can be obtained in a reasonable time. Figure 13.2 shows an optical image of such a test board having four chips on the board for electromigration tests. The layout of the solder joints between the chip and the board in one of the four chips is shown in Figure 13.3. The dimension of the chip is 0.3 × 0.3 cm, which has 36 solder joints. The diameter of the solder joint is 250 μm. The contact opening at under-bump-metallization (UBM) is 200 μm in diameter. The chips are mounted on the printed circuit board using Pb-free solder bump of composition of Sn-1.2% Ag-0.5% Cu. The UBM is a layered thin films of Al/Ni(V)/Cu, in which the thickness of Al, Ni(V), and Cu is 1, 0.3, and 1 μm, respectively. A temperature sensor of Pt serpentine wire on the top surface of the Si chip is shown.

In this chapter, we shall first discuss microstructure changes with or without lattice shift. The definition of lattice shift will be given in the next section.

When the irreversible process is accompanied by lattice shift, the total number of lattice sites is conserved and it is a constant volume process, there is no void or whisker formation, and in turn no failure. When lattice shift is absent, it is a nonconstant volume process, so the excess vacancies

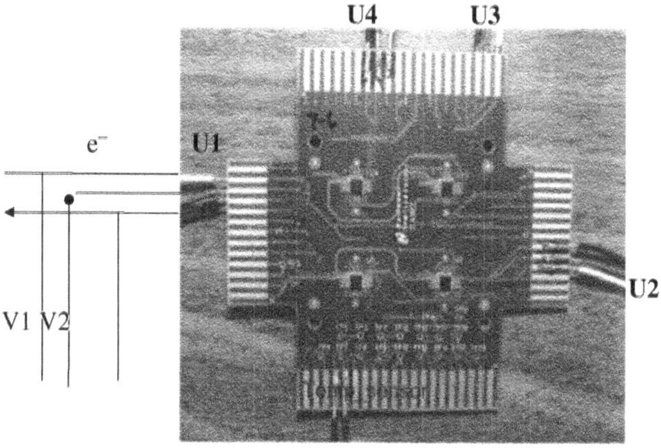


**Figure 13.2** Shows an optical image of such a test board having four chips on a board for electromigration tests.

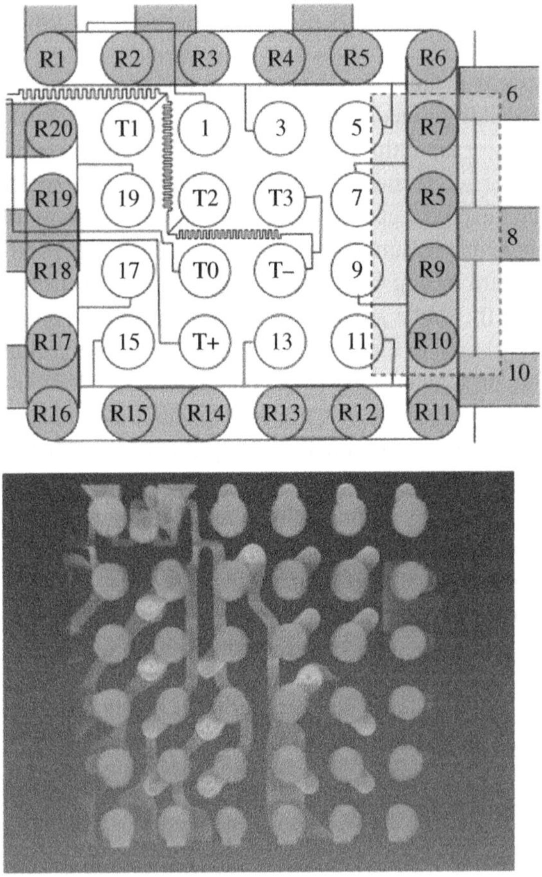


**Figure 13.3** The layout of the solder joints between the chip and the board in one of the four chips is shown here. The dimension of the chip is 0.3 × 0.3 mm, which has 36 solder joints. The diameter of the solder ball is 250 μm.

cannot be absorbed by sinks, and extra lattice sites have to be created for void formation, so failure occurs. Also if the excess atoms cannot be absorbed, they can grow hillocks and whiskers.

Next, we shall discuss MTTF in electromigration, and we shall revisit Black's equation in order to justify the dependence of $j^{-2}$ in the equation. Then, we present a unified model of MTTF, on the basis of entropy production, which can be applied to the failure analysis of electromigration, thermomigration, stress-migration, as well as any other irreversible processes. Finally, several examples of system-level failure study in recent 2.5D IC devices will be presented.

## 13.2 Microstructure Change with or Without Lattice Shift

Phase change in traditional thermodynamics occurs in a closed system between two equilibrium states, usually under the condition of constant temperature and constant pressure. For example, we consider the phase change in a eutectic SnPb solder, going from the molten state to the solid state. We can minimize Gibbs free energy to describe the process and use the equilibrium phase diagram of Sn-Pb to define the composition of the two solid phases in the eutectic microstructure after solidification. Nevertheless, in this special case, it is a near-equilibrium, not the final equilibrium phase change, because we cannot define the lamellar spacing in the eutectic structure.

However, when we have a phase change under a temperature gradient or a pressure (stress) gradient in an open system, we do not have the boundary condition of constant temperature and constant pressure, so the change cannot be described by a minimum of Gibbs free energy. Instead, we have irreversible processes, and the kinetics of phase change is in the domain of nonequilibrium thermodynamics. A classic example is the Soret effect of thermomigration, in which a homogeneous alloy becomes inhomogeneous under a temperature gradient. Because the inhomogeneous state has a higher free energy than the homogeneous state, it is a process of increasing rather than decreasing of free energy.

Nevertheless, irreversible processes may not lead to device failure with lattice shift, if there is no atomic flux divergence. Take electromigration in an Al wire as an example; if the atomic flux driven by electromigration is uniform in the wire and if the cathode and the anode are very large source and sink of Al atoms, respectively, there is no failure because it is a steady state of mass transport of Al atoms from the cathode to the anode, when we define failure as void or whisker formation in the wire. Hence, for failure to occur, we require divergence of the atomic flux in the irreversible processes. For example, void formation has been found in triple points of grain boundaries in Al wires, where atomic flux divergence occurs.

Still, the condition of divergence is necessary but insufficient for void or whisker formation, if it is a constant lattice site or constant volume process. We note that while the total number of atoms is conserved, the total number of lattice sites has to increase when there is void formation or whisker growth. The necessary and sufficient condition of failure is that we must not have lattice shift in the region of flux divergence, so we must have a change of the number of lattice sites or a change of volume. The example below explains what is lattice shift.

In the classical example of Kirkendall effect in the interdiffusion between Cu and CuZn alloy, there is a divergence of flux because the flux of Cu is not equal to that of Zn, and the divergence is balanced by a flux of vacancy. Yet in Darken's analysis of interdiffusion, there is no void formation because the vacancy source and sink are assumed to be effective, so vacancies can be created or absorbed as they are needed in the microstructure; in turn, vacancies can be assumed to be at equilibrium everywhere in the sample. Thus, void cannot nucleate if there is no super-saturation of vacancies. Furthermore, in Darken's analysis, there is no stress issue too.

Figure 13.4 is a schematic diagram to illustrate the idea of lattice shift in interdiffusion, where we assume there is a flux of atoms moving from the right to the left. It is accompanied by a flux of vacancies moving in the opposite direction. On the right hand side of the figure, we assume there is a dislocation loop, which serves as the sink of vacancies, so the dislocation loop will climb to increase its diameter by absorbing vacancies. Also on the left-hand side, we assume there is a dislocation loop, which serves as the source of vacancies, so the loop will climb to decrease its diameter by emitting vacancies. Sooner or later, these motions of dislocation climb will eliminate an atom plane on the right-hand side, and at the same time, it will produce an atomic plane on the left-hand side. Thus, the combined effect will cause all the atomic planes in between the two dislocation loops to shift to the right by one atomic plane thickness. This effect is defined as "lattice shift." If we implant a marker in the middle region, it will move with the lattice shift, so we can measure the marker displacement or marker motion.

In Darken's model of Kirkendall effect, there is no void and no stress. However, in most real cases of interdiffusion and reactions, Kirkendall or Frenkal void formation occurs. Also bending of samples due to stress effect has been found because of insufficient lattice shift. If insufficient

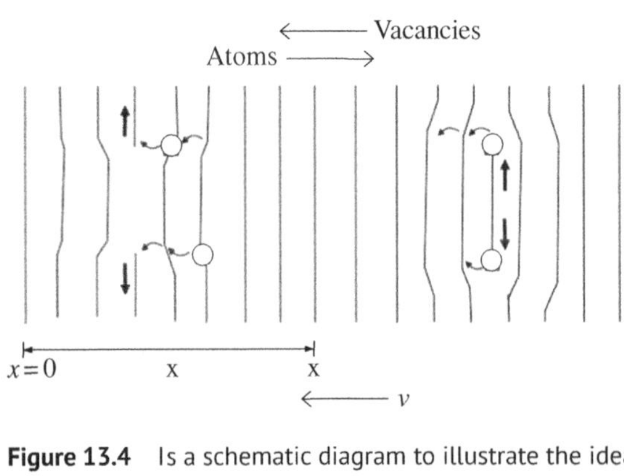


**Figure 13.4** Is a schematic diagram to illustrate the idea of lattice shift.

or no lattice shift occurs, the excess vacancies cannot be absorbed, we have to create lattice sites to accommodate them. If they form voids, the void will take new lattice sites, thus the volume increases. It means we shall have a nonconstant volume process when failure occurs due to void formation.

In the opposite direction, because we have to assume the total number of atoms is conserved, then how about the concept of lattice shift? We can create vacancies, but we cannot create atoms. However, when we have whisker growth, we must allow new lattice sites to be created in order to grow the whisker out of the sample. Thus, the concept of failure requires flux divergence and no lattice shift remains true.

## 13.3 Statistical Analysis of Failure

### 13.3.1 Black's Equation of MTTF for Electromigration

On studying electromigration, we can vary time, temperature, and the applied current density. To obtain MTTF on the basis of electromigration tests, Black has provided an equation as below [1, 2].

$$MTTF = A\left(j^{-n}\right)\exp\left(\frac{E_a}{kT}\right) \tag{13.1}$$

where the time (MTTF) is related to temperature and current density by using three parameters; the pre-factor $A$, current density power factor $n$, and activation energy $E_a$. It is of interest to point out that $n$ is close to 2 in most MTTF data, and $E_a$ tends to relate to the activation energy of atomic diffusion in electromigration. Why $n = 2$ has been a controversial topic and it has been widely discussed [3–8]. In the next section, we shall revisit Black's equation and show that $n = 2$ can be justified.

To determine these parameters from experiments, we need to conduct at the least the electromigration tests under two temperatures and two current densities as a function of time until failure occurs. Figure 13.5 shows the Weibull distribution plots of time to failure of a set of solder joint samples, shown in Figure 13.2, tested at two current densities of $5 \times 10^3$ and $1 \times 10^4$ A/cm$^2$ and at two temperatures of 125 and 150 °C, using the equipment as shown in Figure 13.1. The resistance change, as monitored by voltage change on each sample, was automatically measured by a computer, and the data were recorded. The data were plotted by voltage vs. time, from which we can determine the time to failure for each individual sample.

The Weibull linear plot as shown in Figure 13.5 is important for obtaining MTTF on the basis of Black's equation of failure. However, it is hard to locate the failure or the void formation at a specific site of the test sample, which has caused the resistance increase. This is particularly true in 3D IC devices, because it is not easy to locate the failure site in a 3D structure that led to an open circuit. High-resolution in situ synchrotron radiation X-ray tomographic inspection is needed.

Often, it is asked why we tend to use Weibull distribution in electromigration studies rather than others such as lognormal distribution. It has been found that in Weibull distribution, we can observe the data of early weak-link failure easily. From the point of view of system reliability, the early failure due to weak-links is most undesirable. They must be removed. Another good reason is the correlation to the physical mode of time-dependent failure in electromigration, where the exact similarity between Weibull distribution function and Johnson-Mehl-Avrami equation of phase transformations is intriguing, which will be discussed later.

To find the activation energy in Black's MTTF equation, we use the data at the same current density with two temperatures and the relation below.

$$\frac{MTTF_2}{MTTF_1} = \frac{\exp\left(\dfrac{E_a}{kT_2}\right)}{\exp\left(\dfrac{E_a}{kT_1}\right)}$$

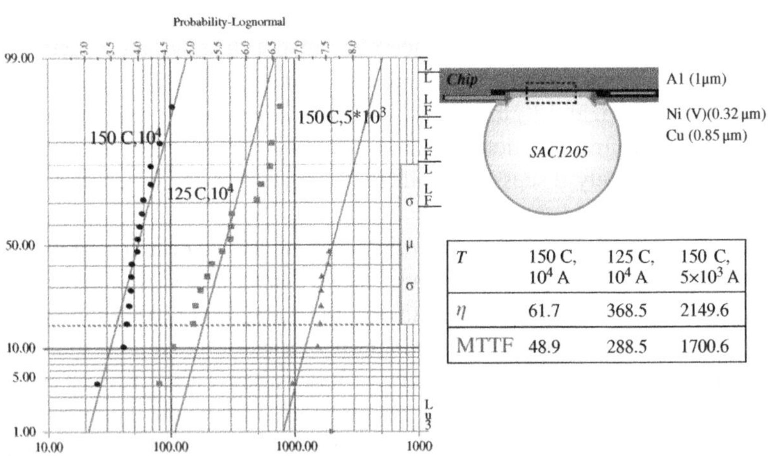


| $T$ | 150 C, $10^4$ A | 125 C, $10^4$ A | 150 C, $5\times10^3$ A |
|---|---|---|---|
| $\eta$ | 61.7 | 368.5 | 2149.6 |
| MTTF | 48.9 | 288.5 | 1700.6 |

**Figure 13.5** Shows the Weibull distribution plots of time to failure of a set of solder joints tested at two current densities of $5 \times 10^3$ and $1 \times 10^4$ A/cm$^2$ and at two temperatures of 125 and 150 °C.

$$E_a = \left(\frac{1}{kT_2} - \frac{1}{kT_1}\right)\log\left(\frac{MTTF_2}{MTTF_1}\right) \quad (13.2)$$

Then, to determine "$n$," we can use the data at the same temperature with two different current densities and the relation below.

$$\frac{MTTF_2}{MTTF_1} = \left(\frac{j_1}{j_2}\right)^n \quad (13.3)$$

However, due to Joule heating, we have to be very careful in applying the above two relations. When we calculate the activation energy using Eq. (13.2), we have to take into account the effect of Joule heating on temperature by changing $kT$ to $k(T + \Delta T)$ where $\Delta T$ is due to Joule heating at the applied current density and it has been measured, so $kT_1$ becomes $k(T_1 + \Delta T)$ and $kT_2$ becomes $k(T_2 + \Delta T)$ in Eq. (13.2). This correction is straight forward.

When we calculate "$n$" using Eq. (13.3), it is harder. Because Joule heating is different at different current densities, the temperature can hardly be assumed to be the same for two different current densities measured at the same annealing temperature. The actual temperature is not the same, so the problem is nontrivial. We need to know the $\Delta T$ increase at different current densities in order to make the temperature correction. We can obtain the temperature calibration by using a temperature sensor of Pt thin film serpentine line deposited on the test sample.

The activation energy was obtained to be 1.15 eV/atom on the basis of the data shown in Figure 13.5. We found $n = 2.08$. Then we can determine the constant $A = 1.5 \times 10^{-1}$ sec-amp$^2$/cm$^4$. On the basis of the obtained MTTF equation, the prediction or extrapolation of lifetime at the condition of field use can be calculated.

### 13.3.2 Weibull Distribution Function and JMA Theory of Phase Transformations

Weibull distribution function is given as,

$$F(t) = 1 - \exp\left[-\left(\frac{t}{\eta}\right)^\beta\right] \quad (13.4)$$

where $F(t)$ is percentage or fraction of the failed sample as a function of time, $\eta$ is the characteristic lifetime, and $\beta$ is the shape factor or the slope of the Weibull plot. It means a larger slope will give a narrower distribution of time to failure.

There is a close similarity in the mathematical form between Weibull distribution function and Johnson-Mehl-Avrami's (JMA) canonical equation of phase transformations [9]. In classical phase transformations, for example the crystallization of an amorphous phase to a crystalline phase without changing composition, the fraction of the transformed volume is expressed by JMA equation as,

$$X_T = 1 - \exp(-X_{ext}) = 1 - \exp\left[-\left(\frac{t}{\lambda}\right)^n\right] \tag{13.5}$$

where $X_T$ is fraction of the volume transformed and $X_{ext}$ is fraction of the extended volume, which is defined as

$$X_{ext} = \int_{\tau=0}^{\tau=t} \frac{4\pi}{3} R_N R_G^3 (t-\tau)^3 \, d\tau \tag{13.6}$$

where $R_N$ is the nucleation rate and $R_G$ is the growth rate of spherical transformed particles, and $t$ is the time of transformation. In essence, the physical meaning of $X_{ext}$ is the sum of the volume of all spherical particles in the period of $\tau = 0$ to $t$ without the consideration of growth impingement and phantom nucleation.

Since the mathematical form of Weibull distribution function, Eq. (13.4), and JMA equation of phase transformations, Eq. (13.5), is similar, it is of interest to correlate them so that we may obtain an explanation of Weibull distribution of MTTF from the point of view of phase transformation of forming voids in a set of interconnects. Electromigration-induced failure of open circuit due to void formation can be regarded as phase transformation in the cathode end of an Al or Cu line or in a flip-chip solder joint where a void nucleates and grows. This is a topic of interdisciplinary, which requires a careful study in the future. In Black's equation of MTTF, the physical meaning of the activation energy has not been defined clearly. It could be the activation energy of nucleation of the void, or the growth of the void, or the sum of both as indicated by $R_N$ and $R_G$ in Eq. (13.6).

## 13.4 A Unified Model of MTTF for Electromigration, Thermomigration, and Stress-Migration

### 13.4.1 Revisit Black's Equation of MTTF for Electromigration

To revisit Black's equation, we consider entropy production in irreversible processes as the basis of microstructure failure. According to Onsager, entropy production rate is equal to

$$\frac{TdS}{Vdt} = JX \tag{13.7}$$

where $T$ is temperature, $dS/dt$ is rate of entropy production, $V$ is volume of the sample in the irreversible process, $J$ and $X$ are the conjugated flux and the conjugated driving force, respectively. Because there are atomic flux, heat flux, and charge flux in irreversible processes, the forces behind the fluxes are correspondingly the conjugated driving forces.

Let $J_e$ be the atomic flux driven by electromigration, we have

$$J_e = C\frac{D}{kT}Z^* e\rho j \tag{13.8}$$

On the conjugated driving force,

$$X_e = Z^* eE = Z^* e\rho j, \tag{13.9}$$

where $E = -d\phi/dx = \rho j$ is electric field.

Now, let $V^*$ be the critical volume of the void formation at the cathode end, which has led to opening failure, and we have $V^* = \Omega J_e At$, where $\Omega$ is atomic volume, $A$ is the cross-section of the diffusion, and $t$ (or MTTF) is time to failure, so

$$t = V^* / \Omega J_e A \tag{13.10}$$

The above equation shows that MTTF is proportional to $1/j$, thus $n = 1$. Yet, some authors have considered that the void formation needs nucleation, and $n$ may equal to 2 when MTTF includes both nucleation and growth of void failure.

Now, to justify that $n = 2$, we consider entropy production. From Eq. (13.7), the total entropy production until failure is, $S_{threshold} = (dS/dt)t$, thus from Eq. (13.7) [10],

$$TS_{threshold} / V = J_e X_e t$$

$$t^{failure} = \frac{TS_{threshold}}{VJ_eX_e} = \frac{TS_{threshold}}{V}\frac{1}{\left(C\frac{D}{kT}Z^* e\rho j\right)\left(Z^* e\rho j\right)}$$

$$= \frac{TS_{threshold}}{V}\frac{1}{C}\frac{kT}{D}\frac{1}{\left(Z^* e\rho\right)^2}j^{-2} = \frac{S_{threshold}}{VC}\frac{kT^2}{\left(Z^* e\rho\right)^2 D_0}\exp\left(\frac{E}{kT}\right)j^{-2} \tag{13.11}$$

$$= A\left(\frac{T}{j}\right)^2 \exp\left(\frac{E}{kT}\right)$$

where $A = \frac{S_{threshold}k}{VCD_0 \left(Z^* e\rho\right)^2}$. The above equation shows that $n = 2$ is justified in Black's equation. Yet, the MTTF dependence is on $(T/j)^2$, not $(1/j)^2$, nor $(1/j)$.

On the other hand, if we consider Joule heating in electrical conduction as presented in Chapter 9, it is

$$\frac{TdS}{Vdt} = j\left[-\frac{d\varphi}{dx}\right] = jE = j^2\rho$$

where $j^2\rho$ is called "Joule heating" per unit volume per unit time. It may be the hidden link of $j^2$ between current density and MTTF in Eq. (13.11).

### 13.4.2 MTTF for Thermomigration

To obtain MTTF for thermomigration, the atomic flux is given as below [10],

$$J_h = C\frac{D}{kT}\frac{Q^*}{T}\left(-\frac{\partial T}{\partial x}\right) \tag{13.12}$$

where $Q^*$ is defined as heat of transport in thermomigration, and $Q^*$ has the same dimension as $\mu$, so it is the heat energy per atom. The conjugated driving force is,

$$X_h = \frac{Q^*}{T}\left(-\frac{dT}{dx}\right) \tag{13.13}$$

We have $J_h X_h t^{failure} = TS_{threshold}/V$

$$MTTF \approx t^{failure} = \frac{TS_{threshold}}{VJ_hX_h} = B\left(-\frac{dT}{dx}\right)^{-2} \exp\left(\frac{E_a}{kT}\right) \tag{13.14}$$

### 13.4.3 MTTF for Stress-Migration

The driving force of stress-migration is given below in Eq. (13.15), where stress potential is defined as $\sigma\Omega$, and $\sigma$ is stress and $\Omega$ is atomic volume.

$$X_S = -\frac{d\sigma\Omega}{dx} \tag{13.15}$$

And the conjugated atomic flux in stress-migration is given as

$$J_S = CMF = C\frac{D}{kT}\left(-\frac{d\sigma\Omega}{dx}\right) = \frac{D}{kT}\left(-\frac{d\sigma}{dx}\right) \tag{13.16}$$

we have $J_S X_S t^{failure} = TS_{threshold}/V$ [10],

$$MTTF \approx t^{failure} = \frac{TS_{threshold}}{VJ_S X_S} = G\left(-\frac{d\sigma}{dx}\right)^{-2} \exp\left(\frac{E_a}{kT}\right) \tag{13.17}$$

### 13.4.4 The Link Among MTTF for Electromigration, Thermomigration, and Stress-Migration

The conjugated forces of electromigration, thermomigration, and stress-migration in Sn or Sn-rich solder have been shown to be nearly the same in the last chapter. This provides the following analysis of MTTF of three of them by equating their driving forces.

$$Z^* e\rho j = 3k\frac{dT}{dx} = -\frac{d\sigma\Omega}{dx} \tag{13.18}$$

Thus, we have

$$j = \frac{3k}{Z^* e\rho}\frac{dT}{dx} = \frac{-1}{Z^* e\rho}\frac{d\sigma\Omega}{dx} \tag{13.19}$$

Then, by substituting $j^{-2}$ into the original Black's equation, we can obtain MTTF for thermomigration and stress-migration, respectively.

### 13.4.5 MTTF Equations for Other Irreversible Processes in Open Systems

The three examples presented in the above showed that for other type of irreversible process in an open system, we can use Eq. 13.7 to obtain MTTF equations. Therefore, for any irreversible process, its MTTF equation can be obtained, provided that the flux and the driving force can be defined. However, the parameters in the MTTF equation must be determined from experimental data.

## 13.5 Failure Analysis in Mobile Technology

In mobile technology, the continuing demand for smaller size, more functionality, lower power consumption, and reduced cost is challenging in manufacturing, as well as in failure analysis. In Chapter 1, we have mentioned that at the moment, the two major trends in electronic packaging are the need of higher and higher number of IO and the demand of lower and lower power or Joule heating. Actually, the idea of 3D IC has occurred in microelectronics industry for more than 10 years. Yet, 3D IC has not been in mass production because of high cost and low reliability. Currently, the use of 5G technology in AI applications has provided the impetus to promote the 2.5D IC into mass production, instead of 3D IC. Thus, the reliability of 2.5D IC devices becomes a major concern.

In 2.5D IC devices, the adding of a Si interposer requires one more redistribution layer (RDL) and one more level of solder joints. This is because of vertical interconnects by using through-Si-vias (TSVs) and micro-bumps. In the dense 2.5D IC packaging, not only Joule heating is serious but also heat dissipation is poor. To enhance heat dissipation, we need to have a high temperature gradient. Unfortunately, the gradient can lead to thermomigration. Furthermore, the use of Si interposer enhances the lateral heat transfer along the interposer, which has been found to have caused an unexpected thermomigration failure.

Thus, thermal management affects all the reliability problems in mobile technology. Electromigration enhanced by Joule heating is the primary concern of failure, especially the early failure. In this section, we shall discuss the synergistic effect of electromigration and Joule heating, and thermal cross-talk on a Si interposer.

### 13.5.1 Joule Heating Enhanced Electromigration Failure of Weak-Link in 2.5D IC Technology

In Chapter 1, Figures 1.4 and 1.5 show, respectively, an SEM cross-sectional image and X-ray tomography images of a 2.5D IC test sample. Basically, there are two pieces of Si chips and three levels of solder joints. The latter are the largest BGA at the bottom, the middle C-4 flip-chip solder joints, and the smallest micro-bumps on the top. To interconnect them, there are two sets of redistribution layer (RDL). The first set of RDL is for the circuit fan-out from BGA to C-4 joints. The second set of RDL is for the circuit fan-out from C-4 joints to micro-bumps. The purpose of fan-out is to increase the number of I/O. This second set of RDL is new in 2.5D IC because it does not exist in 2D IC. It was found that this new RDL, when it is added to a device without enough reliability consideration, becomes the weak-link in electromigration tests [11].

In Figure 1.5a in Chapter 1, a set of arrows was used to indicate that we can pass electric current to study electromigration in the 2.5D IC packaging circuit. One of the BGA balls on the left was used as the cathode, and the electrons will flow through the PTH, flip-chip solder joint, TSV, and a daisy chain of micro-bumps, and return in opposite direction to another BGA ball on the right as the anode. Because the circuit is a packaging circuit, neither transistors nor multilevel Cu damascene interconnects are included. Nevertheless, there are a large number of Cu connections in the TSV and RDL, as well as solder joints in the packaging circuit. This system-level electromigration test is much more complicated than those previous studies of electromigration in a pair of solder joints or a stripe of Cu or Al lines, as discussed in Chapter 10.

The test sample failed in a second when 100 mA passed through a pair of the BGA at 100 °C. Based on this initial finding, three testing conditions of lower current were chosen at 100 °C, 50 mA; 100 °C, 60 mA; and 150 °C, 50 mA. All the failures showed a rather flat resistance change with time until a sudden increase leading to failure, as shown in Figure 13.6. [11]

To locate the failure site, those failed samples were polished layer by layer slowly and carefully, and the images were examined. Figure 13.7a and b shows the failure site images of the sample failed in 164 hours at 100 °C and 50 mA; and Figure 13.7c and d shows the sample failed in 12.25 hours at 150 °C and 50 mA. They were found in the RDL located in the interposer chip. In Figure 13.7a, we can see six periodic holes in the RDL. Focus ion beam (FIB) was used to cut perpendicularly into these holes, and Figure 13.7b shows one of the images. These holes indicate a burn-out type failure, where we see that the Cu lines were molten and the dielectric between them was broken. Figure 13.7c shows another failure site. This

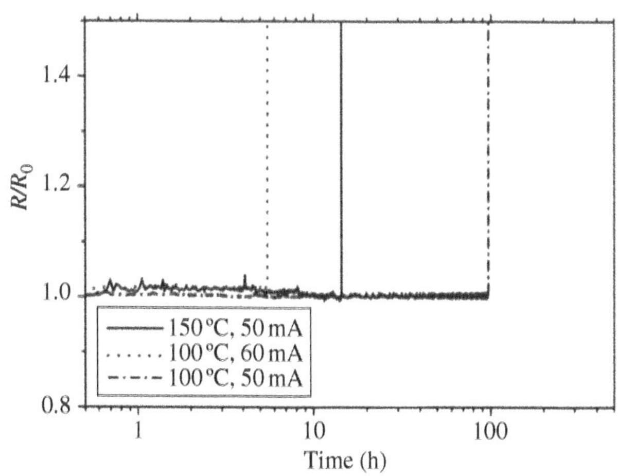


**Figure 13.6** All the failures showed a rather flat resistance change with time until a sudden sharp increase leading to failure.

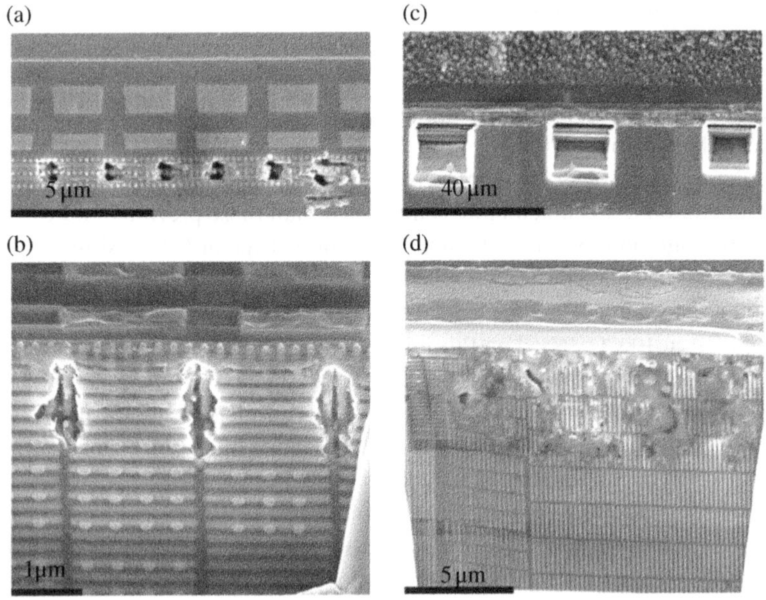


**Figure 13.7** (a) and (b) show the failure site images of the sample failed in 164 hours at 100 °C and 50 mA. And (c) and (d) show the sample failed in 12.25 hours at 150 °C and 50 mA.

failure site is really big, about 80 μm long. FIB was used to cut three holes at the beginning, the middle, and the end of the failure site, respectively, to obtain more clear images. Those images are similar to each other, and one of them is shown in Figure 13.7d. Figure 13.8a shows the location of RDL in the sample. It is near the bottom side of the interposer and below the TSV. Figure 13.8b is a cross-sectional TEM image of the Cu lines in RDL. The cross-section of the Cu line is about 100 nm by 100 nm.

According to Figure 13.8, if we assume 1 mA passes across the cross-sectional area of 100 × 100 nm, the current density will be $1 \times 10^7$ A/cm$^2$, which is very high for electromigration to happen in Cu line at 100 °C. The estimate of current density is not unreasonable. Because from the density change of solder joints (number of joints per unit area) of BGA to C-4 solder joints and to micro-bumps, we expect the high current density when the applied current is 50–60 mA. Furthermore, the density of Cu lines is extremely high in RDL, which means that Joule heating is very serious. Also the heat dissipation is poor, so the temperature can have a time-dependent increase. These situations are very different from the previous electromigration studies carried out in Cu lines, as discussed in Chapter 10.

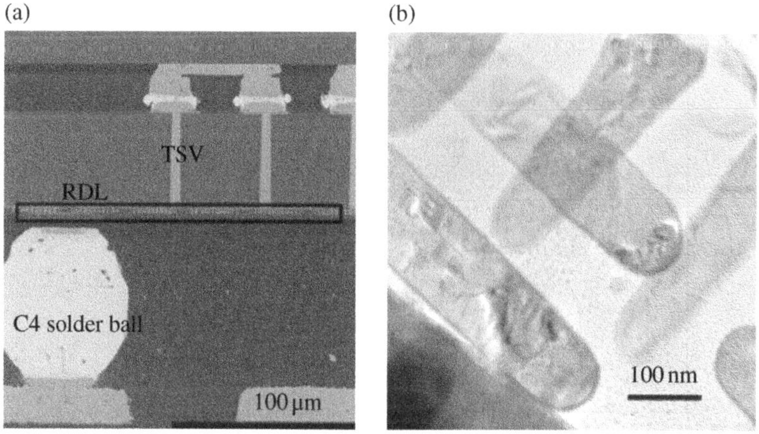


**Figure 13.8** (a) It shows that the location of RDL in the sample is near the bottom side of the interposer and below the TSV. (b) Cross-sectional TEM image of the Cu lines in RDL. The cross-section of the Cu line is about 100 nm by 100 nm.

Because of the time-dependent failure of burn-out, the temperature must be raised above the melting point of Cu, which is 1083 °C. Physically, we note that the thickness of interposer chip is about 50 μm, which is much less than the typical thickness of a Si device chip at 200 μm. Thus, the heat conduction in the Si interposer is much reduced, and this is one of the reasons why heat dissipation is poor. Below, we confirm the time-dependent Joule heating and the positive feedback to enhance electromigration by simulation.

A three-dimensional simulation program was adopted to show the possibility of burn-out in the RDL wires at the test conditions. Since the Cu line in RDL was made similar to Cu damascene structure, we assume that the electromigration damage is also similar to Cu damascene structure by surface diffusion at 100 °C, which will make Cu lines thinner along the current flow direction. Because of the reduced cross-sectional area of current path in Cu line, resistance and Joule heating will increase. Due to poor thermal dissipation, the increased Joule heating will result in temperature increase, which in turn will increase the rate of electromigration. However, the enhanced electromigration will reduce the thickness further and in turn increase resistance and temperature again. This positive feedback process accelerates the thinning of the Cu lines in RDL. Figure 13.9 is the simulation result in three different stressing conditions, from which we can see the synergistic and positive feedback process leading to fast and catastrophic type of burn-out failure.

This synergistic effect of Joule heating and electromigration on burn-out is a new failure mode of reliability. Joule heating will be the most critical

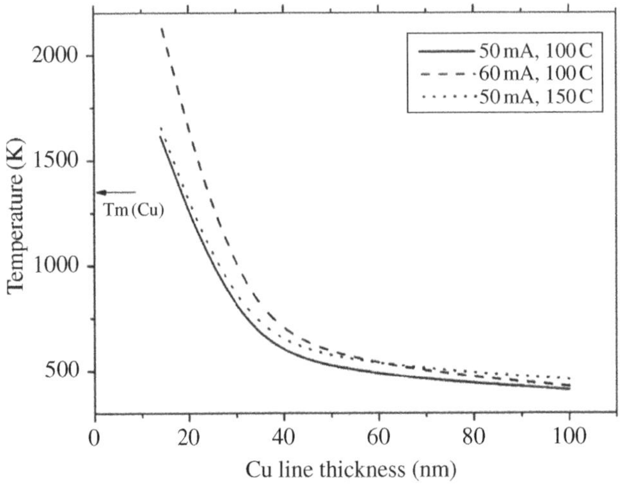


**Figure 13.9** Simulation result in three different stressing conditions, from which the synergistic and positive feed-back process leading to fast and catastrophic type of burn-out failure is clear.

issue as the density of transistor becomes higher while the form factor cannot change much in mobile units. This synergistic effect can couple with mechanical and other failure in the future. Clearly, it needs our further attention and better understanding.

In a short summary, the system-level electromigration in 2.5D IC test samples has shown a time-dependent burn-out type of failure. The failure site has been located in the RDL between the C-4 joints and micro-bumps. Simulation results support the finding that the positive feedback between Joule heating and electromigration can lead to a local temperature increase reaching the melting point of Cu, resulting in a fast and catastrophic burn-out type of failure.

### 13.5.2 Joule Heating Induced Thermomigration Failure Due to Thermal Crosstalk in 2.5D IC Technology

Thermal-crosstalk-induced thermomigration failure in unpowered micro-bumps has been found in 2.5D IC devices [12]. In the test structure, as shown in Figure 13.10, there are two Si chips placed horizontally on a Si interposer. Figure 13.10a and b shows, respectively, an SEM cross-sectional image and an X-ray tomography image of the top view of the test sample. Figure 13.10a shows a stacking of two levels of Si chips; the large bottom Si chip serves as an interposer, and it is connected vertically to the top two smaller Si chips through thermal compression bonded micro-bumps, which are arranged in daisy chains, as shown in Figure 13.10b. The top two Si chips are separated by under-fill materials. Under each of these two chips, there are two daisy chains of micro-bumps with 17 μm in diameter and 30 μm in pitch. The daisy chain circuit of the micro-bumps is shown in Figure 13.10c.

One daisy chain of the micro-bumps under one Si chip (chip 2) in Figure 13.10b was powered under current density of $5.3 \times 10^4$ A/cm$^2$. The testing temperature was 150 and 170 °C. Owing to Joule heating in the daisy chain, the temperature in the powered micro-bumps will have a temperature rise about 30 °C above the test temperature (estimated by simulation). Yet there is little temperature gradient in the powered micro-bumps under chip 2 because of the daisy chair arrangement. So the powered micro-bumps have electromigration but no thermomigration.

However, the unpowered micro-bumps in the neighboring chip (chip 1) in Figure 13.10a were damaged with big holes in the solder layer. This is

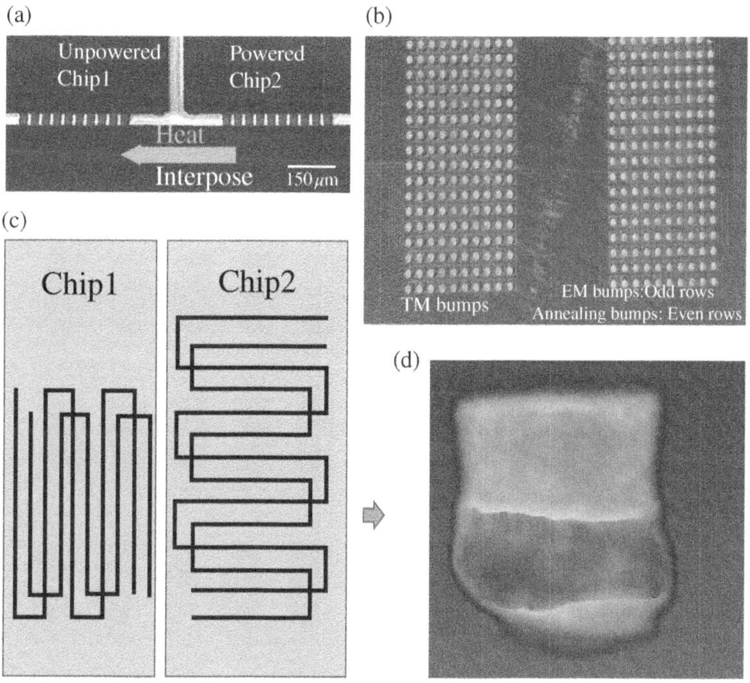


**Figure 13.10** (a) Cross-sectional SEM image showing the test sample, (b) Top view of the sample by Synchrotron Radiation (SR) X-ray micro-tomography, and (c) Schematic of the circuit. (d) One of microbumps from X-ray micro-tomography.

because the Joule heating from the powered micro-bumps was transferred horizontally, through the interposer, to the bottom of the neighboring unpowered micro-bumps, and created a large temperature gradient, in the order of 1000 °C/cm, across the unpowered micro-bumps in the neighboring chip 1, so the latter failed by thermomigration.

Furthermore, as one daisy chain in chip 2 is powered, the Joule heating will be transferred to the unpowered micro-bumps in the neighboring chain in chip 2 through both the top Si chip and the bottom Si interposer. Therefore, the unpowered micro-bumps in chip 2 have no temperature gradient, but they will experience a constant temperature annealing, which is slightly higher than the ambient temperature by 30 °C due to Joule heating, but neither electromigration nor thermomigration.

In short summary, there are three different sets of micro-bumps in the test sample: the micro-bumps under electromigration (the powered micro-bumps in chip 2); the micro-bumps under thermomigration (the micro-bumps in chip1); and the micro-bumps under isothermal annealing (the unpowered micro-bumps in chip 2). Synchrotron radiation tomography was used to compare these three sets of micro-bumps. The results show that the micro-bumps under thermomigration have the largest damage. Figure 13.10d shows the 3D image of one micro-bump in the as-received state by using synchrotron radiation X-ray micro-tomography.

Figure 13.11a shows the micro-bumps under chip 2 that were stressed by electromigration at 120 mA, 170 °C for 72 hours. Because we only powered

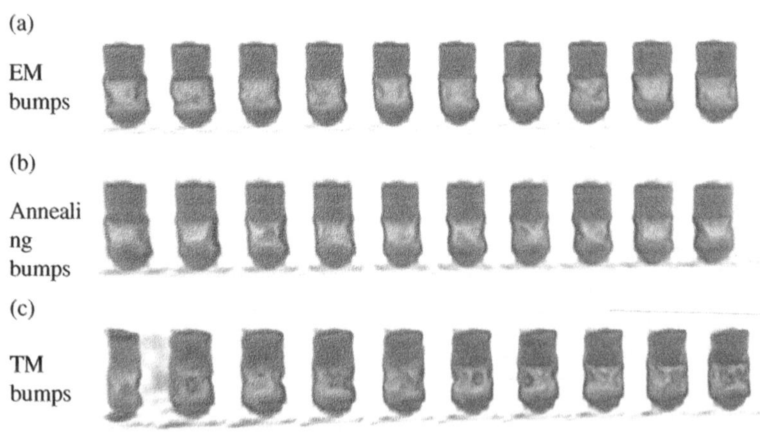


**Figure 13.11** In the test structure, there are two Si chips placed horizontally on a Si interposer. (a) and (b) show, respectively, an SEM cross-sectional image and an X-ray tomography image of the top view of the test sample. (c) shows the daisy chain circuit of the micro-bumps.

the micro-bumps in the odd rows (row 1) under chip 2, the micro-bumps in the even rows (row 2) are unpowered but under isothermal annealing. Figure 13.11b shows the odd row (row 1) micro-bumps under chip 2, under isothermal annealing. Figure 13.11c shows the micro-bumps in Chip 1 that were under thermomigration. In comparison, those micro-bumps under thermomigration have had much more dimples on the surface of the micro-bump, indicating that the surface diffusion of Sn under thermomigration is most serious and has caused more damages. In order to check the damage inside the micro-bump, the samples were analyzed slice by slice in the Avizo software. It was found that even in the middle of the micro-bump, those under thermomigration had at the least twice void volume more than those under electromigration micro-bumps.

## Problems

**1** What is the difference between yield and reliability in microelectronic technology? Please use examples to explain it.

**2** What is mean-time-to failure? Why is it important?

**3** Electromigration has been a major reliability concern in microelectronic devices. How can we have almost never found that our personal computer, laptop computer, and cell phones have failed due to electromigration?

**4** Why Joule heating has been identified to be the key reliability issue in the future 3D IC devices?

**5** In Black's equation of MTTF, why is the sign in front of the activation energy to be positive?

**6** In Black's equation, we have measured the activation energy to $E = 1$ eV/atom, the power factor of current density is $n = 2$, and the pre-exponent factor $A = 10^{-2}$ sec $(\text{amp/cm}^2)^2$. Calculate the MTTF when the device is used at 100 °C with a current density of $2 \times 10^3$ amp/cm$^2$. Then for the next-generation devices, when the current density will be increased to $1 \times 10^4$ amp/cm$^2$, what will be the MTTF?

**7** In flip-chip technology, the line-to-bump configuration is unique, which means that electromigration can occur in the Al line, as well as in the solder bump. Assume that the current density in the Al line is $10^6$ amp/cm$^2$ and the current density in the solder bump is $10^4$ amp/cm$^2$ due to current crowding. Which will fail first at 100 °C due to void formation?

## References


1 Black, J.R. (1967). Mass transport of Al by momentum exchange with conducting electrons. *Proceedings of IEEE International Reliability Physics Symposium*, 144–159.

2 Black, J.R. (1969). Electromigration – a brief survey and some recent results. *IEEE Transactions on Electron Devices* 16: 338–347.

3 Lloyd, J.R. (2007). Black's law revisited – nucleation and growth in electromigration failure. *Microelectronics and Reliability* 47: 1468–1472.

4 de Orio, R.L., Ceric, H., and Selberherr, S. (2010). Physically based model of electromigration: from Black' equation to modern TCAD models. *Microelectronics Reliability* 50: 775–789.

5 Shatzkes, M. and Lloyd, J. (1986). A model for conductor failure considering diffusion concurrently with electromigration resulting in a current exponent of 2. *Journal of Applied Physics* 59: 3890–3893.

6 Kirchheim, R. and Kaeber, U. (1991). Atomistic and computer modeling of metallization failure of integrated circuits by electromigration. *Journal of Applied Physics* 70: 172–181.

7 Korhonen, M.A., Borgesen, P., Tu, K.N., and Li, C.Y. (1993). Stress evolution due to electromigration in confined metal lines. *Journal of Applied Physics* 73: 3790–3799.

8 Clement, J.J. and Thompson, C.V. (1995). Modeling electromigration induced stress evolution in confined metal lines. *Journal of Applied Physics* 78: 900–904.

9 Porter, D.A. and Eastering, K.E. (1992). *Phase Transformations in Metals and Alloys*, 2e, 290. London: Chapman & Hall.

10 Tu, K.N. and Gusak, A.M. (2019). A unified model of mean-time-to-failure for electromigration, thermomigration, and stress-migration based on entropy production. *Journal of Applied Physics* 126: 075109.

11 Liu, Y., Li, M., Kim, D.W. et al. (2015). Synergistic effect of electromigration and Joule heating on system level weak-link failure in 2.5D integrated circuits. *Journal of Applied Physics* 118: 135304.

12 Li, M., Kim, D.W., Gu, S. et al. (2016). Joule heating induced thermomigration failure in un-powered microbumps due to thermal crosstalk in 2.5D IC technology. *Journal of Applied Physics* 120: 075105.

# 14

# Artificial Intelligence in Electronic Packaging Reliability

## 14.1 Introduction

Artificial intelligence (AI) is to mimic human intelligence. In the big data era, many new AI applications have been invented, for example in distance teaching and home office using advanced 5G communication technology and 3D IC devices. Here, we ask what will be the technical innovation in AI apps, which can enhance the study of reliability of electronic packaging technology? After all the analyses presented in the early chapters in this book, we should know the answers on what are the most important issues in reliability and how we can get help from AI?

The most important issue in reliability is "experience." In any company, an employee who has had experience on reliability is valuable. He/she may know roughly the best design, proper integration of different materials, and critical processing steps for a new device, so that he/she can have design-for-reliability (DfR) in packaging technology. The new product can achieve the best compromise among the requirements of reliability, cost-effectiveness, and short time to market. No doubt, innovation is always important.

Fundamentally, reliability failure in electronic devices is due to microstructure change, which leads to void or hillock formation. Our goal of AI is to build a machine to have the experience of the expert on reliability based on having big data, in which every event of reliability failure due to microstructure change is included.

In essence, microstructure change is a time-dependent event, which can lead to void and hillock formation in a device. Because it is time-dependent, we define mean-time-to-failure (MTTF) based on statistical distribution of failures from a set of data of microstructure change which can characterize

*Electronic Packaging Science and Technology*, First Edition. King-Ning Tu, Chih Chen, and Hung-Ming Chen.

the lifetime of the device. Typically, we have to perform tests of a set of samples, one by one in series, and it takes time.

How to "cut-in" to do AI is nontrivial. If you are not an expert in a particular field and if you do not have experience, it is hard to find a point to cut-in. What have been discussed in the previous chapters is to show that our experience in studying electronic packaging reliability can help locate the cut-in point.

## 14.2 To Change Time-Dependent Event to Time-Independent Event

Reliability is a time-dependent event, so the challenge is it takes time to do. For example, in studying electromigration, we have to conduct a large number of tests as a function of time, current density, and temperature, until failure occurs. The purpose is to obtain a distribution of the failure for MTTF analysis. It is a time-consuming event! However, if AI can help change it to a time-independent event, we save time.

There are two key questions in reliability. The first question, asked by any microelectronics company about a new device, is How long is its MTTF? The second question is whether there is any early failure, outside the distribution of MTTF, caused by a weak-link in the system. An early failure is an unacceptable failure, which must be identified and removed.

We recall that MTTF equation can be obtained on the basis of entropy production, as given in Chapter 13. In Section 13.4.5 and Eq. 14.1 in Chapter 14, we have

$$\frac{TdS}{Vdt} = JX \tag{14.1}$$

which means that for any irreversible process, we can obtain its MTTF equation right away, provided that we can define the flux and the driving force. Yet, the parameters in the equation must be calculated based on experimental data, which is time consuming.

Thus, we ask can AI help obtain the parameters in MTTF quickly without the required long and tedious time-dependent tests? To do so, we recognize the advantage of big date is to know the microstructure change of a large number of samples in a single reliability test. We can obtain the distribution of microstructure change at a specific given time and temperature under a specific applied current density for example. In other words, the time-dependent failure distribution of many samples tested one-by-one is replaced by the time-independent distribution of microstructure change of a large set of samples tested once or twice.

We may need all the information about the design, materials, dimension of each component, and processing steps in order to analyze the microstructure of the device. Furthermore, the effect of processing on the stability or change of microstructure should be known, so that we can calculate the electrical resistance, thermal conduction, and stress relaxation of any of the circuit path taken for a reliability test. Indeed, heat treatments will affect interdiffusion and reaction to form alloys or intermetallic compounds, so we will need to know them well in order to calculate the resistance change, Joule heating, heat dissipation, and stress relaxation in the device under a reliability test.

Knowing heat distribution, we will be able to find out where is the hot spot or the poor heat dissipation problem in the circuit. Knowing stress relaxation, we will be able to find out where is the stress center. In turn, we will be able to find the atomic flux divergence and source and sink of vacancies, where a weak-link can occur in the test circuit and may become an early failure in tests.

Experimentally, we need to verify the failure in a 3D IC device by using X-ray graphic processing unit (X-GPU). An example of X-GPU is given below. We go to a hospital to take an X-ray image of our lung to see if we have tuberculosis or lung cancer. A doctor who examines the negative will be able to judge.

In 3D IC devices, we cannot observe an internal void from the surface of the sample, and we need to penetrate the 3D structure. It means we have to examine the internal structure of the device. Thus, we need to use an X-ray graphic processing unit (X-GPU) to do so. We took an X-ray 3D image of a device structure before and after failure, by comparison, we and the machine can detect the failure and its location in the 3D structure. If the X-ray machine has performed a large number of tests and the data and failure images have been stored in the machine, a distribution of microstructure changes can be obtained, and this process is part of machine learning. We can use the mean microstructure change to failure to analyze reliability and obtain the parameters of the needed MTTF.

## 14.3 To Deduce MTTF from Mean Microstructure Change to Failure

We need to prepare the test sample differently. It should have a large number of test samples linked together; for example in a daisy chain of several hundred flip-chip solder joints. After thermal annealing at a given temperature and time, we can obtain the distribution of void formation in the daisy chain solder joints. The void volume can be measured within the resolution of the X-GPU, which has a resolution of 0.7 micron. A distribution of the fraction of void volume to the solder joint volume of all the joints can be obtained. We can define when the fraction is over 20%, it is a failure. By comparing the data before and after test, we can obtain the failure distribution and weak links.

When the machine has performed a large number of tests and has stored all the data and images before and after failure, it has the experience of a reliability expert, as well as the intelligence of the expert. This means when we bring a new device to the machine to take an X-ray image, it can judge whether there is a weak-link in the microstructure. Also it can provide the mean microstructure change to failure (MMTF) of the new device. In turn, the MTTF can be obtained. We note that now the machine can do so without the tedious reliability tests!

## 14.4 Summary

The future study of reliability must have AI in mind because of the trend of strong competition among microelectronic companies, where a large volume of consumer electronic products is produced. We are able to change reliability study from a time-dependent event to a time-independent event.

# Index



*Electronic Packaging Science and Technology*, First Edition. King-Ning Tu, Chih Chen, and Hung-Ming Chen.

## p



## q



## r



## s

*w*



*x*



*y*